Information Security and Cryptography
Texts and Monographs

For further volumes:
http://www.springer.com/series/4752

T0206937

Lars R. Knudsen • Matthew J.B. Robshaw

The Block Cipher
Companion

 Springer

Prof. Lars R. Knudsen
Technical University of Denmark
Dept. of Mathematics
Matematiktorvet
Building 303 S
2800 Kgs. Lyngby
Denmark
lars.r.knudsen@mat.dtu.dk

Dr. Matthew J.B. Robshaw
Orange Labs
rue du Général-Leclerc 38-40
92794 lssy les Moulineaux
France
matt.robshaw@orange-ftgroup.com

Series Editors
Prof. Dr. David Basin
Prof. Dr. Ueli Maurer
ETH Zürich
Switzerland
basin@inf.ethz.ch
maurer@inf.ethz.ch

ISSN 1619-7100
ISBN 978-3-642-27111-3   ISBN 978-3642-17342-4 (eBook)
DOI 10.1007/978-3-642-17342-4
Springer Heidelberg Dordrecht London New York

ACM Codes: E.3

Printed on acid-free paper

Springer is part of Springer Science+Business Media (www.springer.com)

To Mia, Kasper, Sasha, and Heather.

To Lucas, Daniel, and Elke.

Foreword

Block ciphers have played a central role in the development of large-scale commercial cryptology. While some key design principles of block ciphers were already described in Shannon's seminal 1948 paper, open research on block ciphers started at IBM in the late 1960's under the supervision of Feistel. This work culminated in the design of DES, published as the US government FIPS standard for protecting sensitive but unclassified data in 1977. It is hard to overestimate the importance of DES for practical cryptology. DES was widely deployed for commercial and government applications, and even today it survives in its triple-DES variant; DES was also used to construct one-way functions, MAC algorithms, hash functions, and pseudo-random string generators. DES formed an attractive target for academic cryptanalysis; while progress in the first decade was slow, in 1988 differential cryptanalysis was discovered and in 1993 linear cryptanalysis followed. These results taught us that the designers of DES had a very good understanding of the security of block ciphers. These exciting development in cryptanalysis combined with the interest in the design of new software-oriented symmetric algorithms resulted in a thriving research community meeting at Fast Software Encryption from 1993 onwards.

A second milestone in the development of block ciphers is the AES competition; NIST published an open call in 1997, the selection of Rijndael as winner was announced in 2000, and the FIPS standard was published in 2001. The focused attention of the cryptographic community during the competition resulted in substantial progress on the design and cryptanalysis of block ciphers. Since 2004, cryptanalysts have shifted their attention to hash functions, but there is a continued interest in studying the security of block ciphers; during recent years there has been growing interest in developing lightweight block ciphers.

Until the early 1990's, there were only a small number of books on cryptology; the design and application of DES plays a central role in the books of Konheim (1981) and of Meyer and Matyas (1982). Today there are many introductory and more advanced books on cryptology and cryptanalysis, and all of them contain one or more chapters on block ciphers. However, only two books deal exclusively with block ciphers, namely the 1993 book by Biham and Shamir on differential cryptanalysis of DES and the 2002 book by Daemen and Rijmen on the design of the AES standard. In view of this, I am very pleased that the authors of the current volume have done a superb job in writing the first textbook that covers the first four decades of block cipher research. The book is very suitable to learn the field, as the authors restrict themselves in the first seven chapters to the key results; moreover, they make use of simple and well-chosen examples to clearly explain differential and linear attacks. The book offers the right mix between clarity of exposition and mathematical rigor, which is essential to explain some subtle points related to cryptanalytic attacks. Chapter 8 contains an excellent selection of more advanced topics, in which the authors draw on their extensive experience and innovative contributions in cryptanalysis and the design of block ciphers. The readers will also benefit from the description of six prominent block ciphers (in addition to DES and AES) and from the close to 800 references. Overall, the authors have produced an academic tour de force by presenting the essential results on block ciphers in this volume. I hope that you will enjoy reading this book as much as I did.

Leuven, June 2011   *Bart Preneel*

Preface

The Block Cipher Companion has been some considerable time in preparation. We find it hard to believe, but we started work on this book during the AES process in which we both participated. But ongoing changes to our professional and personal lives continually conspired to move the end point just beyond reach.

Our motivation for starting was the lack of an accessible book that gave a compact yet thorough treatment of block ciphers. We're pleasantly surprised to find that, even after ten years during which many excellent textbooks on cryptography have been published, there may still be room for a "companion" for those wanting to learn a little more about block ciphers. In what follows we aim to provide a technically detailed yet readable account of one important component of today's cryptography. This book should be useful to students taking a course in cryptography, to the casual—though admittedly committed—bystander, and to professionals wanting to understand a little more about cryptographic design.

The bulk of the work in the book evolved from course work that was prepared by Lars at the Technical University of Denmark and by Matt when teaching at Royal Holloway, University of London and, previously, when visiting the Universidad de Chile, Santiago. We would like to thank all our colleagues and students at these institutions. We would also like to thank our friends and colleagues who have had to endure block cipher conversations over the past decades and the following people who read parts or all of our manuscript: Mohamed Ahmed Abdelraheem, Julia Borghoff, Praveen Gauravaram, Charlotte V. Miolane, Søren S. Thomsen. Thanks also to Ronan Nugent from Springer for a very pleasant and helpful collaboration.

Our book aims to distill nearly three decades of research into one small handbook. Clearly it is difficult to do full justice to all the work that has taken place in the field. But we hope that we have managed to highlight the talents and insights of our research colleagues who together have made the field of block ciphers one of the richest and most fascinating areas of study. Block ciphers rock!

Denmark and France,
June, 2011

Lars Knudsen
Matt Robshaw

Contents

Chapter 1
Introduction

As our personal and commercial transactions are routinely transmitted around the world, the need to protect information has never been greater. One of the best single line descriptions of cryptography is due to Rivest [636]:

Cryptography is about communication in the presence of an adversary.

Over the past decades the field of cryptography has grown and matured to a remarkable degree. The title of the book you are now reading is sufficient clue to the fact that we are restricting our attention to one very small part of this research, and we will restrict our attention to one particular class of cryptographic algorithm: the *block cipher*. However, we shouldn't feel too discouraged that we will only be covering a small fraction of the cryptographic research. Block ciphers are fundamental to much of today's deployed cryptography and they are, by far, the most widely used cryptographic primitive. Indeed, even though block ciphers are useful in their own right they are also used to build other cryptographic mechanisms. In short, they are particularly versatile objects.

The *block cipher* is the narrow focus of this book and we feel that an overview of the state of the art of block cipher analysis, design, and deployment is timely. The chapters that follow are of one of two types. Several chapters provide descriptions of prominent block ciphers and give an insight into their design. Other chapters consider the role of the cryptanalyst (the "adversary") and provide an overview of some of the most important cryptanalytic methods.

The field of block cipher design was somewhat quiet after the adoption of the *Advanced Encryption Standard (AES)*, a cipher which will be described in Chap. 3. In fact some cryptographic commentators went so far as to suggest that "cryptography was dead", by which they implied that the task of algorithm design was pretty much done and dusted. However, the field of block cipher design and analysis is as strong as it ever was and we believe such work to be vital. Not only do we need to continually validate the security of the techniques we use, but new applications introduce new demands for which existing algorithms might be inadequate. In fact even recent advances in hash function cryptanalysis can be viewed as the application and extension of techniques in block cipher cryptanalysis.

Of course we're not looking at block ciphers just out of curiosity; we want to do something with them. In short we want to solve some security problem, to achieve a security goal. Most books on security and more than a few on cryptography provide a list of the likely security goals for an application. The contents of such lists vary among commentators, but the goals of *confidentiality*, *integrity*, and *availability* are often highlighted. Application designers will attempt to deliver these goals by appealing to a range of *security services* and these can range widely from the technical to the administrative. Cryptography typically forms only a very small component of a security solution, but it has an important role and it is likely to provide some of the following:

- *Confidentiality*: ensuring that an adversary listening to our communication channel cannot gain information about the content of our communications.
- *Data integrity*: ensuring that any adversary with access to our communication channel is unable to manipulate the contents of some communication by unauthorized means.
- *Data origin authentication*: ensuring that any adversary with access to our communication channel is unable to modify and/or misrepresent the true origin of some communication. Some commentators observe that the property of *data origin authentication* directly implies *data integrity* since a message that has been modified has a new source.

Of course there are other services that we haven't explicitly mentioned. For instance, *entity authentication* allows for the corroboration of the identity of an entity such as a person, a computer terminal, or a credit card. We might also consider issues such as *authorization, confirmation, delegation*, and *non-repudiation* along with many others. However, there are so many good treatments of these issues in the broader cryptographic and security literature (*e.g.*, see [493]) that we will not attempt to duplicate those efforts here.

Instead our interest lies in the *cryptographic algorithm* and in how a limited number of bits and bytes might be processed under the action of a secret key. For the most part we won't be interested in how the algorithm is used or in how data or secret-key material is delivered to the algorithm. These are important issues, but they are not our concern in this book.

The popular press seems to relish the opportunity to report on exciting new "security glitches" or "breaks", particularly those involving the Internet. However, it is rare that such security weaknesses are due to the cryptographic algorithm. Of course bad algorithms will give a bad application. However, the overall success of a deployment will depend on all aspects of the engineering effort; the choice of cryptographic algorithm is often a small—but important—part of the solution. When choosing a cryptographic algorithm the two most important concerns will almost certainly be *(i)* security and *(ii)* performance. And while other business-related issues such as licensing might play a part, today we are likely to find standardised algorithms offering the best choice.

It is a testament to the skills of early algorithm designers that we have had trusted algorithms available to us for many years. Nevertheless, anyone reading the cryptographic literature will be struck by the large number of algorithms that have been proposed. What is it that makes a good cryptographic algorithm? What makes one better than another?

1.1 Cryptographic Algorithms

It is customary to classify cryptographic algorithms according to how key material is used. To begin, we will isolate the class of *keyless* algorithms, where we put the cryptographic algorithms that do not use *any* key material. The classic example is the *hash function* or *message-digest* algorithm.

Assuming that key material is used, and further assuming that the communication or interaction we wish to protect has multiple participants, we consider two additional classes of cryptographic algorithms. For some algorithms all participants share the same key material. These form our first type of keyed algorithms and they are referred to as *symmetric* or *secret-key algorithms*. There are three types of algorithms in this category: *block ciphers*, *stream ciphers*, and *message authentication codes (MACs)*. Block ciphers and stream ciphers are encryption primitives while the message authentication code is used for data and data origin authentication. It is worth observing that we can use a block cipher to build both a stream cipher and a message authentication code. In fact we can even build the keyless hash function out of a block cipher and we will explore all of this in Chap. 4.

In contrast to the case where the sender and receiver keys are the same, it is possible to devise cryptographic algorithms where the key used by the receiver cannot be computed from that used by the sender. Such algorithms—where the key used for encrypting a message can be made public—are referred to as *asymmetric* or *public-key algorithms*. Different types of algorithms in this category would include *key agreement algorithms*, *public-key encryption algorithms*, and *digital signature algorithms*. This is a fascinating area of cryptography but not one that we will cover in this book.

The symmetric and keyless algorithms tend to be the workhorses of the cryptographic world. As we move from keyless algorithms to symmetric algorithms, and on to asymmetric algorithms, the algorithms at our disposal tend to be slower. An old rule of thumb used to be that a block cipher might be around 50 times faster than an asymmetric algorithm, a stream cipher might be around twice as fast as a block cipher, and a hash function might be faster still at around three times the speed of a block cipher. However, current design trends in all three fields, block ciphers, stream ciphers, and hash functions, are providing more and more exceptions to this crude comparison and it shouldn't be relied upon too closely. However, it does at least help to set the perspective with regard to symmetric and asymmetric cryptography and it illustrates why we leave the bulk data processing—for both encryption and authentication purposes—to the symmetric and keyless algorithms.

1.2 Block Ciphers

As their name implies, block ciphers operate on "blocks" of data. Such blocks are typically 64 or 128 bits in length and they are transformed into blocks of the same size under the action of a secret key. When using the basic block cipher with the same key, two instances of the same input block will give the same output blocks.

The block cipher encrypts a block of *plaintext* or *message* m into a block of *ciphertext* c under the action of a secret key k. This will typically be denoted as $c = \text{ENC}_k(m)$. The exact form of the encryption transformation will be determined by the choice of the block cipher and the value of the key k. The process of encryption is reversed by decryption, which will use the same user-supplied key. This will be denoted $m = \text{DEC}_k(c)$. Throughout the book, the key input to a block cipher will be indicated with a small circle:

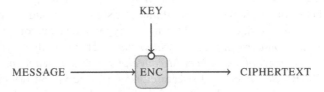

A block cipher has two important parameters:

1. the *block size*, which will be denoted by b, and
2. the *key size*, which will be denoted by κ.

For a given key, a b-bit block cipher maps the set $\mathcal{M}$ of 2^b b-bit inputs onto the same set of 2^b outputs:

$$\mathcal{M} = \{ \overbrace{0\ldots00}^{b}, \overbrace{0\ldots01}^{b}, \overbrace{0\ldots10}^{b}, \overbrace{0\ldots11}^{b}, \ldots\ldots, \overbrace{1\ldots1}^{b} \}.$$

This is done in such a way that every possible output appears once and only once. The mapping is a permutation of the set of inputs and, as we vary the secret key, we obtain different permutations. Thus a block cipher is a way of generating a family of *permutations* and the family is indexed by a secret key k.

For a secure block cipher we expect no exploitable information about the encryption process to leak. Such information might include information about the choice of key, information about the encryption or decryption of as yet unseen inputs, or information about the permutations generated using different keys.

The block size b determines the space of all possible permutations that a block cipher might conceivably generate. The key size κ determines the number of permutations that are actually generated. To appreciate the (difficult) task of the designer it is worth looking at some of the numbers involved. For a block cipher with key size κ there are 2^κ possible keys and each key specifies a permutation of 2^b inputs. There are $(2^b)!$ different permutations on b-bit input blocks which, by using Stirling's approximation, is approximately $2^{(b-1)2^b}$. For typical values of b and κ a block cipher will provide only a tiny fraction of all the available permutations. Furthermore, it will do so in a highly structured way. However, we expect a good block cipher to disguise this and, casually speaking, we expect a randomly chosen key to "select" a permutation seemingly at random from among all $2^{(b-1)2^b}$ possibilities. Being even more demanding, we require that keys that are related in some way yield permutations that have no discernible relation between them.

In the chapters to come we will give detailed descriptions of some prominent block ciphers that accomplish this remarkable feat. In the remainder of this chapter, however, we will explore some basic aims of the designer and the adversary.

1.3 Cryptographer and Cryptanalyst

Much of our work in block ciphers is due to the work of Shannon (1916-2001). In particular, the landmark paper *Communication Theory of Secrecy Systems* of 1949 [680] introduced the twin ideas of *confusion* and *diffusion* for practical cipher design. They are still the most widely used principles in block cipher design.

Shannon was working on a mathematical framework for encryption and he was particularly interested in the encryption of natural languages. It is well known that the letter *e* is the most frequently occurring letter in the English language and frequency tables for single letters and digrams or trigrams of letters can be compiled.[1] A cryptanalyst trying to recover some plaintext in natural English has a great deal of side-information specific to the way English is structured. The plaintext source is said to contain *redundancy*.

Historical or classical ciphers [493] employed simple mechanisms to break up the arrangement of the source plaintext. Yet increasingly complex analysis still allowed the cryptanalyst to use natural language redundancy to decrypt the ciphertext. As well as providing a mathematical framework for the process of encryption using statistically based *Information Theory*, Shannon observed that any good encryption algorithm must disguise redundancy in the source plaintext. To quote Shannon;

> Any significant statistics from the encryption algorithm must be of a highly involved and very sensitive type—the redundancy has been both diffused and confused by the mixing transformation [680].

The idea of confusion is "to make the relation between the simple statistics of the ciphertext and the simple description of the key a very complex and involved one", while in the method of diffusion "the statistical structure of the plaintext which leads to its redundancy is dissipated into long range statistics in the cryptogram" [680]. It should be stressed that the properties of confusion and diffusion are not absolute,

---

[1] The exact distribution will depend on the source. A scientific treatise on zinc may have an inadvertently high number of *z*'s. Similarly the novel *A Void* by George Perec (1938-1982) has a deliberately skewed distribution of letters since the entire text eschews the letter *e* (creating a "void").

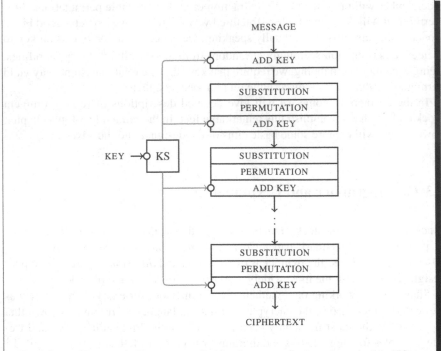

Fig. 1.1 An SP-network. The user-supplied key is processed using a *key schedule* (KS) to derive a set of *round keys*.

quantifiable, concepts. They have been reinterpreted by commentators in different ways and one nice description is given by Massey [469]:

> CONFUSION: The ciphertext statistics should depend on the plaintext statistics in a manner too complicated to be exploited by the cryptanalyst.

> DIFFUSION: Each digit of the plaintext and each digit of the secret key should influence many digits of the ciphertext.

Block ciphers are designed to provide sufficient confusion and diffusion. It is the task of the designer to come up with a judicious mix of components that will give a secure (and efficient) block cipher, though it is sometimes difficult to see exactly what contribution each component makes to the explicit goals of confusion and diffusion. However, these concepts remain useful in highlighting the kind of behaviour we expect from a good block cipher.

We thus have some notion of the properties we're striving for. It turns out that the basic operations of *substitution* and *permutation* are particularly important in achieving these goals. Most, if not all, block ciphers will contain some combination of substitution and permutation, though the exact form of the substitution and the permutation may vary greatly.

Substitution is often used as a way to provide confusion within a cipher. Such substitution might be designed around an arithmetic function such as *integer addition* or *integer multiplication*. More typically, substitution is achieved with a suitably designed *lookup table*, *substitution box*, or what we simply refer to as an *S-box*. S-boxes can be carefully designed to have specific security properties and they can be a quick operation in practice. One downside is that S-boxes typically need to be stored, thereby embodying some memory constraints. Provided the boxes are not too large, this is unlikely to cause a significant problem. However, in demanding hardware implementations space can be at a premium [116] while in software implementations interactions between the S-box and on-processor cache can lead to timing variations that potentially key-related information [41, 584].

Permutations are often used to contribute to the good diffusion in a cipher. Very often the permutation is performed at the bit level, by which we mean that individual bits are moved into a new ordering and DES (see Chap. 2) is the obvious example of this approach. However, bit-level permutations have important performance implications. While they are easy to achieve in hardware, where implementation is simply a matter of aligning wires, software is suited to operations that operate on words of data. Manipulating individual bits is not especially natural and this can slow down the performance of the cipher. We will see in a more modern cipher such as the AES (see Chap. 3) that diffusion is provided by byte-level operations with some optimisations exploiting 32-bit words, and this leads to a flexible performance profile.

The mix of substitution and permutation is an important component of most block cipher designs. Indeed, an important class of block ciphers illustrated in Fig. 1.1 is designed using just these operations. Called *substitution/permutation networks* or simply *SP-networks*, these ciphers consist of the repeated application of a carefully chosen substitution, a carefully chosen permutation, and the addition of key material. In Fig. 1.1 we introduce something called the *key schedule*. This is an important feature of block ciphers that rely on several rounds of computation.[2] While the user supplies the encryption key, it is a good design principle to reuse as much of that key material as often as possible throughout the encryption process. It is the role of the key schedule to present a series of *round keys* to each round of encryption and these round keys are computed from the user-supplied encryption key. Designers have different approaches to good key schedules; some key schedules are computationally lightweight whereas others are very complex.

So far we have considered the role of the cryptographer; now we turn our attention to the cryptanalyst. What is it that the attacker is trying to do? In the most extreme case the cryptanalyst will want to recover the user-supplied secret key. However, the attacker may be satisfied with much less. With this in mind it is possible to establish a hierarchy of possible attacks [382]:

1. TOTAL BREAK: The attacker recovers the user-supplied key k.
2. GLOBAL DEDUCTION: The attacker finds an algorithm A that is functionally equivalent to either $\text{ENC}_k(\cdot)$ or $\text{DEC}_k(\cdot)$.

---

[2] When each round is the same the cipher is sometimes called an *iterated cipher*.

3. LOCAL DEDUCTION: The attacker can generate the message (or ciphertext) corresponding to a previously unseen ciphertext (or message).
4. DISTINGUISHING ALGORITHM: The attacker can effectively distinguish between two black boxes; one contains the block cipher with a randomly chosen encryption key while the other contains a randomly chosen permutation.

An attacker achieving a total break can achieve all the other goals. Indeed these attacks have been ordered so that achieving any given goal automatically achieves those that follow. Taking the converse argument, if an attacker is unable to distinguish between the implementation of a block cipher and a randomly chosen permutation then we have, in some sense, achieved an ideal block cipher. This is the basis for an argument often used within the *provable security* research community where a great deal of work is done on "proving" the security of a construction that uses a block cipher on the assumption that the underlying block cipher offers some "ideal behaviour", *i.e.*, that it cannot be efficiently distinguished from a randomly chosen permutation.

For our analysis we always take it for granted that the cryptanalyst knows the details of the encryption technique. In 1883, Kerckhoffs (1835-1903) laid out six requirements for a usable field cipher [345]. These have been reinterpreted by commentators and the term *Kerckhoffs' Assumption* or *Kerckhoffs' Principle* is now used to refer to the assumption that the cryptanalyst knows every detail of the encryption mechanism except the user-supplied secret key.

We might also assume that the cryptanalyst has access to different types of data. Indeed, as we look at attacks in more detail it will become clear that we need to specify exactly the kind of data required in an attack. Not all kinds of data are equally useful, nor equally available, and attacks can be classified according to the type of data that they require:

1. CIPHERTEXT ONLY: The attacker intercepts the ciphertext generated by the encryption algorithm. Here the attacker will rely on some knowledge of the plaintext source; *i.e.*, some form of redundancy.
2. KNOWN PLAINTEXT/MESSAGE: The attacker is able to intercept a set of n ciphertexts $c_0 \ldots c_{n-1}$ corresponding to some known messages $m_0 \ldots m_{n-1}$ with $c_i = \text{ENC}_k(m_i)$ for $0 \leq i \leq n-1$.
3. CHOSEN PLAINTEXT/MESSAGE: The attacker is able to request the encryption of a set of n messages $m_0 \ldots m_{n-1}$ of the attackers' choosing[3] and intercepts the ciphertexts $c_i = \text{ENC}_k(m_i)$ for $0 \leq i \leq n-1$.
4. ADAPTIVELY CHOSEN PLAINTEXT/MESSAGE: The attacker obtains the encryption of new chosen messages m_i for $i \geq n$ in an interactive way, perhaps after seeing an original pool of n chosen message/ciphertext pairs $c_i = \text{ENC}_k(m_i)$ for $0 \leq i \leq n-1$.

---

[3] This might appear unlikely but there are situations where this is a real possibility. During the Second World War the British RAF would drop mines at specific chosen grid references (a process referred to as "gardening") so that their opponents, on discovering the mines, would generate warning messages exploited by the cryptanalysts at Bletchley Park [296].

5. CHOSEN AND ADAPTIVELY CHOSEN CIPHERTEXT: The attacker recovers the decryption of ciphertexts of his own choosing. As in the case of a chosen message attack, there is an adaptive version.

With each escalation in the type of data available, the attacker has more control over the analysis of the block cipher and can devise increasingly sophisticated attacks. However, at the same time collecting data of a given type becomes more demanding as we move down the list. Many of the attacks we consider will be differentiated according to how much data is required as well as the type of data required.

The success of different types of cryptanalysis will be measured according to the resources they consume. We have already seen that the amount, and type, of data is important. We can add some additional resources that are likely to be of interest below:

1. TIME: The time, or work effort, required to mount the attack is usually the first thing analysts and commentators consider in a new attack. Sometimes—and therefore potentially erroneously—it is the only requirement considered.
2. MEMORY: The amount of memory, or storage, for an attack is very important. Sometimes the amount of memory is so great that it creates an insurmountable bottleneck and the attack remains impractical. One frequently cited motto is that "time is cheaper than memory" [288]. While it might not provide an eternal truth, it nicely captures the idea that the time to perform, say, 2^{40} encryption operations is easier to accumulate than the memory to store the 2^{40} results.
3. DATA: We have already seen that the type of data for an attack is important, but so is the amount of data. For instance, an attack might not require much time but it might require an enormous amount of data. If the time required to generate the data far exceeds normal usage patterns then the practical impact of the attack is limited.

When considering a cryptographic scheme in isolation, the data types and the resources listed above are typically the most appropriate way to quantify a cryptanalytic attack. However, when we turn to cryptographic implementation and deployment, a whole range of new attacks that can be more powerful than classical cryptanalysis become vital. These are attacks that allow the attacker to exploit the leakage of physical information, for instance, the encryption time [411] or the power consumption [412], during the implementation of an algorithm. This area is termed *side-channel analysis*, and the design of attacks and countermeasures is a vast and constantly evolving field. Regrettably, however, including such work would take us out of the scope we have already established.

Instead we turn our attention to the simplest and most widely applicable method of cryptanalysis, that of guessing the secret encryption key.

Table 1.1 A broad headline assessment of the security offered by different key lengths, in the absence of any cryptanalysis weakness.

κ (bits)	search time (operations)	Current Status (2010)	
40	2^{40}	easy to break	no security
64	2^{64}	practical to break	poor security
80	2^{80}	not currently feasible	reasonable security
128	2^{128}	very strong	excellent security
256	2^{256}	exceptionally strong	astronomical levels of security

1.4 Security

To use a block cipher we must choose a secret key k. The easiest attack an adversary can mount is to simply try and guess the value of the key being used. If the key is κ bits long there are 2^{κ} alternatives. The probability of correctly guessing the key at the first attempt is $2^{-\kappa}$. Adding an additional bit to the length of the key halves the probability that the key is correctly guessed.

A more systematic approach is to exhaustively search for the secret key. No matter how clever or secure the algorithm, this type of attack is always available to the cryptanalyst and the only way to protect against it is to have a sufficiently large key. The time required to exhaust all possible keys (the *keyspace*) is proportional to the time required to perform 2^{κ} encryption operations. Having to exhaust the entire keyspace before finding the correct key would be very unlucky, just as being correct on the first guess would be very lucky. So often we refer to the *expected time* to recover a κ-bit key which is $2^{\kappa-1}$ operations.

In translating this to a physical time, testing a key depends on the time to encrypt a block of message and the time to run the key schedule. Recall that for a multi-round block cipher we need to extract a set of round keys from the user-supplied key. Normally the cost of running the key schedule is not so significant. It occurs once at the start of encryption and any time penalty is spread or amortised over the whole encryption process. However, during exhaustive search we are changing keys often and the cost of the key schedule can be significant. For example, it is estimated that testing an RC5 key takes four times longer than a DES key [650].

In Chap. 5 we will consider some sophisticated variants of exhaustive search. These use memory to reduce the time required to find the secret key. But in its raw form exhaustive key search has no memory requirements and has very light data requirements. To identify the correct key we would rely on redundancy in the message source so that we can recognise when the decrypted plaintext is correct. Of course, the best way to do this would be with a known message.

Depending on the block and key size, it is possible that several keys map one particular message to the same ciphertext. For a good b-bit block cipher the probability that a key maps a given message to a specific ciphertext should be 2^{-b}. Considering the entire keyspace we would therefore expect $2^\kappa \times 2^{-b}$ keys to provide a match for a single message/ciphertext pair. Depending on the sizes of κ and b we might therefore need additional message/ciphertext pairs to eliminate any false alarms from our exhaustive key search.

It is interesting to note that changing encryption keys does not necessarily provide much additional protection against exhaustive key search. To illustrate, we suppose that κ is not that large and that we can exhaustively test the keyspace in a year. This means that in the course of any single day we can search $\frac{1}{365} \approx 0.3\%$ of all possible keys. Suppose now that the key is changed every day. Then the probability that we will recover the key that is used on a given day is around 0.3%. If we were to search over the course of an entire year then we would expect to be successful at least once. While the attacker cannot predict which of the 365 keys he will recover during the year, he should get one of them. Thus there is still some risk even though keys are being changed frequently. It is however slightly better than the case of a key that remains unchanged. In this case the attacker would be guaranteed to be successful in recovering an active key during that year of searching.

So how do we know when a key is large enough? Much will depend on the risks facing the application and no definitive answer can be given. However, it seems that there is reasonable consensus in the cryptographic community and several benchmark key sizes have been established; see Table 1.1. To provide a physical comparison to these enormous numbers, the universe is less than 2^{80} microseconds old and it is estimated to contain around 2^{250} protons.

1.5 Summary

During our opening discussion on exhaustive search we have identified a *necessary* condition for the practical security of a block cipher, namely that the encryption key k be sufficiently long. Note, however, that this is not a *sufficient* condition and the field of cryptography is littered with poor algorithms whose only claim to security is a large key!

Indeed it might be tempting to ensure that every block cipher—even those of high-quality design—have an enormously long secret key. However, one problem in the deployment of cryptography is that of generating and managing secret keys. We really don't want to have to deal with more key material in a system than we need to. There are also significant concerns about how one might go about designing a cryptosystem that uses an enormously long key without any attendant compromise in performance. But perhaps most importantly from the authors' perspective, it is just not very elegant.

The rest of the book, therefore, is intended to help us identify some block ciphers that achieved this careful balancing act. But what is a *good* block cipher? In our opinion a good block cipher is one for which the best attack is an exhaustive search. Such a cipher performs according to the designers' intentions. However, a good block cipher might not necessarily be a practically secure block cipher; for instance, the key length could be shorter than we would like and exhaustive search could therefore pose more of a threat than we would like. In the next chapter we will encounter such a cipher.

Chapter 2
DES

In this chapter we will describe one particular block cipher, the *Data Encryption Algorithm (DEA)* or the *Data Encryption Standard (DES)*. The importance of DES cannot be overstated and it has been said that "DES trained a generation of cryptographers". When DES was first published in 1975, it was anticipated that the cipher would remain acceptable for ten years. Yet today, 30 years on, it remains one of the most widely implemented ciphers in the world and, in one guise or another, it is likely to remain in widespread use for considerable time to come.

The story of DES is a fascinating one. In May 1973 the US *National Bureau of Standards (NBS)*[1] issued a call for proposals for a block cipher suitable for federal use. Initial results were disappointing and so, in August 1974, a second call was made. A block cipher, based on an earlier internal design called *Lucifer*, was submitted by IBM. The algorithm was published for comment in March 1975, accepted as a standard in August 1976, and published as FIPS 46 in January 1977 [545].

Early reactions to DES were mixed and two public workshops were held by the NBS to provide a forum for discussions. As we will see, DES uses a 56-bit key and even in the 1970s it was unclear just how much security this really offered. To some commentators, while the economic investment required to break DES seemed to be more than most could afford, it didn't seem to be out of reach for everyone. Early estimates by Diffie and Hellman [215] suggested that a machine could be built for around $20 million that would find a DES key in a day. NBS responded that such a machine would "cost over $70 million to build between 1977 and 1990, be 256 feet long, [and] draw millions of watts of power" [539]. Other discussions revolved around specific design features and in particular they focused on the S-boxes (internal components on which the security of DES relies). While analysis soon confirmed that the S-boxes had been carefully designed, the design criteria were not made public. Some even speculated that there might be a *trapdoor* in the algorithm; an additional piece of information that would allow unauthorized decryption. For the rest of the 1970s and most of the 1980s, academics attacked DES without success. The

---

[1] The NBS has been superseded by the *National Institute of Standards and Technology (NIST)*. This body is charged with establishing standards to satisfy U.S. Government requirements. These are typically published as *Federal Information Processing Standards (FIPS)*.

clever design features of DES became increasingly clear and when coupled with an increasingly long list of abandoned cryptanalysis, the net result was a massive level of trust being conferred upon the cipher.

In 1990 an attack called *differential cryptanalysis*, see Chap. 6, was described that could recover a DES key, though with vast quantities of data. Additional analysis appeared to show that DES had in some ways been optimised to resist this attack. This was confirmed in 1994 by Coppersmith, one of the original DES designers [167]. Then, in 1993, Matsui described a new cryptanalytic technique called *linear cryptanalysis*, see Chap. 7, and in 1994 experimentally recovered a DES key [475]. While a dramatic development, the practical impact was somewhat limited by the vast amounts of data required. In Matsui's experiment just generating the data required 40 days of continuous encryption! Nevertheless, it was the first time that any DES key had been successfully recovered in the public domain.

Impressive as all this cryptanalysis was, perhaps the most significant development over these years was the relentless increase in computational power. Estimates for the cost of exhaustive key search were dropping and becoming increasingly worrisome. Slowly, but inevitably, a DES exhaustive key search was becoming a reality [744, 745]. In 1997 a distributed search effort using the Internet accumulated sufficient spare computing cycles to recover a DES key as part of a challenge launched by RSA Laboratories [650]. Then, in 1998, a dedicated DES key search machine called *DES Cracker* was built that revealed in the most dramatic way the true practicality of hardware-based DES exhaustive key search [226].

In both 1988 and 1993, DES had been reaffirmed as a federal standard for five additional years. Finally, in 1998, this came to an end and NIST finally confirmed the withdrawal of DES at the *Crypto* conference in August 2004. Cryptographically speaking DES, as a stand-alone cipher, is at the end of its useful life. However, this is not the end of the DES story and a cipher called *Triple-DES* remains a federal standard [549], see Sect. 2.4. Triple-DES is, without argument, the most trusted block cipher available today.

2.1 DES Description

DES is a block cipher that operates on 64-bit blocks. It uses a 56-bit key, but this is often embedded within a 64-bit block where one bit in eight is used as a parity bit. When describing DES we will denote the leftmost bit position of the 64-bit block as bit 1, and label the bits as we move to the right, making the rightmost bit 64. This is the convention that was established in the original FIPS documentation and it is the one that we will use here.[2]

DES is an example of a *Feistel cipher*. A b-bit Feistel cipher consists of the repetition of r rounds of an identical structure. This repeated structure consists of a *round function* and a *swap*. The round function maps a $\frac{b}{2}$-bit input into a $\frac{b}{2}$-bit output

---

[2] Today we would not label the bits $\{1 \ldots 64\}$ but would most likely use $\{63 \ldots 0\}$ to reflect modern software techniques. However, we stay with the established notation.

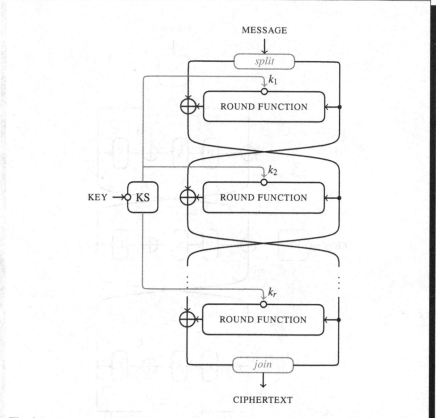

MESSAGE

split

k_1

ROUND FUNCTION

k_2

KEY → KS

ROUND FUNCTION

k_r

ROUND FUNCTION

join

CIPHERTEXT

Fig. 2.1 An r-round Feistel network with the user-supplied key being processed to give *round keys $k_1, \ldots, k_r$*.

under the action of a set of *round keys $k_1, \ldots, k_r$*. This $\frac{b}{2}$-bit output is used to modify one half of the text being encrypted. The two halves of text are then swapped and the round is repeated. The sole exception to the application of round function *and* swap is in the last round of the cipher, where there is no swap. The general form of a Feistel network is illustrated in Fig. 2.1. At each round we use a round key which is derived from the κ-bit user-supplied key k using a *key schedule*.

More formally, we write the message input m as $m = (u_0 \| v_0)$ where u_0 and v_0 are 32-bit strings and we use $\|$ to denote bitwise concatenation. The successive intermediate values u_i and v_i for $1 \le i \le r$ are generated as follows, where $f(\cdot, \cdot)$ denotes the round function and the ciphertext is given by $c = (u_r \| v_r)$:

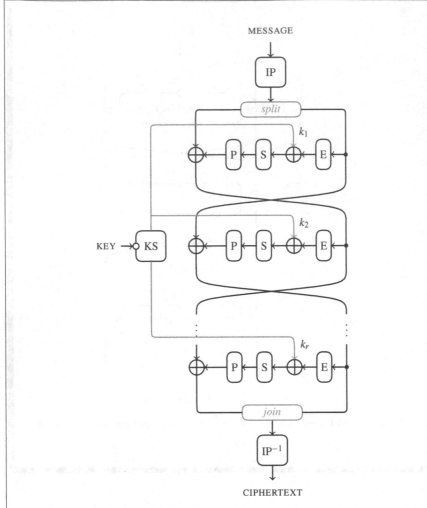

Fig. 2.2 DES with round function components; the bit expansion E, the S-boxes S, and the bit permutation P.

$$\left.\begin{array}{l} v_i = u_{i-1} \oplus f(v_{i-1}, k_i) \\ u_i = v_{i-1} \end{array}\right\} \text{ for } 1 \le i \le r-1,$$

$$u_r = u_{r-1} \oplus f(v_{r-1}, k_r), \text{ and}$$

$$v_r = v_{r-1}.$$

One advantage of the Feistel network is that the decryption process is identical to encryption provided the round keys are taken in reverse order. While there

Table 2.1 The DES initial permutation IP and its inverse. To interpret the table, consider the fifth entry (reading from left to right) of the first row in IP. This entry is 26 and the table indicates that the bit in position 26 of the DES input is moved to position 5. Similarly, the third entry of the second row in IP is 44, which tells us that the bit in position 44 of the DES input is moved to position 11.

IP	IP^{-1}
58 50 42 34 26 18 10 2	40 8 48 16 56 24 64 32
60 52 44 36 28 20 12 4	39 7 47 15 55 23 63 31
62 54 46 38 30 22 14 6	38 6 46 14 54 22 62 30
64 56 48 40 32 24 16 8	37 5 45 13 53 21 61 29
57 49 41 33 25 17 9 1	36 4 44 12 52 20 60 28
59 51 43 35 27 19 11 3	35 3 43 11 51 19 59 27
61 53 45 37 29 21 13 5	34 2 42 10 50 18 58 26
63 55 47 39 31 23 15 7	33 1 41 9 49 17 57 25

are several different designs with similar properties the Feistel network happens to be the one used in DES. This has spurred much theoretical work on the "ideal behaviour" of block ciphers that are built in this way (see Sect. 8.6), but despite such efforts, few recent ciphers adopt a similar approach. Regarding DES there are sixteen rounds and the user-supplied key is 56 bits in length and so $\kappa = 56$. For DES an *initial permutation (IP)* of the 64-bit input block is added to the start of the Feistel network. To maintain the property that the encryption network can be reused for decryption, we require the inverse operation IP^{-1} to be applied to the output of the network. The DES initial permutation and its inverse are given in Table 2.1.

2.1.1 The Round Function

Fig. 2.2 gives an overview of DES and we now consider the components in more detail. Since DES operates on 64-bit blocks, each iteration of the round function takes a 32-bit input and returns a 32-bit output. The second input to the round function, the DES round key, is a 48-bit quantity derived using the key schedule and the round function has four components:

1. The 32-bit input is expanded to a 48-bit intermediate value by a bitwise *expansion* typically denoted by E. This is done by duplicating particular bits of the input.
2. Using bitwise exclusive-or the 48-bit intermediate value is combined with the round key.
3. The 48 bits that result are split into eight groups of six, and used as inputs to eight different *S-boxes*. Each S-box returns four bits which, when concatenated together, give a 32-bit intermediate quantity.

Table 2.2 The bitwise expansion E and bitwise permutation P in DES. The tables should be interpreted as those for IP and IP^{-1} in that the first bit of the output of E is taken from the 32nd bit of the input while the first bit of the output of P is taken from the 16th bit of the input.

		E			
32	1	2	3	4	5
4	5	6	7	8	9
8	9	10	11	12	13
12	13	14	15	16	17
16	17	18	19	20	21
20	21	22	23	24	25
24	25	26	27	28	29
28	29	30	31	32	1

			P				
16	7	20	21	29	12	28	17
1	15	23	26	5	18	31	10
2	8	24	14	32	27	3	9
19	13	30	6	22	11	4	25

4. A bit-level *permutation*, typically denoted by P, is then applied to the 32 bits coming out of the S-boxes and provides a 32-bit output from the round function.

2.1.1.1 Expansion E and permutation P

The bitwise expansion E is given in Table 2.2. The mapping takes a 32-bit input and returns a 48-bit output and expansion is achieved by duplicating 16 bits so that they appear twice in the output. The remaining 16 bits appear only once in the output. One effect of the particular expansion chosen is that any two neighbouring S-boxes share two input bits. The bitwise permutation P used in the DES round function is given in Table 2.2. Fig. 2.4 presents the DES round function showing the wirings of the E expansion and of the P permutation. We shall discuss some of the properties of E and P in Sect. 2.2.

2.1.1.2 The Round Key Addition

While addition of the round key is a simple operation, the process by which the 48-bit round keys are derived from the user-supplied key is a little complicated at first sight. Cryptographically speaking, DES uses a 56-bit key though the user-supplied key is represented as a 64-bit quantity. Each byte of the 64-bit representation of the key consists of seven bits of cryptographic key followed by an extra bit used as a parity bit. These parity bits are in positions 8, 16, ..., 64 of the 64-bit representation of the key.

The first thing that happens in the key schedule is that these parity bits are removed (see Fig. 2.3). This is done by what is referred to as *Permuted Choice 1*

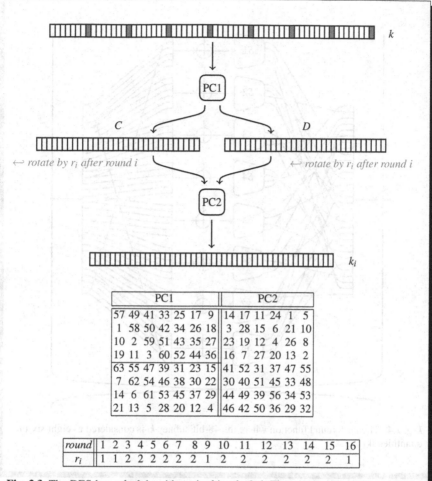

Fig. 2.3 The DES key schedule with parity bits shaded. The permutations PC1 and PC2 are given where in PC1 the top half gives register C while the lower half gives register D. We see that bit 1 in register C is bit 57 of the user-supplied key while bit 1 in register D is taken from bit position 63. The round-dependent rotation values r_i are also given.

(PC1) which, conceptually, loads the 56 key bits into two 28-bit key registers referred to as C and D. PC1 is described in Fig. 2.3 and since it extracts 56 bits from 64 the table for PC1 has eight rows and seven columns. The parity bits numbered 8, 16, ..., 64 are ignored. To generate a new round key we rotate both the C and D registers by one or two bit positions to the left. The number of bit positions depends on the round, and the correspondence is listed in Fig. 2.3. Then, 48 bits of the round key are extracted from the two registers C and D using a table known as *Permuted Choice 2 (PC2)* and this is also described in Fig. 2.3. Looking closely at the table

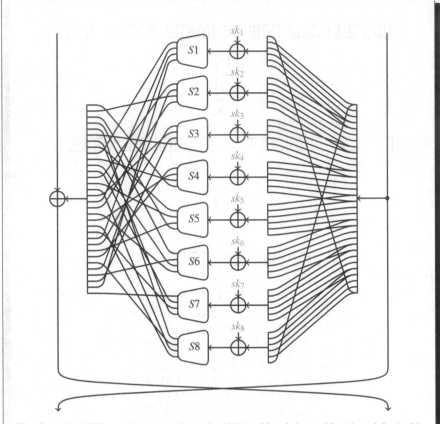

Fig. 2.4 The DES round function where the 48-bit subkey k_i is considered as eight six-bit quantities $\text{sk}_1 \ldots \text{sk}_8$.

we see PC2 extracts 48 bits from the full set of 56 bits held in both registers. So the first bit of a round key is taken from bit position 14; this is the current value at bit position 14 of register C. Bit 25 of a round key is taken from bit position 41; this is the current value of bit 13 (41 minus 28) of register D. Since PC2 extracts 48 bits from 56 the table has eight rows and six columns. Each bit in a round key corresponds to a bit from the user-supplied key and Table 2.3 lists the sixteen DES round keys and gives their composition in terms of the bits in the user-supplied key.

Table 2.3 The bit composition of the DES round keys k_1, ..., k_{16} in terms of their bit position in the user-supplied key $k = \ell_1\ell_2 \ldots \ell_{63}\ell_{64}$.

k_1	k_2
10 51 34 60 49 17 33 57 2 9 19 42	2 43 26 52 41 9 25 49 59 1 11 34
3 35 26 25 44 58 59 1 36 27 18 41	60 27 18 17 36 50 51 58 57 19 10 33
22 28 39 54 37 4 47 30 5 53 23 29	14 20 31 46 29 63 39 22 28 45 15 21
61 21 38 63 15 20 45 14 13 62 55 31	53 13 30 55 7 12 37 6 5 54 47 23

k_3	k_4
51 27 10 36 25 58 9 33 43 50 60 18	35 11 59 49 9 42 58 17 27 34 44 2
44 11 2 1 49 34 35 42 41 3 59 17	57 60 51 50 33 18 19 26 25 52 43 1
61 4 15 30 13 47 23 6 12 29 62 5	45 55 62 14 28 31 7 53 63 13 46 20
37 28 14 39 54 63 21 53 20 38 31 7	21 12 61 23 38 47 5 37 4 22 15 54

k_5	k_6
19 60 43 33 58 26 42 1 11 18 57 51	3 44 27 17 42 10 26 50 60 2 41 35
41 44 35 34 17 2 3 10 9 36 27 50	25 57 19 18 1 51 52 59 58 49 11 34
29 39 46 61 12 15 54 37 47 28 30 4	13 23 30 45 63 62 38 21 31 12 14 55
5 63 45 7 22 31 20 21 55 6 62 38	20 47 29 54 6 15 4 5 39 53 46 22

k_7	k_8
52 57 11 1 26 59 10 34 44 51 25 19	36 41 60 50 10 43 59 18 57 35 9 3
9 41 3 2 50 35 36 43 42 33 60 18	58 25 52 51 34 19 49 27 26 17 44 2
28 7 14 29 47 46 22 5 15 63 61 39	12 54 61 13 31 30 6 20 62 47 45 23
4 31 13 38 53 62 55 20 23 37 30 6	55 15 28 22 37 46 39 4 7 21 14 53

k_9	k_{10}
57 33 52 42 2 35 51 10 49 27 1 60	41 17 36 26 51 19 35 59 33 11 50 44
50 17 44 43 26 11 41 19 18 9 36 59	34 1 57 27 10 60 25 3 2 58 49 43
4 46 53 5 23 22 61 12 54 39 37 15	55 30 37 20 7 6 45 63 38 23 21 62
47 7 20 14 29 38 31 63 62 13 6 45	31 54 4 61 13 22 15 47 46 28 53 29

k_{11}	k_{12}
25 1 49 10 35 3 19 43 17 60 34 57	9 50 33 59 19 52 3 27 1 44 18 41
18 50 41 11 59 44 9 52 51 42 33 27	2 34 25 60 43 57 58 36 35 26 17 11
39 14 21 4 54 53 29 47 22 7 5 46	23 61 5 55 38 37 13 31 6 54 20 30
15 38 55 45 28 6 62 31 30 12 37 13	62 22 39 29 12 53 46 15 14 63 21 28

k_{13}	k_{14}
58 34 17 43 3 36 52 11 50 57 2 25	42 18 1 27 52 49 36 60 34 41 51 9
51 18 9 44 27 41 42 49 19 10 1 60	35 2 58 57 11 25 26 33 3 59 50 44
7 45 20 39 22 21 28 15 53 38 4 14	54 29 4 23 6 5 12 62 37 22 55 61
46 6 23 13 63 37 30 62 61 47 5 12	30 53 7 28 47 21 14 46 45 31 20 63

k_{15}	k_{16}
26 2 50 11 36 33 49 44 18 25 35 58	18 59 42 3 57 25 41 36 10 17 27 50
19 51 42 41 60 9 10 17 52 43 34 57	11 43 34 33 52 1 2 9 44 35 26 49
38 13 55 7 53 20 63 46 21 6 39 45	30 5 47 62 45 12 55 38 13 61 31 37
14 37 54 12 31 5 61 30 29 15 4 47	6 29 46 4 23 28 53 22 21 7 63 39

Table 2.4 Using the DES test vector given in [493], the message 4e6f772069732074 is encrypted to the ciphertext 3fa40e8a984d4815 using the 64-bit user-supplied key 0123456789abcdef. For this encryption, the 16 round keys of 48 bits and the 16 intermediate round outputs are as follows.

round	round output	round key
1	00fe1327c9efe379	0b02679b49a5
2	c9efe379c225d717	69a659256a26
3	c225d7171efc7384	45d48ab428d2
4	1efc738476f2b3de	7289d2a58257
5	76f2b3de10d55380	3ce80317a6c2
6	10d55380e90739fd	23251e3c8545
7	e90739fd572337f0	6c04950ae4c6
8	572337f0cd9968e4	5788386ce581
9	cd9968e4256a96b9	c0c9e926b839
10	256a96b98049c24c	91e307631d72
11	8049c24ca1663aa6	211f830d893a
12	a1663aa6b714e099	7130e5455c54
13	b714e099a3eb2c46	91c4d04980fc
14	a3eb2c46b94da965	5443b681dc8d
15	b94da9651a037d0d	b691050a16b5
16	6091a7a11a037d0d	ca3d03b87032

2.1.1.3 The S-Boxes

The security of DES depends on the S-boxes since these are the only nonlinear components of the cipher. The eight S-boxes are different, but they each take a six-bit input and return a four-bit output. To illustrate their action consider S-box 5 given (using hexadecimal notation) in the following table.

	0	1	2	3	4	5	6	7	8	9	a	b	c	d	e	f
p_0	2	c	4	1	7	a	b	6	8	5	3	f	d	0	e	9
p_1	e	b	2	c	4	7	d	1	5	0	f	a	3	9	8	6
p_2	4	2	1	b	a	d	7	8	f	9	c	5	6	3	0	e
p_3	b	8	c	7	1	e	2	d	6	f	0	9	a	4	5	3

The S-box consists of four rows labeled p_0 through to p_3. Each row has 16 entries and the numbers 0 through to 15 occur once, and only once. Therefore each row represents a permutation of the numbers $\{0, \ldots, 15\}$. The six-bit input is split into two parts. The outer two bits are used to choose a row of the S-box and the inner four bits are used to pick a column of the S-box. The entry identified in this way gives the four bits of output from the S-box. As an example, suppose that the six-bit input to S-box 5 is 001101 in binary notation. Using the two outer bits first we get (in binary notation) the number 01 and so we use the permutation (or row) p_1. We

Table 2.5 The DES S-boxes given in hexadecimal notation.

S1	0 1 2 3 4 5 6 7 8 9 a b c d e f		S2	0 1 2 3 4 5 6 7 8 9 a b c d e f
p_0	e 4 d 1 2 f b 8 3 a 6 c 5 9 0 7		p_0	f 1 8 e 6 b 3 4 9 7 2 d c 0 5 a
p_1	0 f 7 4 e 2 d 1 a 6 c b 9 5 3 8		p_1	3 d 4 7 f 2 8 e c 0 1 a 6 9 b 5
p_2	4 1 e 8 d 6 2 b f c 9 7 3 a 5 0		p_2	0 e 7 b a 4 d 1 5 8 c 6 9 3 2 f
p_3	f c 8 2 4 9 1 7 5 b 3 e a 0 6 d		p_3	d 8 a 1 3 f 4 2 b 6 7 c 0 5 e 9

S3	0 1 2 3 4 5 6 7 8 9 a b c d e f		S4	0 1 2 3 4 5 6 7 8 9 a b c d e f
p_0	a 0 9 e 6 3 f 5 1 d c 7 b 4 2 8		p_0	7 d e 3 0 6 9 a 1 2 8 5 b c 4 f
p_1	d 7 0 9 3 4 6 a 2 8 5 e c b f 1		p_1	d 8 b 5 6 f 0 3 4 7 2 c 1 a e 9
p_2	d 6 4 9 8 f 3 0 b 1 2 c 5 a e 7		p_2	a 6 9 0 c b 7 d f 1 3 e 5 2 8 4
p_3	1 a d 0 6 9 8 7 4 f e 3 b 5 2 c		p_3	3 f 0 6 a 1 d 8 9 4 5 b c 7 2 e

S5	0 1 2 3 4 5 6 7 8 9 a b c d e f		S6	0 1 2 3 4 5 6 7 8 9 a b c d e f
p_0	2 c 4 1 7 a b 6 8 5 3 f d 0 e 9		p_0	c 1 a f 9 2 6 8 0 d 3 4 e 7 5 b
p_1	e b 2 c 4 7 d 1 5 0 f a 3 9 8 6		p_1	a f 4 2 7 c 9 5 6 1 d e 0 b 3 8
p_2	4 2 1 b a d 7 8 f 9 c 5 6 3 0 e		p_2	9 e f 5 2 8 c 3 7 0 4 a 1 d b 6
p_3	b 8 c 7 1 e 2 d 6 f 0 9 a 4 5 3		p_3	4 3 2 c 9 5 f a b e 1 7 6 0 8 d

S7	0 1 2 3 4 5 6 7 8 9 a b c d e f		S8	0 1 2 3 4 5 6 7 8 9 a b c d e f
p_0	4 b 2 e f 0 8 d 3 c 9 7 5 a 6 1		p_0	d 2 8 4 6 f b 1 a 9 3 e 5 0 c 7
p_1	d 0 b 7 4 9 1 a e 3 5 c 2 f 8 6		p_1	1 f d 8 a 3 7 4 c 5 6 b 0 e 9 2
p_2	1 4 b d c 3 7 e a f 6 8 0 5 9 2		p_2	7 b 4 1 9 c e 2 0 6 a d f 3 5 8
p_3	6 b d 8 1 4 a 7 9 5 0 f e 2 3 c		p_3	2 1 e 7 4 a 8 d f c 9 0 3 5 6 b

then consider the inner four bits, which gives us 0110. This is the binary notation of six and so we should look at column six to identify the number that is output. In this particular case (for S-box 5) the entry in p_1 for column six is d in hexadecimal notation and so the output from the S-box (in binary representation) is 1101.

$$S5[001101] = 1101.$$

Each S-box consists of four permutations of the numbers $\{0, \ldots, 15\}$ and across all S-boxes the 32 permutations are different. The S-boxes were designed carefully and we will discuss some of the design features in Sect. 2.2. The full list of DES S-boxes is given in Table 2.5.

2.2 Design Features

The initial design of DES stretched over a period of five years [498]. But the design principles were only published many years later, leading to much speculation and some criticism in the meantime. An early study of DES revealed some peculiarities [289], one of which is the so-called *complementation* property, see Sect. 2.3.1. In reply the NBS observed that this feature could be useful for implementation testing [539]. Others wondered whether there might not be a hidden trapdoor. From [498]: "To answer the criticism the US Senate Committee on Intelligence conducted an investigation into the matter". Some results of the investigation were that DES was free of any known weakness and that the NSA (National Security Agency) had not tampered with the design. However, the investigation failed to satisfy all the critics at that time.

During the late 1970s and throughout the 1980s several studies of DES were published, including [145, 194, 209, 679]. Some of the issues raised included regularities and irregularities in the S-boxes and properties of the P permutation. The fourth S-box was highlighted as suspicious in [289] while [679] singled out the fifth. But it was not until the invention of differential cryptanalysis, see Chap. 6, that real progress was made in reverse-engineering the design principles of DES. Let us start at the beginning.

In response to questions raised at one of the DES workshops, the NBS revealed some of the design criteria of the S-boxes [124].

1. No S-box is a linear or affine function of the input.
2. Changing one bit in the input to an S-box results in changing at least two output bits.
3. The S-boxes were chosen to minimise the difference between the number of 1's and 0's when any single bit is held constant.
4. For any S-box S, it holds that $S[x]$ and $S[x \oplus 001100]$ differ in at least two bits.
5. For any S-box S, it holds that $S[x] \neq S[x \oplus 11rs00]$ for any binary values r and s.

In 1994 Coppersmith [167] confirmed that these were indeed some of the original design criteria for the DES S-boxes. These requirements on the S-boxes result in the following important property of the DES S-boxes.

- If two different 48-bit inputs to the ensemble of eight S-boxes result in equal outputs, then there must be different inputs to at least three neighbouring S-boxes.

The importance of this property will become clear in Chap. 6. Coppersmith also gave another design criterion:

- For any S-box it holds for any nonzero 6-bit value α, and for any 4-bit value β, that the number of solutions (for x) to the equation $S[x] \oplus S[x \oplus \alpha] = \beta$ is at most 16.

The following properties of the P permutation are also known.

1. The four bits output from an S-box are distributed so that they affect six different S-boxes in the following round (four boxes directly and two via the expansion mapping).
2. If an output bit from S-box i affects one of the two middle input bits to S-box j (in the next round), then an output bit from S-box i cannot affect a middle bit of S-box i.

3. The middle six inputs to two neighbouring S-boxes (those not shared by any other S-boxes) are constructed from the outputs from six different S-boxes in the previous round.
4. The middle ten input bits to three neighbouring S-boxes, four bits from the two outer S-boxes and six from the middle S-box (*i.e.*, those not shared by any other S-boxes), are constructed from the outputs from all S-boxes in the previous round.

Coppersmith [167] lists the first two of these as design criteria for the P permutation. They have been identified before, for instance in [194]. The last two are given in [381].

After the publication of differential cryptanalysis [81] it became clear that both the S-boxes as well as the expansion E and the permutation P were designed to increase the resistance of DES to this attack. This was confirmed in [167]. However, it is not clear whether the DES designers were aware of linear cryptanalysis [476, 475], see Chap. 7, which gives the fastest analytical attacks on DES.

There were other design criteria than those given above for the S-boxes, the P permutation, and the E expansion. It is worth observing that if we were to pick the contents of the DES S-boxes at random, we would almost certainly have a weaker cipher. In fact, even if we were to consider the $8! = 40,320$ possible orderings of the S-boxes (with the contents of the S-boxes being unchanged) for the vast majority of orderings the resultant cipher would be weaker than DES. Thus it is not only the contents of the S-boxes that matter, but also their position within the algorithm. Other criteria considered the avalanche of change and the ciphertext bits depend on all plaintext bits and on all key bits after five rounds of encryption [498]. A range of statistical tests was also conducted [414].

In addition to the cryptographic constraints posed on the design of DES, implementation issues were also considered. Since DES was designed for hardware implementation, it was deemed important that the cipher components was chosen so that the number of logical circuits would be kept at a minimum. Early in the design process a pool of possible S-boxes was generated according to the design criteria developed thus far and these boxes were then ranked according to how efficiently they could be implemented. As the process evolved and more constraints were put on the S-boxes, the pool of boxes shrunk and the complexity of the most efficient ones increased. In the end there were three S-boxes with the same most-efficient implementation and seven S-boxes with the same second most-efficient implementation. It seems that the eight S-boxes for DES were chosen from this set of ten [498]. Interestingly, this is an issue that continues to have some significance in contemporary applications [116, 438].

We might also consider the initial and final permutations. Given a ciphertext, an attacker can undo the action of IP^{-1}, and, given a plaintext, he can account for the action of initial permutation (through IP). Thus, in a chosen text attack, IP and IP^{-1} have no cryptographic value. However, the system was designed for hardware implementation, and the purpose of the initial permutation was to provide a suitable ordering of bits when loaded from the input device to the DES chip [498].

Table 2.6 The number of times each bit from the user-supplied key appears in some round key of DES.

Frequency	Bit of the user-supplied key
0	8,16,24,32,40,48,56,64
12	3,42,52,58
13	7,10,12,14,19,23,26,28,29,33,36,38,39,43,45,49,54,55,59,61
14	1,4,5,6,9,11,13,15,17,20,21,22,25,27,30,35,46,51,62,63
15	2,18,31,34,37,41,44,47,50,53,57,60

Turning to the key schedule, the design principles are not public and it is probably fair to say that it is still not fully understood (at least publicly). However, a few properties are recognised.

For instance, the sum of the rotation amounts $r_1, \ldots, r_{16}$ for the C and D registers is equal to 28. This is no coincidence and at the end of an encryption the registers C and D are back at their initial state. The registers are ready for the next encryption. It is also interesting to note that the key schedule can be reversed for decryption, with the register rotations being applied in the opposite directions (to the right).

There is a somewhat irregular appearance to the rotations by one or two bit positions. In particular it may seem odd that the rotation amount in the ninth round is 1. The plausible explanation is that this irregularity avoids the existence of so-called *related keys*; see Sect. 8.5. If, for example, all rotations in the key schedule of DES were set to 2, then the pair of keys k and k^*, where k^* is equal to k rotated by two positions, would have many round keys in common [46, 45]. If, for example, the rotation amounts for r_9 and r_{15} were swapped then there would also be many pairs of keys having many round keys in common [385].

Clearly, each bit in a round key corresponds to a bit from the user-supplied key. Table 2.3 listed the 16 round keys in terms of the bits in the user-supplied key. Note, however, that the key bits of the user-supplied key do not appear equally often in the set of round keys. The frequencies of bits is given in Table 2.6 and we see that some bits occur in only 12 round keys, whereas others are used in 13, 14, and even 15 round keys. Since there are 48 bits in a DES round key, a round key cannot depend on all of the user-supplied key. It is interesting to observe that, for any pair of round keys, at least 54 bits of the (effective) key bits appear in one or both round keys. However, inspection of Table 2.3 reveals that all 56 (effective) key bits appear in either the first, or the last round key, or in both. This can have some consequence for a more advanced form of cryptanalysis, see Chap. 8, and it is a property that does not hold for any other pair of round keys.

2.3 Structural Properties

It is well-known that DES displays some basic structural phenomena. Today these properties would be unwelcome in a block cipher design and, indeed, the AES design requirements were drafted to explicitly rule such properties out, see Chap. 3. That said, while such properties are unfortunate they are not too drastic when DES is used for encryption. They must nevertheless be kept in mind when DES is used as a component in some construction, perhaps in a hash function or within a cryptographic protocol.

2.3.1 The Complementation Property

In this section we let $\overline{m}$ denote the bitwise complement of a binary input block m. We also let $\text{DES}_k(m)$ represent the encryption of message m with DES under the action of key k. Then the *DES complementation property* [289] can be written as

$$\overline{\text{DES}_k(m)} = \text{DES}_{\overline{k}}(\overline{m}).$$

In words, suppose c denotes the ciphertext obtained by encrypting message m with DES and the key k. Then if we were to encrypt the complement of m using the complement of k as the key, then the resulting ciphertext would be the complement of c. In terms of its impact on encryption, this property doesn't buy too much for the cryptanalyst. Nevertheless, there are immediate implications for constructions or protocols where an adversary might have more control over the choice or use of a key. It is therefore always important that the complementation property be considered when DES is used in some higher-level design, see Chap. 4.

While the impact to encryption is limited, it is interesting to observe that the complementation property can be used to halve the cost of exhaustive key search in a chosen message attack [289]. To see this, assume that an attacker knows a message m and its encryption c under an unknown key k that he is trying to recover. Then, since this is a chosen message attack, the attacker can also request the encryption of the message $\overline{m}$, and we will denote the resulting ciphertext c^*. So the attacker knows m, c, and c^* and he is trying to recover k where

$$c = \text{DES}_k(m) \quad \text{and} \quad c^* = \text{DES}_k(\overline{m}).$$

Suppose the attacker tries a value ℓ for k, so he computes $d = \text{DES}_\ell(m)$. He can immediately check whether ℓ is a valid guess by checking to see if $d = c$. However, he should also check whether $\overline{d} = c^*$. By the DES complementation property the cryptanalyst knows that

$$\overline{d} = \text{DES}_{\overline{\ell}}(\overline{m})$$

and so if we have $\overline{d} = c^*$ then we must have that $\overline{\ell}$ is a candidate for k. Since bitwise complementation is much more efficient than encryption, with each DES computa-

Table 2.7 The four weak keys of DES presented in hexadecimal notation with parity bits included. For these keys w, along with any input m, we have that $\text{DES}_w(\text{DES}_w(m)) = m$.

```
0101010101010101      fefefefefefefefe

1f1f1f1f1f1f1f1f      e0e0e0e0e0e0e0e0
```

tion the attacker can rule out two values for the encryption key. We would therefore expect to find the encryption key k after 2^{54} steps instead of the 2^{55} steps we might expect on average during an exhaustive key search.

2.3.2 Weak and Semi-Weak Keys

DES has another unfortunate property. There are four keys for which encrypting twice with the same key leaves the plaintext unchanged! These are typically referred to as *DES weak keys*. More formally, we can describe a DES key w as being weak if the following relationship holds for any input m:

$$\text{DES}_w(\text{DES}_w(m)) = m.$$

There are four DES keys that allow this to happen and these are given in Table 2.7. The explanation is straightforward. In effect, a weak key occurs when encryption is exactly the same operation as decryption. When this occurs, encrypting twice means that all plaintexts will be unchanged since the second encryption undoes the action of the first. The only way encryption and decryption differ for DES is in the order in which the round keys are taken. It is possible that a key k leads to a set of round keys for which reversing their order leaves the actual values used, unchanged. So the round key for round 1 will be the same as the round key for round 16, the round key for round 2 will be equal to the round key for round 15, and so on and so forth. As a result encryption with the key k will be the same as decryption.

Such weak keys have other properties. Each weak key w has 2^{32} fixed points m where $\text{DES}_w(m) = m$. In other words, encrypting a fixed point with a weak key gives exactly the same point (as the name implies). To see this, consider the 2^{32} ciphertexts after eight rounds of encryption for which the two 32-bit halves are equal. For a weak DES key, the round keys of the eighth and ninth rounds are equal. Thus the intermediate texts after seven rounds and after nine rounds of encryption will be equal. Since the round keys of the seventh and tenth rounds are equal, the intermediate texts after six and after ten rounds of encryption are equal. Continuing,

Table 2.8 The six pairs of semi-weak keys for DES presented in hexadecimal notation with parity bits included. For any corresponding pair of keys w_1 and w_2 and any input m we have that $\mathrm{DES}_{w_1}(\mathrm{DES}_{w_2}(m)) = m$.

```
01fe01fe01fe01fe        1fe01fe01fe01fe0        01e001e001e001e0
fe01fe01fe01fe01        e01fe01fe01fe01f        e001e001e001e001

1ffe1ffe1ffe1ffe        011f011f011f011f        e0fee0fee0fee0fe
fe1ffe1ffe1ffe1f        1f011f011f011f01        fee0fee0fee0fee0
```

it follows that the plaintext will be equal to the ciphertext. By similar reasoning it can be shown that these are the only fixed points for DES used with a weak key.

There are also six pairs of *semi-weak* keys. These are defined to be any pair of keys w_1 and w_2 such that for any input m

$$\mathrm{DES}_{w_1}(\mathrm{DES}_{w_2}(m)) = m.$$

There are six possible key pairs that allow this to happen and these are presented in Table 2.8. In other words we have found a pair of keys for which encryption with one key followed by encryption with the other key leaves the plaintext unchanged. If encryption with one key is decryption with another key, then the two sets of round keys generated must have a reverse symmetry. Thus the round key for round 1 with the first key will be the same as the round key for round 16 with the second key. The round key for round 2 with the first key will be equal to the round key for round 15 with the second key, and so on and so forth.

It is interesting to consider how much these structural features matter. In terms of a practical implementation of DES for encryption, such structural features are not that significant. For instance, there are 2^{56} possible keys and only four weak keys, so the chance of picking a weak key at random is 2^{-54} (which is very unlikely). But while the presence of weak keys might not be too important for encryption, just as in the case with the complementation property, their existence can be vital when DES is used within some construction. Indeed, since it is so easy to test for a weak key—after all, the four possibilities are known—the cost of testing is slight and weak keys can easily be prevented from being used in practice.

2.4 DES Variants

There have been many suggestions on how to strengthen DES by modifying components or by using DES in some hybrid construction. Here we restrict ourselves to proposals that are believed to offer good alternatives to the original.

Table 2.9 Common proposals for triple encryption using a generic block cipher.

	Shorthand	Description
2-key ENC-DEC-ENC	EDE_2	$c = \text{ENC}_{k_1}(\text{DEC}_{k_2}(\text{ENC}_{k_1}(m)))$
2-key ENC-ENC-ENC	EEE_2	$c = \text{ENC}_{k_1}(\text{ENC}_{k_2}(\text{ENC}_{k_1}(m)))$
3-key ENC-DEC-ENC	EDE_3	$c = \text{ENC}_{k_3}(\text{DEC}_{k_2}(\text{ENC}_{k_1}(m)))$
3-key ENC-ENC-ENC	EEE_3	$c = \text{ENC}_{k_3}(\text{ENC}_{k_2}(\text{ENC}_{k_1}(m)))$

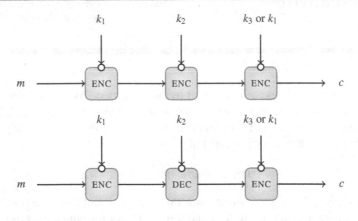

2.4.1 Triple-DES

As the name implies, when we use triple-DES (sometimes referred to in the literature as *TDES*, *TDEA*, or *3DES*) we use DES three times. In principle this means that we could be using an encryption key with an effective cryptographic strength of 168 bits, but the commonly deployed version of triple-DES uses a key of 112 bits. Either way, it is clear that we are now avoiding the problem of the short length of the DES encryption key. However, we are making the issue of software performance even more pressing and we will see that the AES provides an alternative way of balancing these demands.

In Table 2.9 we summarise some different approaches to triple-DES. Perhaps the most common deployment is as *two-key EDE* (EDE_2) where EDE stands for *encrypt-decrypt-encrypt*. Note that the NIST standard [549] specifies the more general form *three-key EDE* (EDE_3) and allows three different keying options for the *bundle* of keys (k_1, k_2, k_3):

1. Choose all three keys k_1, k_2, and k_3 independently.
2. Choose the two keys k_1 and k_2 independently and set $k_3 = k_1$.
3. After choosing k_1 set $k_1 = k_2 = k_3$.

At first sight, using ENC-DEC-ENC as the basis for triple encryption might be a bit surprising and the form ENC-ENC-ENC appears more natural. However, a form based on ENC-DEC-ENC rather than ENC-ENC-ENC allows backward compatibility with single encryption. For instance, if we take EDE_2 or EDE_3 and set all keys equal to the same value k_1 then we have that

$$c = \text{ENC}_{k_1}(\text{DEC}_{k_1}(\text{ENC}_{k_1}(m))) = \text{ENC}_{k_1}(m).$$

We have achieved single encryption using the same circuitry as that used for triple encryption and so the same hardware/software modules could be used for implementations of both DES and triple-DES.

Note that this backward compatibility of EDE with single encryption is independent of the underlying block cipher. In the specific case of DES, however, we can also obtain backward compatibility with a single encryption in the case of EEE_3. For example, if we wish to have $\text{ENC}_{k_3}(\text{ENC}_{k_2}(\text{ENC}_{k_1}(m)))$ being equivalent to $\text{ENC}_{k_1}(m)$ then we could choose $k_3 = k_2 = w$ where w is one of the four DES weak keys. In this case we would have that

$$c = \text{ENC}_{k_3}(\text{ENC}_{k_2}(\text{ENC}_{k_1}(m))) = \text{ENC}_w(\text{ENC}_w(\text{ENC}_{k_1}(m))) = \text{ENC}_{k_1}(m).$$

This clearly depends on having weak keys of the right form and it is, therefore, a DES-specific phenomenon.

One obvious question is why have we moved to triple encryption without considering double encryption; *i.e.*, an encryption of the form EE_2? The short answer is that double encryption does not offer us the gain in security that we might expect (though contrary to many claims it does offer *some* increased security) and a move to triple encryption is preferable. We will leave an analysis of the attacks on double and triple encryption until Chap. 5 but in Table 2.10 we summarise the performance of various attacks on double- and triple-DES. The conclusion is that triple-DES is a very secure cipher that builds upon the good cryptanalytic strength of DES and provides improved resistance to exhaustive search. Triple-DES is supported in a range of industry standards and is very widely deployed.

The main drawback to triple-DES is its performance since, at a naïve level, it is three times slower than DES. However, there are techniques that can be used to speed things up. Suppose that we have a hardware implementation of triple-DES consisting of three independent DES encryption units placed serially. Then instead of waiting for the third unit to finish before starting with the next block of encryption, it is possible to ensure that all three encryption units are active during encryption and we start encryption of a second message that is interleaved with the first. This requires some moderately sophisticated scheduling of the messages, but depending on the mode of use of triple-DES and the application, it may well be a suitable implementation.

Finally, we mention a body of work [348] from the 1980s that attempted to answer one very important question for triple-DES: is the set of DES-generated permutations closed under functional composition? Put a different way, is encrypting twice with DES under different keys equivalent to a single DES encryption under

Table 2.10 Some selected attacks on double- and triple-DES encryption. More details are presented in Chap. 5.

Variant	Time	Precomputation	Memory	Messages
EE_2	2^{112}	–	–	2 known
EE_2	2^{56}	2^{56}	2^{56}	2 chosen
EDE_2	2^{112}	–	–	2 known
EDE_2	2^{56}	2^{56}	2^{56}	2^{56} chosen
EDE_2	2^{120-n}	2^n	2^n	2^n known
EEE_2	2^{112}	–	–	2 known
EDE_3	2^{168}	–	–	3 known
EDE_3	2^{112}	2^{56}	2^{56}	3 chosen
EEE_3	2^{168}	–	–	3 known
EEE_3	2^{112}	2^{56}	2^{56}	3 chosen

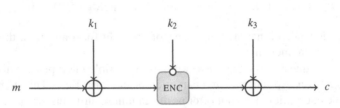

Fig. 2.5 Encryption of message m with DESX under the keys k_1, k_2, and k_3.

a third key? If so, this would imply that multiple encryption with DES would be equivalent to a single encryption. There would also be implications for single encryption since if DES *were* closed in this way then on average one would be able to find a DES key in approximately 2^{28} operations. However, it has been shown that this is not the case and the result was proved using some of the weak-key properties of DES [133, 166] along with a range of ingenious experiments [348].

2.4.2 DESX

DESX was proposed by Rivest [639] and offers a simple and elegant way to increase the security of DES against exhaustive key search. For DESX, additional key material is supplied by the user and this provides what is sometimes referred to as *pre-whitening* and *post-whitening*. As the name suggests, this additional key material is used to disguise the input to, and the output from, the DES operation. Encryption of message m with DESX can be written as

$$\text{DESX}_{(k_1,k_2,k_3)}(m) = k_3 \oplus \text{DES}_{k_2}(m \oplus k_1).$$

Backward compatibility with single-DES can be provided by setting $k_1 = k_3 = 0$.

While such a pre- and post-whitening technique offers only a moderate improvement against advanced cryptanalytic techniques such as differential and linear cryptanalysis, see Chaps. 6 and 7, it does provide an effective way to increase the resistance of the cipher to exhaustive key search. Further, since the computational cost of exclusive-or is negligible compared to the cost of DES encryption, DESX is effectively the same speed as DES and offers a very compelling alternative to multiple encryption. The only slight complication for DESX is the unusual key length of $64 + 56 + 64$ bits. In implementations, however, it is common for the 184 bits to be derived from a more application-friendly key of, say, 128 bits.

It is worth noting that pre- and post-whitening can be used on any block cipher. It has been shown [370, 371] that if F is an idealised block cipher then

$$FX_{k_1,k_2,k_3}(m) = k_3 \oplus F_{k_2}(m \oplus k_1)$$

has an improved resistance against exhaustive key search when compared to F itself. If κ is the key size of F and b is the block size, FX has an effective key length of $\kappa + b - \log m$ when the results of m encryptions (using FX) are available to an attacker. This result also applies to the variant where $k_1 = k_3$. So, for example, if we assume that DES is an idealised block cipher and that 2^{32} encryptions are available to an attacker, then a search to recover the key would have a complexity of at least $2^{120-32} = 2^{88}$ operations.

2.5 DES in Context

Over the years there have been many proposals to modify internal components of DES. This is widely viewed as a dangerous thing to do and it is very hard to come up with a variant of DES that is as strong as the original. Over several iterations Kim proposed a set of eight S-boxes that could be used as a replacement for the originals. Early versions [374, 377] were weaker than DES with respect to differential cryptanalysis [381] though later versions [375] added new design criteria to resist the under-recognised attack of Davies and Murphy [195]. There is other work on providing alternatives to DES [53] and it is an issue that some continue to consider even today [438]. However, generally speaking, such work has had a mixed history and it seems to be very hard to come up with a "better DES than DES".

DES has now been around for more than 30 years. Over that time the cryptographic community has learned to trust the cipher and those, like triple-DES, that are built around it. There are, however, some drawbacks. The key size is insufficient for modern applications and there can be problems in having a block size of only

64 bits, see Chap. 4. Furthermore DES doesn't offer the performance that we often require in software and reflects a time (mid-1970s) before the great explosion in software applications.

During the 1990's many other ciphers, see Chap. 9, adopted different components and architectures as a way of getting a block cipher with much improved software performance. This period of invention and analysis fueled a decade of block cipher development and gave us the successor to DES: the *Advanced Encryption Standard* or, more succinctly, the AES.

2.6 Getting to the Source

Here we give a selection of references to the work of many who have contributed to an understanding of this remarkable cipher. This list is far from exhaustive, but it should provide a good range of pointers to work in the field:

Descriptions and surveys	[414, 435, 545, 691]
Early cryptanalysis	[106, 145, 161, 193, 194, 199, 201, 209, 231, 238]
	[290, 348, 511, 620, 621, 622, 629, 660, 761]
Mature cryptanalysis	[48, 54, 55, 77, 78, 81, 82, 110, 195, 381, 385]
	[394, 420, 475, 476, 477, 617, 669, 682]
Design insights	[46, 124, 128, 129, 133, 166, 167, 196, 347, 679]
	[738]
Variants	[42, 49, 53, 138, 139, 375, 438, 521, 549, 639]
Implementation aspects	[5, 160, 200, 208, 223, 266, 267, 291, 307, 358]
	[441, 711, 712, 726, 727, 746, 751]

Chapter 3
AES

The influence of DES on the development of secret-key cryptography is immense. Its success inspired many closely related designs, though improving on DES was always a rather lofty ambition. Indeed, very few of these DES-relatives offered any serious advantages and many ended up being weaker than the original.

The development of the Advanced Encryption Standard (AES) allowed the cryptographic community to take a major step forward in block cipher design. Just as DES trained a generation of cryptographers, the AES process provided a massive spur to the design, analysis, and implementation of block ciphers. Ironically, the establishment and standardisation of the AES tended to dampen, at least for a while, any new research on block cipher designs. While some researchers continued to analyse the AES, there was little real purpose (or indeed desire) to design alternative block ciphers. The situation has now changed somewhat and new designs for new applications are now being proposed. However, unless there is a catastrophic weakness in the AES, this is the cipher that will be used almost everywhere for the next 20 to 30 years.

Before we describe today's most important cipher, it is worth reviewing the process by which the block cipher *Rijndael* [185] was finally chosen as the AES on October 2, 2000. As we will see in Chap. 9, the 1990s witnessed an explosion in block cipher proposals. Many of these were never taken up, but a fair number were deployed in products and some are in use today. Several of these ciphers contained interesting architectural features that later found their way into other block cipher designs as well as into some of the ciphers in the final round of the AES competition.

Between January 1997 and November 2001, when FIPS 197 was finally published [551], the National Institute of Standards and Technology (NIST) coordinated the search for a DES replacement. Designers had a clean slate in terms of putting together a cipher matching the NIST-specified requirements, and these requirements were few but clear:

1. The cipher should be a single block cipher.
2. The cipher should be available royalty-free worldwide.
3. The cipher should have a public and flexible design.
4. The cipher should offer the security of two-key triple-DES as a minimum.

Some specific parameters were set for the new AES. Among them was the stipulation that a block size of 128 bits should be supported and that any proposal should be able to support at least three key sizes of $\kappa = 128$, 192, and 256 bits.[1]

The AES process was run as an open competition and there were entries from all over the world. Out of 21 entries, 15 were accepted for first round consideration. This first round of evaluation lasted around a year and depended for the most part on input from the cryptographic community. At the end of the first round, five finalists were chosen for second round consideration and, after DES, they are perhaps the most scrutinized block ciphers available today. All finalists had positive and negative attributes, all had distinctive (and elegant) design attributes, and yet they were radically different. The best way to get an appreciation of the full range of issues involved in the AES process—and the enormous amount of work undertaken by NIST, the submitters, and analysts alike—is to look at the final NIST report [550]. The AES winner was the block cipher Rijndael proposed by Joan Daemen and Vincent Rijmen. An elegant cipher, Rijndael offered the most appealing performance profile of the AES finalists and this made it a popular choice among implementors.

In the coming sections we will describe the AES/Rijndael. First though, we need to decide how we are going to refer to the algorithm since there is a subtle difference between them. The submission Rijndael was a more flexible cipher than the final standardised AES since Rijndael accommodates multiple block lengths and multiple key lengths. However, the AES is explicitly restricted to a block length of 128 bits and a key length of 128, 192, or 256 bits. Since there are good technical reasons to have a standardised block cipher with a block length of 256 bits (see Chap. 4), some view this as a missed opportunity[2]. Others, however, focus on the issue of interoperability and stress that it is one thing to have applications negotiate a key length but another, more complicated issue to have applications that support multiple block lengths.

3.1 AES Description

The AES is an SP-network and so there will be clearly defined layers of S-box substitution and diffusion. Perhaps the most striking feature of the AES is the very regular byte structure that is maintained throughout the cipher. Every operation takes place on bytes and this makes it very versatile for implementation. Indeed, encryption with the AES is best described with reference to a square array of bytes. A second notable feature of the AES is that whenever a component can be reused, it is reused, and whenever alternatives are available, the simplest and most efficient option is taken. The net result is a cipher that is very attractive to implementors.

It might not be immediately obvious from the description that follows, but the security of the AES depends on viewing a single byte of data in two different ways.

---

[1] Initial calls from NIST also discussed the requirement for block sizes $b = 128$, 192, and 256 bits.
[2] In fact FIPS 197 [551] mentions (Sect. 6.3) the possibility of supporting different block sizes in future versions of the standard.

The first way is to view a byte as a string of eight bits. This is the representation of a byte that we have used most of the time so far. For those who prefer a mathematical formalization, we are regarding eight bits as an eight-dimensional vector over $GF(2)$. A second way to treat a byte is to view it as representing an element of the field $GF(2^8)$ of 256 elements. Operations that are easy to analyse in one representation can be difficult to cope with (as a cryptanalyst) in the other representation. By mixing operations that combine bytes in different ways a very strong cipher can be derived. More background on this and related issues is provided in [159].

3.1.1 Arithmetic in $GF(2^n)$

Entries in $GF(2^n)$ can be expressed as n-bit strings. Equivalently, they can be viewed as polynomials of degree up to $n - 1$. In this way all 2^n members of $GF(2^n)$ can be written as a polynomial

$$c_{n-1}X^{n-1} + c_{n-2}X^{n-2} + \ldots c_2X^2 + c_1X + c_0$$

where c_0 to c_{n-1} are coefficients that take the values 0 or 1.

The arithmetic defined over $GF(2^n)$ allows us to both add and multiply these values. Adding elements is straightforward. Two polynomials of degree $n - 1$ are added termwise, with the resultant coefficients being computed modulo 2. Thus

$$\begin{aligned}
&(c_{n-1}X^{n-1} + c_{n-2}X^{n-2} + \ldots c_2X^2 + c_1X + c_0) \\
+ &(d_{n-1}X^{n-1} + d_{n-2}X^{n-2} + \ldots d_2X^2 + d_1X + d_0) \\
= &(c_{n-1} \oplus d_{n-1})X^{n-1} + (c_{n-2} \oplus d_{n-2})X^{n-2} + \ldots \\
&+ (c_2 \oplus d_2)X^2 + (c_1 \oplus d_1)X + (c_0 \oplus d_0).
\end{aligned}$$

For multiplication we need to specify the *irreducible polynomial* of degree n that defines the representation of the field $GF(2^n)$. We might denote this polynomial by R_p and we have

$$R_p = X^n + y_{n-1}X^{n-1} + y_{n-2}X^{n-2} + \ldots y_2X^2 + y_1X + y_0.$$

Now when we multiply two entries in $GF(2^n)$ we can view this as multiplying two polynomials of degree $n - 1$. Since the coefficients are all $\{0, 1\}$ the polynomial that results will have $\{0, 1\}$ coefficients. If the degree of this product is $n - 1$ or less, then we are done and we have our entry in $GF(2^n)$. Otherwise, the product contains some nonzero coefficients for the terms X^n or higher. If the largest power of X that appears in the intermediate product is the term X^{n+i} for some $i \geq 0$, then we add the polynomial $R_p \times X^i$ to the intermediate result. This will eliminate the leading term X^{n+i}, while likely changing the values of the lower coefficients. We then look for the next greatest power of X that remains and repeat the process. This is, in effect, a method to reduce the polynomial product modulo R_p.

For the purposes of the AES we will be operating on bytes and so we have $n = 8$. The field representation used for the AES is defined by what is now referred to as the *Rijndael polynomial* where

$$R_p = X^8 + X^4 + X^3 + X + 1.$$

3.1.2 Encryption with the AES

The AES is defined for 128-bit blocks and keys of length 128, 192, and 256 bits. We will refer to these three variants as AES-128, AES-192, and AES-256. The essential differences between these three variants are the number of rounds used for encryption and some slight changes to the key schedule [551].

To begin, the 128-bit input can be written in bits as $b_0 b_1 \dots b_{127}$. These are divided into 16 bytes $B_0 B_1 \dots B_{15}$ where we set $B_i = b_{8i} b_{8i+1} \dots b_{8i+7}$. We know that this byte can be viewed as an element of $GF(2^8)$ and so we can equivalently write byte B_i, for $0 \le i \le 15$, as a polynomial of degree 7:

$$b_{8i} X^7 + b_{8i+1} X^6 + b_{8i+2} X^5 + b_{8i+3} X^4 + b_{8i+4} X^3 + b_{8i+5} X^2 + b_{8i+6} X + b_{8i+7}.$$

Conceptually, we then place these bytes in a square array as follows:

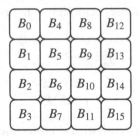

The AES round function uses the following operations: SubBytes transforms each byte of the array using a nonlinear substitution box; ShiftRows ensures that each row of the array is moved by a different number of byte positions; MixColumns mixes each column of the array while AddKey mixes each byte of the array with a byte of subkey material. We will now look at each of these operations in turn.

3.1.2.1 SubBytes

Each byte of the array is transformed by the AES S-box. This takes in an eight-bit quantity and gives an eight-bit quantity as output. Only one S-box is used throughout the cipher and, as we will see, it has been carefully constructed. So an array $m_0, \dots, m_{15}$ is transformed into an array $n_0, \dots, n_{15}$ using the following identities (which we illustrate for $i = 0$ and $i = 11$):

$$n_i = S[m_i] \text{ for } 0 \leq i \leq 15.$$

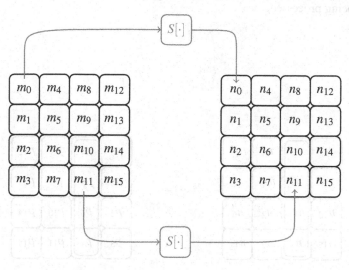

3.1.2.2 ShiftRows

Each row of the array is moved to the left by a different number of byte positions. More precisely, row i is moved by i byte positions for $0 \leq i \leq 3$.

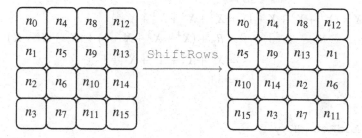

3.1.2.3 MixColumns

Each column of the array is mixed together. To do this, we view the column as a (4×1) column vector of entries in $GF(2^8)$ and we pre-multiply this column vector by the (4×4) $GF(2^8)$-matrix M where

$$M = \begin{pmatrix} 02 & 03 & 01 & 01 \\ 01 & 02 & 03 & 01 \\ 01 & 01 & 02 & 03 \\ 03 & 01 & 01 & 02 \end{pmatrix}.$$

The output will be a (4×1) column vector of entries in $\mathrm{GF}(2^8)$ which replaces the column being processed.

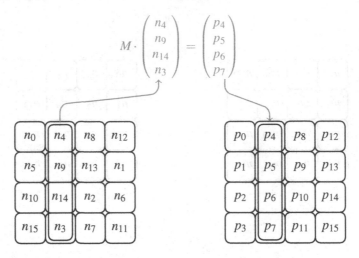

The product of two polynomials for the `MixColumns` operation can be easily described as regular multiplication of polynomials, but with any term of degree greater than 7 being reduced modulo the Rijndael polynomial R_p; see Sect. 3.1.1. Thus, as an example, the product of the two bytes 02 and $a3$ can be computed as

$$
\begin{aligned}
02 \times a3 &= \\
X \times (X^7 + X^5 + X + 1) = X^8 + X^6 + X^2 + X \\
&= R_p + (X^4 + X^3 + X + 1) + (X^6 + X^2 + X) \\
&= X^6 + X^4 + X^3 + X^2 + 1 \\
&= 5d.
\end{aligned}
$$

3.1.2.4 AddRoundKey

At each round, the array of bytes (the *state array*) is combined with a similarly sized array of subkey material. This is derived from the user-supplied key using a key schedule, that will be described in Sect. 3.1.4.

If we denote the current state of the byte array by $p_0 \ldots p_{15}$ and the subkey array as $k_0 \ldots k_{15}$ then the result of `AddRoundKey` is to combine these arrays in a bytewise fashion to give an array $q_0 \ldots q_{15}$ where $q_i = p_i \oplus k_i$ for $0 \le i \le 15$. We illustrate this for $i = 12$.

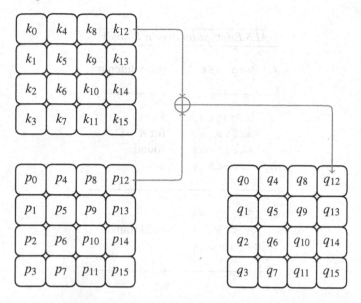

3.1.2.5 Overall Structure for Encryption

We have seen the form of a basic AES round. There are a couple of additional details
to consider. The first is that encryption begins with an iteration of `AddRoundKey`
where the round key in question is part of the user-supplied key material. We will
see this more when we consider the key schedule in Sect. 3.1.4. The second detail
is that the final round of the AES does not use the `MixColumns` operation and this
allows us to write the decryption process in a way that is very similar to encryption.
So, in summary, we have the following form to encryption over n rounds of the AES.
This is also illustrated in Fig. 3.4 on page 49. For AES-128 we have $n = 10$ rounds,
for AES-192 we have $n = 12$ rounds, and for AES-256 we must use $n = 14$ rounds.

AES Encryption over n Rounds	
AddRoundKey	pre-whitening
SubBytes ShiftRows MixColumns AddRoundKey	↑ for $n-1$ rounds ↓
SubBytes ShiftRows AddRoundKey	final round

3.1.3 Decryption with the AES

For each of the operations in a round, it is straightforward to define the corresponding inverse operation.

Encryption	Decryption
SubBytes: use S-box $S[\cdot]$	InvSubBytes: use S-box $S^{-1}[\cdot]$
ShiftRows: rotate to the left	InvShiftRows: rotate to the right
MixColumns: multiply by M	InvMixColumns: multiply by M^{-1}
AddRoundKey	AddRoundKey

To decrypt, just use these operations in the correct order so as to undo the encryption process, making sure to reverse the order of the round keys. Noting that the operation AddRoundKey is self-inverse (*i.e.*, an *involution*), we have the following form to decrypt over *n* rounds.

AES Decryption over n Rounds

```
AddRoundKey
InvShiftRows        undo final round
InvSubBytes
```

```
AddRoundKey         ↑
InvMixColumns       for n − 1
InvShiftRows        rounds
InvSubBytes         ↓
```

```
AddRoundKey         undo pre-whitening
```

Since the operations `InvSubBytes` and `InvShiftRows` operate in a byte-wise fashion they commute with one another. Further, if we make some slight adjustments to the decryption key schedule we can reverse the order of the `AddRoundKey` and `InvMixColumns` operations. This leads us to the following description of the AES decryption process which emphasises a similarity between the forms of encryption and decryption.

Alternative form to AES Decryption over n Rounds

```
AddRoundKey         pre-whitening
```

```
InvSubBytes         ↑
InvShiftRows        for n − 1
InvMixColumns       rounds
AddRoundKey         ↓
```

```
InvSubBytes
InvShiftRows        final round
AddRoundKey
```

For completeness, $S[\cdot]$ and $S^{-1}[\cdot]$ are given in Table 3.1 while the matrices M and M^{-1} are given by

Table 3.1 The AES substitution and inverse substitution boxes $S[\cdot]$ and $S^{-1}[\cdot]$. The value of $S[\mathtt{ab}]$ is given by the entry in row $\mathtt{a}$ and column $\mathtt{b}$.

$S[\cdot]$															
	0	1	2	3	4	5	6	7	8	9	a	b	c	d	e	f
0	63	7c	77	7b	f2	6b	6f	c5	30	01	67	2b	fe	d7	ab	76
1	ca	82	c9	7d	fa	59	47	f0	ad	d4	a2	af	9c	a4	72	c0
2	b7	fd	93	26	36	3f	f7	cc	34	a5	e5	f1	71	d8	31	15
3	04	c7	23	c3	18	96	05	9a	07	12	80	e2	eb	27	b2	75
4	09	83	2c	1a	1b	6e	5a	a0	52	3b	d6	b3	29	e3	2f	84
5	53	d1	00	ed	20	fc	b1	5b	6a	cb	be	39	4a	4c	58	cf
6	d0	ef	aa	fb	43	4d	33	85	45	f9	02	7f	50	3c	9f	a8
7	51	a3	40	8f	92	9d	38	f5	bc	b6	da	21	10	ff	f3	d2
8	cd	0c	13	ec	5f	97	44	17	c4	a7	7e	3d	64	5d	19	73
9	60	81	4f	dc	22	2a	90	88	46	ee	b8	14	de	5e	0b	db
a	e0	32	3a	0a	49	06	24	5c	c2	d3	ac	62	91	95	e4	79
b	e7	c8	37	6d	8d	d5	4e	a9	6c	56	f4	ea	65	7a	ae	08
c	ba	78	25	2e	1c	a6	b4	c6	e8	dd	74	1f	4b	bd	8b	8a
d	70	3e	b5	66	48	03	f6	0e	61	35	57	b9	86	c1	1d	9e
e	e1	f8	98	11	69	d9	8e	94	9b	1e	87	e9	ce	55	28	df
f	8c	a1	89	0d	bf	e6	42	68	41	99	2d	0f	b0	54	bb	16

$S^{-1}[\cdot]$															
	0	1	2	3	4	5	6	7	8	9	a	b	c	d	e	f
0	52	09	6a	d5	30	36	a5	38	bf	40	a3	9e	81	f3	d7	fb
1	7c	e3	39	82	9b	2f	ff	87	34	8e	43	44	c4	de	e9	cb
2	54	7b	94	32	a6	c2	23	3d	ee	4c	95	0b	42	fa	c3	4e
3	08	2e	a1	66	28	d9	24	b2	76	5b	a2	49	6d	8b	d1	25
4	72	f8	f6	64	86	68	98	16	d4	a4	5c	cc	5d	65	b6	92
5	6c	70	48	50	fd	ed	b9	da	5e	15	46	57	a7	8d	9d	84
6	90	d8	ab	00	8c	bc	d3	0a	f7	e4	58	05	b8	b3	45	06
7	d0	2c	1e	8f	ca	3f	0f	02	c1	af	bd	03	01	13	8a	6b
8	3a	91	11	41	4f	67	dc	ea	97	f2	cf	ce	f0	b4	e6	73
9	96	ac	74	22	e7	ad	35	85	e2	f9	37	e8	1c	75	df	6e
a	47	f1	1a	71	1d	29	c5	89	6f	b7	62	0e	aa	18	be	1b
b	fc	56	3e	4b	c6	d2	79	20	9a	db	c0	fe	78	cd	5a	f4
c	1f	dd	a8	33	88	07	c7	31	b1	12	10	59	27	80	ec	5f
d	60	51	7f	a9	19	b5	4a	0d	2d	e5	7a	9f	93	c9	9c	ef
e	a0	e0	3b	4d	ae	2a	f5	b0	c8	eb	bb	3c	83	53	99	61
f	17	2b	04	7e	ba	77	d6	26	e1	69	14	63	55	21	0c	7d

$$M = \begin{pmatrix} 02 & 03 & 01 & 01 \\ 01 & 02 & 03 & 01 \\ 01 & 01 & 02 & 03 \\ 03 & 01 & 01 & 02 \end{pmatrix} \quad \text{and} \quad M^{-1} = \begin{pmatrix} 0e & 0b & 0d & 09 \\ 09 & 0e & 0b & 0d \\ 0d & 09 & 0e & 0b \\ 0b & 0d & 09 & 0e \end{pmatrix}.$$

3.1.4 AES Key Schedule

The key schedule for the AES is computationally lightweight and efficient in terms of memory and performance. It takes as input a user-supplied key of 16, 24, or 32 bytes and returns what is termed an `ExpandedKey` of 16×11, 16×13, and 16×15 bytes respectively. Descriptions of the generation of `ExpandedKey` can be found in [189, 551] while we illustrate the process in Figs. 3.1, 3.2, and 3.3.

For the most part we will be able to describe the key schedule in terms of 32-bit words or four-byte columns of the state array of the AES. Since there are four 32-bit words in each round subkey in the AES, which together give a 128-bit round key, we will denote these words at round i, for $0 \le i \le n$, as $k_{(i,0)} \ldots k_{(i,3)}$, and the subkey for round i is given by $k_{(i,0)} \| k_{(i,1)} \| k_{(i,2)} \| k_{(i,3)}$. So $k_{(i,0)}$ is used as the first column of the round-key array, and so on through to $k_{(i,3)}$, which is used as the last column. For $i = 0$ the first subkey is derived directly from the user-supplied key. For AES-128 we have $n = 10$, for AES-192 $n = 12$, and for AES-256 $n = 14$.

The AES key schedule works in an iterative manner and at each iteration the next t words of key material are computed as a function of the current t words where $t = \frac{\kappa}{32}$ for AES-κ. Thus, for AES-128 we have $t = 4$, for AES-192 we have $t = 6$, while for AES-256 $t = 8$. In all cases, as we have already mentioned, the first t columns of the AES round keys are derived directly from the user-supplied key. For AES-128 this is straightforward and the round key $k_{(0,0)} \| k_{(0,1)} \| k_{(0,2)} \| k_{(0,3)}$ in the first `AddRoundKey` operation is identical to the user-supplied key. For AES-192, four columns of the user-supplied key are used in the first round subkey $k_{(0,0)} \| k_{(0,1)} \| k_{(0,2)} \| k_{(0,3)}$ while the remaining two words form the first two columns $k_{(1,0)}$ and $k_{(1,1)}$ of the second round subkey. For AES-256, the entirety of both the first two round keys are provided by the user-supplied key.

The subkeys for later rounds are derived in different ways depending on the length of the user-supplied key. For AES-128 the subkey $k_{(i,0)} \| k_{(i,1)} \| k_{(i,2)} \| k_{(i,3)}$ is derived from the subkey in the previous round $k_{(i-1,0)} \| k_{(i-1,1)} \| k_{(i-1,2)} \| k_{(i-1,3)}$. At each round there is a simple operation on one 32-bit word (viewed as a set of four bytes). This consists of a byte rotation, an S-box operation (using the same S-box as for encryption), and the addition of an iteration constant r_i, for $i \ge 1$, which is given by $r_i = 02^{i-1}$ in $GF(2^8)$.

For AES-192 the subkey generation is very similar. Since the user-supplied key is larger than the state of the AES, we find for $1 \le i \le 11$ that two successive values of the key registers, each t words long, hold three successive round keys, see Fig. 3.2.

For AES-256 the subkey generation is a bit more involved. It still uses all the same basic operations as those in the key schedules for AES-128 and AES-192, but there is an additional set of operations included. Since the key registers now contain eights words, two successive values provide four successive round keys. For this reason, each iteration of the key schedule for 256-bit keys involves a bit more computation, see Fig. 3.3.

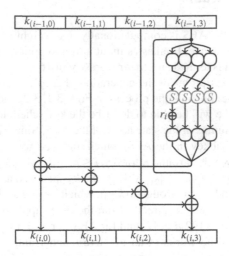

Fig. 3.1 A diagrammatic representation of the key schedule for AES-128. At times the 32-bit words are treated as four bytes, as indicated. The constant r_i for $i \geq 1$ is given by $r_i = 02^{i-1}$ in $GF(2^8)$.

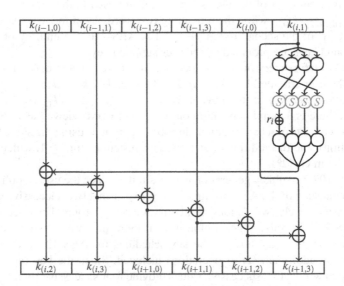

Fig. 3.2 A diagrammatic representation of the key schedule for AES-192. At times the 32-bit words are treated as four bytes, as indicated. The constant r_i for $i \geq 1$ is given by $r_i = 02^{i-1}$ in $GF(2^8)$.

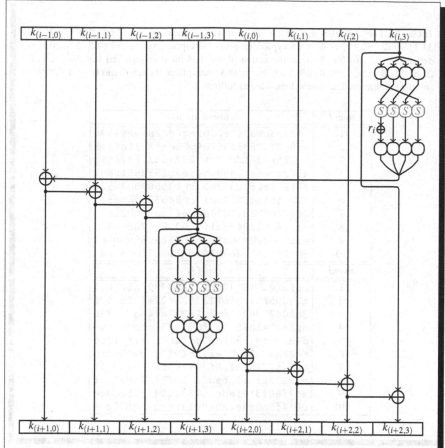

Fig. 3.3 A diagrammatic representation of the key schedule for AES-256. At times the 32-bit words are treated as four bytes, as indicated. The constant r_i for $i \geq 1$ is given by $r_i = 02^{i-1}$ in GF(2^8).

Table 3.2 For the AES test vector given in [551], message 3243f6a8 885a308d 313198a2 e0370734 is encrypted to give the ciphertext 3925841d 02dc09fb dc118597 196a0b32 under the action of the 128-bit user-supplied key 2b7e1516 28aed2a6 abf71588 09cf4f3c. For this encryption, the ten round keys of 128 bits and the ten intermediate round inputs are as follows.

round	round input
1	193de3bea0f4e22b9ac68d2ae9f84808
2	a49c7ff2689f352b6b5bea43026a5049
3	aa8f5f0361dde3ef82d24ad26832469a
4	486c4eee671d9d0d4de3b138d65f58e7
5	e0927fe8c86363c0d9b1355085b8be01
6	f1006f55c1924cef7cc88b325db5d50c
7	260e2e173d41b77de86472a9fdd28b25
8	5a4142b11949dc1fa3e019657a8c040c
9	ea835cf00445332d655d98ad8596b0c5
10	eb40f21e592e38848ba113e71bc342d2

round	round key
1	a0fafe1788542cb123a339392a6C7605
2	f2c295f27a96b9435935807a7359f67f
3	3d80477d4716fe3e1e237e446d7a883b
4	ef44a541a8525b7fb671253bdb0bad00
5	d4d1c6f87c839d87caf2b8bc11f915bc
6	6d88a37a110b3efddbf98641ca0093fd
7	4e54f70e5f5fc9f384a64fb24ea6dc4f
8	ead27321b58dbad2312bf5607f8d292f
9	ac7766f319fadc2128d12941575c006e
10	d014f9a8c9ee2589e13f0cc8b6630ca6

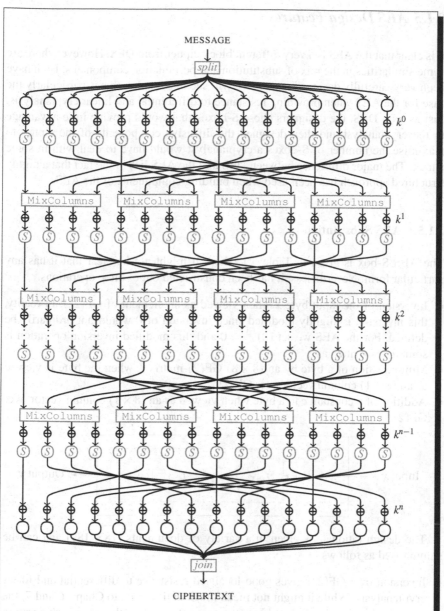

Fig. 3.4 An overview of encryption using the AES. The user-supplied key k is processed using a *key schedule* (*KS*) to derive a set of $n + 1$ *round keys* $k^0 \ldots k^n$.

3.1.5 AES Design Features

It is clear that the AES is a very different block cipher from DES. However, there are some similarities in the mix of substitution and permutation components. Both have been very carefully designed and, as we will see in Sect. 3.2, this is particularly the case for the AES when providing resistance to differential and linear cryptanalysis. Just as with DES, the designers chose S-boxes that would provide little advantage to the cryptanalyst and then designed the diffusive components of the cipher to maximise the number of S-boxes a cryptanalyst would have to deal with in some attack. The major difference between DES and the AES lies in the fact that a highly structured approach has been used when building components of the AES.

3.1.5.1 AES Substitution

The AES S-box is given in Table 3.1 where it might not be clear that it has any particular form. Nevertheless, it has been constructed out of three operations.

- Inversion of the input byte x when viewed as an element of $GF(2^8)$. In reality, this inversion is slightly modified since inversion of 0 would not ordinarily be defined. For the AES we set $(0)^{-1} = 0$ and this modified inversion operation is sometimes written as $x^{(-1)}$.
- Multiplication of a byte by an (8×8) $GF(2)$-matrix L when the byte is viewed as an (8×1) column vector over $GF(2)$.
- Addition of a constant c to a byte when viewed as an (8×1) column vector over $GF(2)$.

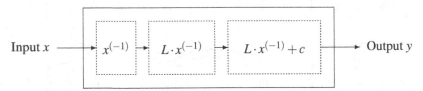

The design rationale is given in a variety of documents [185, 189] and can be summarised as follows.

- Inversion over $GF(2^8)$ gives good localised resistance to differential and linear cryptanalysis. While it might not mean much until we get to Chaps. 6 and 7, the maximum probability for a difference propagation across the inversion operation is 2^{-6} while the maximum bitwise correlation between the input to and the output from the inversion operation is 2^{-3} [185]. Together with the carefully designed diffusion layer, this is going to give excellent security against these attacks. Many of the theoretical foundations for using such algebraic operations within a block cipher construction were established in [568, 564, 569].
- On its own, the inversion operation is algebraically simple. For instance, even though we know that the bitwise difference does not propagate well across the

$GF(2^8)$-inversion operation we do know that if $y = x^{(-1)}$ for some $x, y \in GF(2^8)$ then $yx = 1$ except when $x = y = 0$. To hinder such properties being exploited by an adversary, the designers of the AES immediately follow the inversion operation with a bitwise operation, namely, a transformation of the byte by a $GF(2)$-matrix that is typically denoted by L. This has no adverse effect on the differential and linear properties of the (modified) inversion operation and yet should provide a barrier to viewing the S-box entirely as an operation over $GF(2)$ or $GF(2^8)$. It is in the design of the S-box that we see the conflict between operations in $GF(2^8)$ and $GF(2)$ most pronounced. The $GF(2)$-matrix L is given by[3]

$$L = \begin{pmatrix} 1 & 0 & 0 & 0 & 1 & 1 & 1 & 1 \\ 1 & 1 & 0 & 0 & 0 & 1 & 1 & 1 \\ 1 & 1 & 1 & 0 & 0 & 0 & 1 & 1 \\ 1 & 1 & 1 & 1 & 0 & 0 & 0 & 1 \\ 1 & 1 & 1 & 1 & 1 & 0 & 0 & 0 \\ 0 & 1 & 1 & 1 & 1 & 1 & 0 & 0 \\ 0 & 0 & 1 & 1 & 1 & 1 & 1 & 0 \\ 0 & 0 & 0 & 1 & 1 & 1 & 1 & 1 \end{pmatrix}.$$

- The final operation in the S-box is the addition of a constant c to the byte value that results from the $GF(2)$-matrix multiplication. The value of the constant is set to $01100011 = 63_x$. Why is there a constant? The designers of the AES expressed a concern that without this addition, a zero input to the S-box would provide a zero output. Thus we would have a fixed point. It is unclear how important this would be for security in the overall cipher, but it was felt that it would be better if all fixed points (and any anti-fixed points for which $S[a] = \overline{a}$) were removed [185].

In Sect. 3.2 we will consider the implications of this S-box design. With regards to established attacks such as differential and linear cryptanalysis it is particularly successful. With some other forms of analysis, however, we will see that this construction opens up some interesting opportunities for the cryptanalyst.

3.1.5.2 AES diffusion

Diffusion within the AES is provided by the `ShiftRows` and `MixColumns` operation. Clearly, the two operations complement each other; `ShiftRows` ensures that any pattern or relationship between elements within the same column is disrupted, while `MixColumns` ensures not only that relationships within a column are difficult to track, but also that elements in the same row will be modified differently. Of course, this is the AES, and so while `ShiftRows` is a fairly routine operation, the designers have designed `MixColumns` so that it functions in a particularly useful

---

[3] Note that a different bit ordering means that the specification of L in FIPS 197 [551] differs from that provided in *The Design of Rijndael* [189].

way. Recall that `MixColumns` can be described as pre-multiplication of a (4×1)
$GF(2^8)$-column vector by a (4×4) $GF(2^8)$-matrix M given by

$$M = \begin{pmatrix} 02 & 03 & 01 & 01 \\ 01 & 02 & 03 & 01 \\ 01 & 01 & 02 & 03 \\ 03 & 01 & 01 & 02 \end{pmatrix}.$$

This matrix is derived from the parity-check matrix of an MDS or *maximally distance separable* code. A discussion of MDS codes would take us too far off-topic so more information can be found in [185, 189, 715]. The important feature of the matrix used in the AES is that we can associate what is termed a *branch number* β with the matrix. The branch number β describes the following property. Suppose we have eight bytes $x_0, x_1, x_2, x_3, y_0, y_1, y_2,$ and y_3 satisfying

$$\begin{pmatrix} 02 & 03 & 01 & 01 \\ 01 & 02 & 03 & 01 \\ 01 & 01 & 02 & 03 \\ 03 & 01 & 01 & 02 \end{pmatrix} \begin{pmatrix} x_0 \\ x_1 \\ x_2 \\ x_3 \end{pmatrix} = \begin{pmatrix} y_0 \\ y_1 \\ y_2 \\ y_3 \end{pmatrix}.$$

Then either all x_i and all y_i are zero $(0 \le i \le 3)$ or there are at least β nonzero values among the x_i and y_i $(0 \le i \le 3)$.

It might not be immediately clear why this is important. However, we can at least motivate the use of an MDS matrix in terms of the *avalanche of change*. Suppose we have two inputs to the matrix M that are given by

$$\begin{pmatrix} x_0 \\ x_1 \\ x_2 \\ x_3 \end{pmatrix} \text{ and } \begin{pmatrix} x_0' \\ x_1' \\ x_2' \\ x_3' \end{pmatrix}.$$

Further, imagine that these two inputs are so similar that $x_1 = x_1'$, $x_2 = x_2'$, and $x_3 = x_3'$. We then have that

$$\begin{pmatrix} 02 & 03 & 01 & 01 \\ 01 & 02 & 03 & 01 \\ 01 & 01 & 02 & 03 \\ 03 & 01 & 01 & 02 \end{pmatrix} \begin{pmatrix} x_0 \\ x_1 \\ x_2 \\ x_3 \end{pmatrix} = \begin{pmatrix} y_0 \\ y_1 \\ y_2 \\ y_3 \end{pmatrix} \text{ and }$$

$$\begin{pmatrix} 02 & 03 & 01 & 01 \\ 01 & 02 & 03 & 01 \\ 01 & 01 & 02 & 03 \\ 03 & 01 & 01 & 02 \end{pmatrix} \begin{pmatrix} x_0' \\ x_1 \\ x_2 \\ x_3 \end{pmatrix} = \begin{pmatrix} y_0' \\ y_1' \\ y_2' \\ y_3' \end{pmatrix}.$$

Since multiplication by M is a linear operation over $GF(2)$ we can write

$$\begin{pmatrix} 02\ 03\ 01\ 01 \\ 01\ 02\ 03\ 01 \\ 01\ 01\ 02\ 03 \\ 03\ 01\ 01\ 02 \end{pmatrix} \begin{pmatrix} x_0 + x'_0 \\ x_1 + x_1 \\ x_2 + x_2 \\ x_3 + x_3 \end{pmatrix} = \begin{pmatrix} y_0 + y'_0 \\ y_1 + y'_1 \\ y_2 + y'_2 \\ y_3 + y'_3 \end{pmatrix} \text{ and so}$$

$$\begin{pmatrix} 02\ 03\ 01\ 01 \\ 01\ 02\ 03\ 01 \\ 01\ 01\ 02\ 03 \\ 03\ 01\ 01\ 02 \end{pmatrix} \begin{pmatrix} x_0 + x'_0 \\ 0 \\ 0 \\ 0 \end{pmatrix} = \begin{pmatrix} y_0 + y'_0 \\ y_1 + y'_1 \\ y_2 + y'_2 \\ y_3 + y'_3 \end{pmatrix}.$$

Since the branch number is 5, this means that all the values $y_0 + y'_0, y_1 + y'_1, y_2 + y'_2$, and $y_3 + y'_3$ must be nonzero and that the two outputs generated from two very closely related inputs must differ in every byte.

Of course, we could have deduced this simple case by observing that there are no zero entries in the first column of the matrix M. However, even in the slightly more complicated case of two nonzero input bytes we can be sure that at least three of the output bytes must be nonzero. Thus, when two inputs to the MixColumns layer are very similar in terms of the bitwise difference between them, the outputs will be very different and this should help to amplify even a small amount of "change" or "difference" over successive rounds of the cipher.

Together the ShiftRows and MixColumns operations help to ensure that the number of what are termed *active* S-boxes is either large to begin with, or becomes large very soon. This forms the basis for the very sound security offered by the AES against differential and linear cryptanalysis.

3.1.5.3 AES Performance

It is worth considering the performance of the AES. We have seen that it is heavily structured and all operations are essentially byte operations. Even the bitwise operations within the S-box are hidden from view since the entire S-box substitution can be accomplished with a 256-byte lookup table. This will make the cipher suitable for old, eight-bit processors, and the opportunities for parallelism will mean that performance on more modern 32- and 64-bit processors needn't be hampered. Note that even the coefficients chosen in the MDS matrix M are very simple and suited for implementation in limited environments, though we should note that the coefficients are not quite so nice for M^{-1} in the decryption operation.[4] If there is the luxury of a reasonable amount of memory (as is typical these days) then there are tricks that allow much of the encryption (and decryption) process to be performed by table lookups. All in all, it is a very nice cipher for the implementor and it has a very smooth performance profile across a wide range of platforms.

---

[4] Junod and Vaudenay [341] suggest that different choices of MDS matrix might be better in this respect.

3.2 AES State of the Art

At the time of writing no serious weaknesses have been found in the AES. Certainly with regards to the more traditional forms of cryptanalysis such as differential and linear cryptanalysis—along with their variants and extensions—there seems to be little hope in compromising the cipher.

And while the exceptionally clean structures of the AES—the bytewise operations and opportunities for parallelism—provide some interesting structural opportunities for the cryptanalyst, there are still no attacks that begin to threaten the AES. One exciting avenue for attack came to be termed *algebraic attacks*. But while some first crude attempts were formulated, the community is undecided as to whether such approaches will ever be fruitful.

3.2.1 Differential and Linear Cryptanalysis

This will be a short section since the AES is wonderfully protected against these classic methods of cryptanalysis. For those not familiar with differential and linear cryptanalysis, the details that follow might not be immediately useful without first referring to Chaps. 6 and 7.

In [185] it is shown that any four-round differential trail holds with probability less than 2^{-150} while any four-round linear trail holds with a correlation less than 2^{-75}. Already these are sufficient to thwart basic differential and linear cryptanalysis (see Chaps. 6 and 7).

Since the publication of the Rijndael proposal there has been work in trying to refine the true extent of differential and linear cryptanalytic techniques [155, 306, 359, 361, 362, 589, 588]. But refinement is the operative word and there appears to be little hope for the cryptanalyst in this direction.

In Chap. 8 we will look at some extensions of the basic differential attack. Among these will be attacks that use what are termed *impossible differentials* and *boomerangs*. However, neither impossible differentials [148, 604, 605, 606] nor boomerang attacks [87] make much headway in breaking the cipher.

3.2.2 Structural Attacks

The very clean structure of the AES has been noted by many. And at the time of this writing these provide the best existing attacks on the AES. Nevertheless, they fall far short of compromising the cipher. The best structural attack (perhaps the best attack on the AES in terms of balancing data, memory, and time requirements) is what is known as the *Square attack*. This attack was known to the AES designers [185], but it was first described in [183] where it was used against the block cipher Square [183], a forerunner of the AES.

3.2.2.1 Square Attack

Here we illustrate the basic points of the Square attack and leave the details of the full attack to [183]. Imagine we have a set of 256 plaintexts for which the first byte takes all possible values while the remaining 15 bytes in each text can take any value provided the same value is used in a given byte position across all 256 texts. We will describe a set of texts that have this property as a Λ-set. Imagine that we begin an AES round with a Λ-set. In the following diagram we denote the variable byte-position with $\star$. Consider the actions of SubBytes, ShiftRows, and MixColumns.

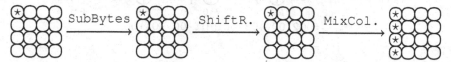

Since the AddRoundKey operation adds the same round key to all of the 256 texts, any Λ-set before AddRoundKey will remain one afterwards. We can therefore consider the next round of transformations.

So at the moment, we know that in each byte position, every possible value is taken once, and only once, as we move through all 256 texts. Now consider the next round.

It looks like we're unable to predict what will happen across the MixColumns operation but this is not quite the case. To illustrate, consider the application of MixColumns in the first column and recall that the $\star$ symbol indicates that every byte value occurs once, and only once, across our 256 texts. Labeling the contents of the first column for the ith text as x_0^i, x_1^i, x_2^i, and x_3^i, for $0 \le i \le 255$, and the outputs after MixColumns as y_0^i, y_1^i, y_2^i, and y_3^i, for $0 \le i \le 255$, we have that

$$\begin{pmatrix} 02 & 03 & 01 & 01 \\ 01 & 02 & 03 & 01 \\ 01 & 01 & 02 & 03 \\ 03 & 01 & 01 & 02 \end{pmatrix} \begin{pmatrix} x_0^i \\ x_1^i \\ x_2^i \\ x_3^i \end{pmatrix} = \begin{pmatrix} y_0^i \\ y_1^i \\ y_2^i \\ y_3^i \end{pmatrix}.$$

If we then add up all the values in the first byte position after the MixColumns operation we note that $y_0^0 \oplus \cdots \oplus y_0^{255}$ is given by

$$y_0^0 \oplus \cdots \oplus y_0^{255} = (02 \cdot x_0^0) \oplus (03 \cdot x_1^0) \oplus (01 \cdot x_2^0) \oplus (01 \cdot x_3^0)$$
$$\oplus (02 \cdot x_0^1) \oplus (03 \cdot x_1^1) \oplus (01 \cdot x_2^1) \oplus (01 \cdot x_3^1)$$
$$\vdots$$
$$\oplus (02 \cdot x_0^{255}) \oplus (03 \cdot x_1^{255}) \oplus (01 \cdot x_2^{255}) \oplus (01 \cdot x_3^{255}).$$

This can be re-written as

$$y_0^0 \oplus \cdots \oplus y_0^{255} = 02 \cdot \bigoplus_{i=0}^{255} x_0^i \oplus 03 \cdot \bigoplus_{i=0}^{255} x_1^i \oplus 01 \cdot \bigoplus_{i=0}^{255} x_2^i \oplus 01 \cdot \bigoplus_{i=0}^{255} x_3^i$$
$$= 02 \cdot 00 \oplus 03 \cdot 00 \oplus 01 \cdot 00 \oplus 01 \cdot 00$$
$$= 00.$$

This means that after the application of `MixColumns` every byte position will be balanced; that is, if we exclusive-or all 256 values in any single byte position after three rounds of AES encryption, we will get zero as a result. This property holds over three rounds of the AES and holds with probability 1. Note that it is entirely independent of the key. This three-round structure can be summarised in the following figure, where $\mathscr{R}$ denotes a full round of computation and s indicates that the sum of the texts in a particular byte can be determined (and in this case is equal to zero).

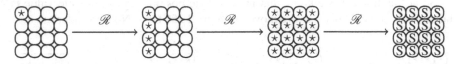

Note that the byte that takes all 256 values in the first round of the three-round structure can be any of the 16 bytes. Consequently, the following structure has similar properties to the one above.

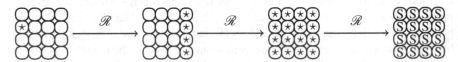

3.2.2.2 Recovering Key Information in the Square Attack

The structures we have identified can be used to attack six rounds of the AES, where the first round consists of the pre-whitening `AddRoundKey` and the last round consists of the final round without `MixColumns`.

The three-round structure is used over rounds 2 to 4. By guessing four key bytes in the first round, four key bytes in the final application of `AddRoundKey` and one key byte in the second last application of `AddRoundKey`, in total nine key bytes, one can compute a candidate value for the sum of the texts in one byte position after four rounds of encryption. For a structure of 256 plaintexts of the form above, this sum is known to be zero. In fact there will be values of the nine key bytes which will return the value of the sum as zero by chance. To eliminate false alarms, the attack needs to be repeated a few times to uniquely determine the correct key bytes. Once the nine key bytes have been found it is relatively easier to find the remaining twelve key bytes of the final application of AddRoundKey, after which the user-selected key can easily be derived.

In total this attack succeeds using roughly 2^{32} plaintexts in time equivalent to 2^{72} encryptions and 2^{32} words of memory [183, 185].

3.2.2.3 Extensions to the Square attack

At first glance it seems that these three-round structures cannot be extended; the next application of SubBytes will destroy the property of the bytes being balanced and there is little that we can say about structural properties as we move forwards through the cipher. But by moving backwards we can gain a further advantage [240, 183].

Consider a group of 2^{32} texts for which the bytes in the first column of the plaintexts take all 32-bit values exactly once when viewed as a four-byte vector. In each of the other 12 byte positions the same constant value is used. These texts can be divided into 2^{24} structures each consisting of 2^8 texts. Each of these structures yields our integral property after three rounds of encryption, that is, the sum of the 2^8 texts in each byte position is zero. But this property also holds over the sum of all 2^{32} texts, which is depicted in the following figure.

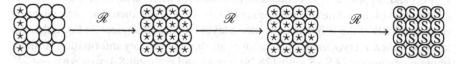

Using the properties of ShiftRows to our advantage, we might consider a structure with 2^{32} texts where the four bytes in the diagonal of the state matrix take each 32-bit value exactly once. It follows that after one round of encryption, the four bytes in the first column will take each 32-bit value exactly once and this is illustrated in the following figure.

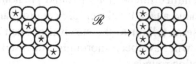

It remains to observe that the structures in the previous two figures concatenate and the resultant four-round structure can be used to attack six rounds of AES more efficiently than the basic Square attack. Instead of guessing nine key bytes, it suffices to guess the value of five key bytes. For each of these values one processes all 2^{32}

texts, but this can be done efficiently. In a clever implementation of this attack one can find the secret key using 6×2^{32} chosen plaintexts in time equivalent to 2^{44} encryptions and 2^{32} words of memory [240, 457].

Some further extensions to the basic Square attack have been noted [240, 457]. These appear to gain some extra rounds for the cryptanalyst, but do so at the expense of a massive increase in work effort. They remain, at best, academic observations on versions of the AES with a larger key size. In fact, while the cryptanalyst might gain an extra round or two by attacking AES-192 and AES-256, he is in fact worse off since AES-192 and AES-256 have two and four additional rounds respectively.

3.2.2.4 Other Structural Attacks

While we will only look at one attack in any specific detail—the Square attack—we note that there are several other attacks that take advantage of the simple and elegant structure of the AES.

A second major style of structural attacks are sometimes referred to as *collision* or *bottleneck* attacks. These require only around 2^{32} plaintexts and exploit a structural phenomenon that manifests itself over three rounds [259, 501]. With additional techniques, this observation can be embedded within an attack on seven rounds of AES-128, though the work effort is commensurate with exhaustive search, and seven rounds of AES-192, with a work effort of 2^{144} operations. Currently the attack doesn't seem to extend.

There is also a form of cryptanalysis that is more generally applicable than the Square attack, but it doesn't reach so far into the cipher and so it is not that successful against the AES. This attack has been explicitly termed *structural cryptanalysis* [97] and is applied to so-called *SASAS* structures where a layer of S-boxes (S) is alternated with an affine diffusion layer (A). In [97] it is shown that even when all specific details of the S-box and the affine layer are key-dependent and/or unknown to the attacker, a cryptanalyst can recover all the relevant key information from an implementation of SASAS with 128-bit blocks and eight-bit S-boxes with just 2^{16} chosen plaintexts and in negligible time. However, again, the attack doesn't extend.

In general, structural attacks seem to be very effective against the AES up to a point. They work exceptionally well over three, four, five, even six rounds. But at six rounds we seem to hit a wall. What is it about the six or seven round boundary that appears to be so difficult to overcome? One way or another these attacks all depend on an easily identified three- or four-round phenomenon being embedded within a five- and six-round attack by clever key counting arguments. And three or four rounds makes sense as a limit to the survival of such strong structures since most localised changes will have been thoroughly diffused across the entire AES array over found encryption rounds.

If there's going to be a weakness in the AES then we will need to find some property that fits with the design of the AES rather than fights against it; for some months the topic of the next section seemed to be a particularly interesting possibility.

Table 3.3 Summary of structural attacks on AES-128, AES-192, and AES-256.

	Rounds	*Texts*	*Time*	*Type*	*Reference*
AES-128	5	2^{30}	2^{31}	Impossible	[74]
AES-128	6	2^{32}	2^{72}	Square	[185]
AES-128	6	2^{92}	2^{122}	Impossible	[148]
AES-128	7	$2^{128} - 2^{119}$	2^{120}	Square	[240]
AES-128	7	2^{32}	$\approx 2^{128}$	Bottleneck	[259]
AES-192	7	2^{32}	2^{144}	Bottleneck	[259]
AES-192	7	2^{92}	2^{186}	Impossible	[605]
AES-256	7	2^{93}	2^{251}	Impossible	[605]
AES-256	8	$2^{128} - 2^{119}$	2^{144}	Square	[240]
AES-256	8	2^{32}	2^{194}	Square	[457]

3.2.3 Algebraic Analysis

The elegant structure and careful design of the AES provide wonderful protection against the established forms of block cipher cryptanalysis. But a significant body of work has examined the algebraic foundations of the cipher with the hope of turning its elegant design into an Achilles' heel.

3.2.3.1 Different Representations

One implication of the clean structure of the AES is the ability to find alternative representations of the AES. Why should these be interesting? There are three major reasons:

1. An alternative representation might offer performance or implementation advantages.
2. An alternative representation might offer new insights into the behaviour of the AES.
3. An alternative representation might provide new perspectives that facilitate cryptanalysis.

Since the AES was established, there have been many papers discussing alternative representations. Some early papers highlighted this approach with the representation of AES encryption using a continued fraction evaluation [241]. Others considered a regrouping of the internal components [533] of the AES so as to highlight certain diffusion properties.

Other representation work has considered the concept of a *dual cipher* [26]. Two ciphers E and E' are said to be dual ciphers if there exist transformations $f(\cdot)$, $g(\cdot)$, and $h(\cdot)$ such that for all messages m and keys k we have

$$f(E_k(m)) = E'_{g(k)}(h(m)).$$

The implications of dual ciphers are explored in [26], where opportunities for cryptanalysis in special situations are described along with the opportunities to provide improved resistance to side-channel cryptanalysis. Among the ciphers dual to the AES were the 240 that are identified [26] by a change in the representation of $GF(2^8)$. This approach was extended in [626] and the potential, but in this case unsuccessful, impact on algebraic cryptanalysis was noted.

The value of these approaches has always been a matter of some debate [186, 533, 532]. But some of these observations [533] have proved to be useful in at least making possible some new types of cryptanalysis such as *algebraic cryptanalysis*.

3.2.3.2 Algebraic Cryptanalysis

The algebraic cryptanalysis of block ciphers is a developing field. With this in mind, we don't intend to go much beyond the basic motivation for this kind of analysis. It should be noted that algebraic approaches to cryptanalysis are typically the only approaches available within the field of public key cryptanalysis. And while such approaches for block ciphers have been hinted at in the literature, for instance, Schaumüller-Bichl [660] proposed a method of formal coding for the cryptanalysis of DES, it was the work of Courtois and Pieprzyk [172] that first brought the possibilities of this style of analysis to a broader audience. Somewhat ironically, progress on the algebraic cryptanalysis of block ciphers has been slow in recent years.[5] One reason behind this stall could be the new mathematical skills and tools that are required to make much headway. Or it could just be a research dead-end. Either way, a greater understanding of algebraic cryptanalysis will develop as researchers consider the full implications of this approach.

While we won't go into details it is easy to give a flavour of what lies behind algebraic cryptanalysis of the AES. First, consider the inversion operation that forms the first component of the AES S-box (see Sect. 3.1.5). The nonzero input a and output b of the inversion operation satisfy the equation $ab = 1$ over $GF(2^8)$. As described in [172], if we write the same equation in terms of the bits of the inputs $a = a_7 a_6 \ldots a_0$ and $b = b_7 b_6 \ldots b_0$ and use polynomials over $GF(2)$ then, provided $a \neq 0$, we would have that

$$(a_7 X^7 + a_6 X^6 + a_5 X^5 + a_4 X^4 + a_3 X^3 + a_2 X^2 + a_1 X + a_0)$$
$$\times (b_7 X^7 + b_6 X^6 + b_5 X^5 + b_4 X^4 + b_3 X^3 + b_2 X^2 + b_1 X + b_0)$$
$$= 1.$$

---

[5] Certainly when we compare this to the field of stream ciphers for which algebraic cryptanalysis seems to be better suited.

If we multiply out the polynomials, reducing the product modulo the Rijndael poly-nomial $X^8 + X^4 + X^3 + X + 1$, then we obtain a polynomial of degree 7 (with eight coefficients) where each coefficient is some combination of the values of $\{a_0 \ldots a_7\}$ and $\{b_0 \ldots b_7\}$. Such terms will be no more complicated than quadratic terms and so the expressions we derive will consist of the sum of products of no more than two starting coefficients. By equating the left-hand and right-hand sides of our equation we can derive a set of eight *multivariate quadratic* equations over GF(2) which connect the input to the output of the inversion operation in a bitwise fashion. For example, we might consider the value of the constant term and we will find that for any nonzero input

$$a_0 b_0 + a_7 b_1 + a_6 b_2 + a_5 b_3 + a_4 b_4 + a_3 b_5$$
$$+ \, a_2 b_6 + a_1 b_7 + a_7 b_6 + a_6 b_7 + a_7 b_5 + a_6 b_6 + a_5 b_7$$
$$= 1. \tag{3.1}$$

By equating coefficients in other bit positions we can derive another seven equations that hold across the inversion operation.

Other observations [172] allow us to increase the number of equations that can be derived across the inversion operation, and hence across the S-box. For instance, across the inversion operation we have that $ab = 1$ over GF(2^8) for input a and in-verse b. But we also have the equation $a^2 b = a$ over GF(2^8), which gives another eight equations. In this way other equations can be harnessed to give us 39 multi-variate quadratic equations holding with probability 1 and Equation (3.1) holding with probability $\frac{255}{256}$.

Since the rest of the encryption process in a round is easily described in a bitwise fashion, for instance see [533], we know that the bits that are input to the inversion operation in one round can be simply described in terms of the bits output from the previous inversion operation. With the equations over the inversion operation itself, we can derive a (large) set of quadratic equations over GF(2) that fully describe the AES encryption process. In [172] it is claimed that recovering the key for AES-128 in a known plaintext setting is equivalent to solving a system of 8,000 quadratic equations in 1,600 variables over GF(2).

When looking at the bitwise equations for the inversion operation in [172], one is struck by their complexity. This is not surprising since inversion is most naturally described over GF(2^8) and we are trying to represent its behaviour in terms of bit-wise operations. While this bitwise perspective is fine for the rest of the S-box and the diffusion layer, it is not ideal across inversion. To avoid this, one way of repre-senting an AES encryption with operations that take place exclusively over GF(2^8) is given in [534]. Considering each of the primitive operations within the AES, ev-erything is byte-oriented and amenable to representation over GF(2^8) except for the map $L(\cdot)$ which is used within the S-box. However, this need not pose too great a problem [534] since we can interpolate the action of the map $L(\cdot)$ by the following polynomial over GF(2^8):

$$L(X) = \mathtt{8f}X^{2^7} + \mathtt{b5}X^{2^6} + \mathtt{01}X^{2^5} + \mathtt{f4}X^{2^4} + \mathtt{25}X^{2^3} + \mathtt{f9}X^{2^2} + \mathtt{09}X^{2^1} + \mathtt{05}X.$$

The reader is referred to [534] for more details about this representation of the AES. The net result is that encryption with the AES can be represented by a large system of $GF(2^8)$ multivariate quadratic equations. The advantage of this system appears to be that we now have a much sparser and more structured system of equations than is derived over $GF(2)$. Whether this system might be more amenable to cryptanalysis than the equivalent system over $GF(2)$ is not known. But it does give an interesting alternative for ongoing and future research.

At the time of this writing noone knows how the equations systems described in this section can be practically solved. While solution methods have been claimed—most famously the XSL approach of [172] which extends the XL method [171, 378]—these have yet to be properly analysed and currently can be viewed as no more than speculation [535]. Algorithms such as Buchberger's algorithm and Faugère's F4 and F5 for Gröbner bases reduction [173, 232, 233] feature when discussing attempts to find an AES key by equation solving [157]. There is considerable overlap and interdependence between many of these algorithms [17] and so the implications of the algebraic approach are not always so clear. However, a better understanding and its possible application to the cryptanalysis of the AES will develop with time. For more details of this particular aspect of AES analysis, the reader is referred to [159].

3.2.4 Related-Key Cryptanalysis

Related-key cryptanalysis allows an adversary to examine the encryption of data under different keys. Some variants allow the attacker to choose the relation between the keys while other variants require that the adversary "only" knows the relation between keys. As an example, one can imagine an attack using encryptions computed using the keys k_0 and $k_1 = k_0 \oplus \alpha$, where the attacker is allowed to pick α, but where k_0 remains secret. There might also be attacks where α is not chosen by the attacker but is known. See Sect. 8.5 for more context to related-key attacks.

In 2000, before Rijndael was chosen as the AES, Ferguson et al. devised a related-key attack which could potentially break AES-256 up to nine rounds [240]. In the years that followed a series of attacks followed this line of research on reduced-round versions of AES variants [68, 243, 324, 305, 763, 372]. Then in 2009 Biryukov and Khovratovich [94] discovered related-key attacks that were able to recover the secret key of (the full) AES-256 using 2^{119} chosen plaintexts and a running time equivalent to the time to compute 2^{119} encryptions using 2^{77} words of memory. This attack requires the adversary to receive encryptions with four different related keys. A similar attack has been devised on AES-192 but with higher complexity. No variant of the attack is known for AES-128. In [93] other variants of this attack are given for reduced-round versions of the two ciphers.

3.2.5 Side-Channel Cryptanalysis

During the AES process Rijndael was often viewed as one of the more promising
AES finalists in terms of resistance to side-channel cryptanalysis. However, since
the publication of the AES various results have shown the AES to be just as vulnerable to side-channel cryptanalysis as other cryptosystems. As such, appropriate
precautions—or a good understanding of the threats—are best advised for any software or hardware implementation of the AES. A good overview of some of the
work in the area can be obtained from the *AES Lounge*, which is maintained by
T. U. Graz [714]. Other interesting work on the impact of using large lookup tables
in fast software implementations is due to Bernstein [41] and Osvik *et al.* [584].
Interestingly the support of low-level AES instructions in new generation Intel processors [270] won't just provide speed advantages; it will also help mitigate certain
forms of side-channel cryptanalysis.

3.3 AES in Context

While the construction of Rijndael appeared to be new to some commentators during the AES process, earlier versions of the cipher had appeared at academic conferences. There are common themes to these AES predecessors—a *Flemish school*
of block cipher design perhaps—and the reader is referred to the designers' own
words [189] in tracing that influence from the cipher Shark [632] in 1995 through
to the AES itself. Unsurprisingly, the success of the AES inspired a range of other
block ciphers [28, 187, 120, 29, 709] and the AES has also been used as the foundation of several hash function designs [40, 253, 415, 559, 369, 312, 272, 298, 60, 141]
that were submitted to the NIST SHA-3 competition [541].

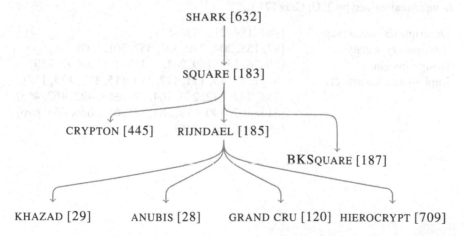

Looking to the future, the purpose behind the AES initiative was to agree upon
a single block cipher for the twenty-first century; one that offered at least as much

security as two-key triple-DES but far better performance (in software). Migrating to a new block cipher is not that easy from an installed base. Further, in the case of the AES and when moving from DES and/or triple-DES, we need to account for differences in the block size as well as the key size. Certainly, most new applications will use the AES. However, interoperability with older applications may require the continued use of other ciphers, particularly triple-DES. However, this is not such a bad outcome and triple-DES is very well trusted and widely used, for example, in the banking sector.

The view of NIST is interesting in this regard. At the fourth AES conference in 2004, Kelsey from NIST outlined the following projected timescale [363] in getting the world to use the AES:

- Single DES is no longer supported even for legacy applications.
- Two-key triple-DES is only supported to 2010.
- Three-key triple-DES is supported to 2020.

In clarifying that NIST finds both triple-DES and the AES equally suitable for federal purposes, publication SP 800-67 [555] makes the following statement:

> Through the year 2030, Triple DES (TDEA) and the FIPS 197 Advanced Encryption Standard (AES) will coexist as FIPS approved algorithms — thus allowing for a gradual transition to AES.

3.4 Getting to the Source

To close this chapter we give a selection of the growing number of references to description and analysis of Rijndael and the AES. A good resource for selected references on implementation of the AES and side-channel cryptanalysis is the *AES Lounge* maintained by T. U. Graz [714].

Descriptions and surveys	[185, 189, 217, 436, 551]
Selected cryptanalysis	[87, 158, 204, 240, 305, 457, 501, 605]
Design insights	[25, 26, 159, 189, 241, 341, 361, 626, 692, 739]
Implementation aspects	[5, 41, 43, 85, 111, 112, 114, 115, 117, 135, 152]
	[153, 237, 242, 263, 264, 275, 368, 422, 462, 465]
	[513, 467, 490, 519, 584, 585, 655, 656, 657, 696]
	[693, 706, 707, 710]

Chapter 4
Using Block Ciphers

A block cipher on its own is rather limited. It takes a b-bit string and outputs a b-bit string under the action of a secret key. But what happens when the message we wish to encrypt is 37 bits long and is too small to fill an entire block? What happens if the message is too long for a single block?

When a block cipher is standardised for widespread use, we need to know how it should be used. Published shortly after the publication of DES, four modes of operation for the DES were specified in FIPS 81 [546]. The modes described in that document are often viewed as the standard modes of block cipher operation:

- Electronic Code Book Mode (ECB).
- Cipher Block Chaining Mode (CBC).
- Cipher Feedback Mode (CFB).
- Output Feedback Mode (OFB).

Similarly, after establishing the AES in 2000 [551], NIST published guidance on how the cipher should be used in special publication SP800-38A [552]. These recommendations included the four standard modes of block cipher operation that are given in [546] and added a fifth:

- Counter Mode (CTR).

The five modes contained in [552] can be used with any block cipher and are likely to cover most block cipher encryption applications and environments.

Actually, the modes of operation of a block cipher provide more than just a practical resolution to the problem of encrypting long messages. Different modes have different security properties and they contribute in different ways to the secure use of a block cipher. Further, they have very different performance attributes that make a block cipher suitable for encryption in a variety of environments.

In Sects. 4.1 and 4.2 we consider different ways of using a block cipher to provide encryption. In the opening chapters of the book we hinted at how useful and versatile block ciphers would turn out to be. In Sects. 4.4 and 4.6 we will explore this issue more and consider how the block cipher might be used as the basis of a message authentication code and a hash function. First, however, we continue to explore how block ciphers can be used for encrypting messages of different sizes.

4.1 Block Encryption

Throughout the following sections we let m_1, m_2, ..., m_n be the n blocks of the message we wish to encrypt.

4.1.1 Electronic Code Book Mode

Electronic Code Book (ECB) is the native mode and one message block is encrypted independently of the encryptions of other blocks. The encryption of multi-block messages is straightforward. For message block m_i and ciphertext block c_i encryption and decryption are given by

$$c_i = \text{ENC}_k(m_i) \quad \text{and} \quad m_i = \text{DEC}_k(c_i) \quad \text{for} \quad 1 \le i \le n.$$

In this description we assume that the final message block m_n is, in fact, a full b-bit block. Questions on how to pad a message to achieve this are addressed in Sect. 4.3.

The ECB mode provides an obvious way to use a block cipher. It might be used for the encryption of keys of fixed length that fit within a single block. However, it is not well suited to the encryption of longer messages and there is always the serious risk of information leakage since equal message blocks are encrypted into identical ciphertext blocks under the same key. One possible way to alleviate this, provided there is spare capacity within the block, is to add some random characters to the data being encrypted. This would help to provide different ciphertext blocks even if the same core data were encrypted. However, such randomisation must be removed in an unambiguous way for successful decryption.

If one does want to use the ECB mode for longer messages then the encryption process is parallelisable; one can encrypt a later block without having to wait for the encryption of an earlier block. Furthermore the mode offers the property of *random access* since any block can be decrypted independently of any other.

When receiving the ciphertext for an ECB-encrypted message there might be two kinds of errors. First, bits of the ciphertext might be incorrect due to a transmission error that has caused a 0 to change to 1 or *vice versa*. Second, the occasional bit might have been dropped altogether. The first type of error in a ciphertext block would mean that the entire corresponding plaintext block would be garbled after decryption (that is each bit would be correct with probability $\frac{1}{2}$). However, the encryption of any other block would be unaffected and so we have error propagation that is contained within the block. For the second type of error, all plaintext will be garbled after decryption until the correct block alignment has been recovered, most

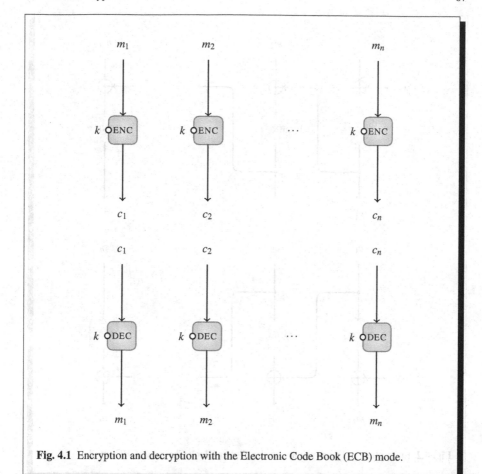

Fig. 4.1 Encryption and decryption with the Electronic Code Book (ECB) mode.

typically by recourse to some form of external block synchronisation. Clearly, a lost ciphertext block results in a lost plaintext block.

4.1.2 Cipher Block Chaining Mode

In the *Cipher Block Chaining (CBC)* mode the encryption of a block depends on the encryptions of previous blocks. The ciphertext blocks c_i are given by

$$c_1 = \mathrm{ENC}_k(m_1 \oplus \mathrm{IV}) \quad \text{and} \quad c_i = \mathrm{ENC}_k(m_i \oplus c_{i-1}) \quad \text{for} \quad 2 \leq i \leq n,$$

where IV is some initial value. The decryption of ciphertext block c_i is given by

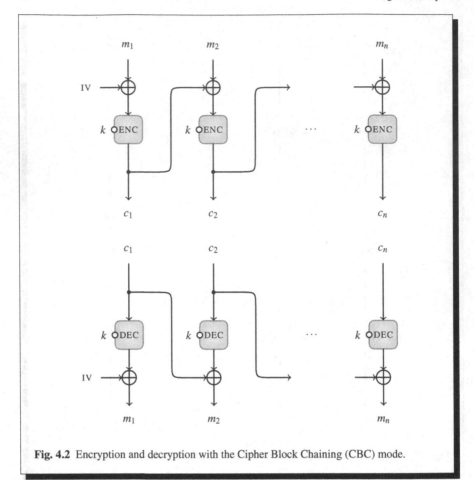

Fig. 4.2 Encryption and decryption with the Cipher Block Chaining (CBC) mode.

$$m_1 = \text{DEC}_k(c_1) \oplus \text{IV} \quad \text{and} \quad m_i = \text{DEC}_k(c_i) \oplus c_{i-1} \quad \text{for} \quad 2 \leq i \leq n,$$

where IV is the same initial value that was used by the sender. In this description we assume that the final message block m_n is, in fact, a full b-bit block. Questions of how to pad a message and how to choose an IV are addressed in Sect. 4.3.

One problem with the ECB mode was that information about the plaintext could easily leak into the ciphertext. Matching ciphertext blocks immediately told the cryptanalyst that matching plaintext blocks were used as input. This provided the motivation to randomise the input, and one convenient way to add unpredictable data to the input is to use the ciphertext that has just been generated. This is the basic idea behind the CBC mode, though for the first plaintext block we are restricted to using a chosen IV. More details on choosing the IV are given in Sect. 4.3. Note that the CBC encryption of any single plaintext block depends on all previous ciphertext blocks and so the encryption process cannot be parallelised; there is no easy way to provide random access or to jump ahead to later parts of the encryption process.

Clearly, a lost ciphertext block or an incorrectly received bit results in a lost plaintext block. For error propagation, an error in the single ciphertext block c_i will ensure that the recovered value of m_i is garbage and after that the recovered value of m_{i+1} will be in error. However, the error pattern in m_{i+1} will be identical to the error pattern in c_i. After this, provided two consecutive ciphertext blocks are received without error and correctly aligned, the second ciphertext block can be correctly decrypted. Thus CBC mode has a self-synchronising property.

One motivation to move away from the ECB mode was the possibility of information leakage. However, the CBC mode also exhibits a limited form of information leakage. Suppose that we have $c_i = c_j$ for some $1 \leq i, j \leq n$ with $i \neq j$. Then we know that

$$c_i = c_j \Rightarrow \text{ENC}(m_i \oplus c_{i-1}) = \text{ENC}(m_j \oplus c_{j-1}),$$
$$\Rightarrow m_i \oplus c_{i-1} = m_j \oplus c_{j-1},$$
$$\Rightarrow m_i \oplus m_j = c_{i-1} \oplus c_{j-1}.$$

Thus, when two blocks of ciphertext are observed to have the same value, we can recover the value of the exclusive-or of two unknown plaintext blocks and information is leaked. For a b-bit block cipher, the birthday paradox (see Fig. 4.11) means we might expect to see such a repetition in the value of a ciphertext block after around $\sqrt{2^b}$ blocks. While this problem is not as immediate as for the ECB mode, it is an issue that motivated calls for a larger block size when moving from DES to the AES.

4.2 Stream Encryption

Three modes of operation of a block cipher effectively turn the block cipher into a stream cipher. While there is a substantial literature on the design of stream ciphers, see [643] for some contemporary approaches, it remains useful to be able to use a block cipher in the kind of environments where stream cipher properties might be preferred. One motivation for considering a stream cipher mode might where we need to encrypt single characters, say, instead of whole blocks. For this purpose the CFB, OFB, and CTR modes would all be suitable.

4.2.1 Cipher Feedback Mode

We will first consider is the *Cipher Feedback (CFB)* mode. In this mode, t bits are encrypted with each call to the block cipher where $1 \leq t \leq b$. Here we will use MSB_t and LSB_t to denote the t most and least significant bits, respectively, of a block. In the description of the CFB mode we assume that each message and ciphertext block is t bits in length. The issues of padding m_n and the choice of an initial value are addressed in Sect. 4.3. The encryption of block m_i is given by

$$c_i = m_i \oplus \mathrm{MSB}_t(\mathrm{ENC}_k(x_i)) \quad \text{and} \quad x_{i+1} = \mathrm{LSB}_{b-t}(x_i)\|c_i \quad \text{for} \quad 1 \leq i \leq n$$

where x_1 is some chosen initial value IV. Decryption has a similar form to encryption. The decryption of block c_i for $1 \leq i \leq n$ is given by

$$m_i = c_i \oplus \mathrm{MSB}_t(\mathrm{ENC}_k(x_i)) \quad \text{and} \quad x_{i+1} = \mathrm{LSB}_{b-t}(x_i)\|c_i \quad \text{for} \quad 1 \leq i \leq n$$

where x_1 is the same IV as that used by the sender.

The generation of t bits of ciphertext requires one block cipher invocation, so when $t < b$ the encryption rate of CFB will be slower than that achieved in the ECB or CBC modes. Further, there are no opportunities to parallelise the encryption process for greater throughput. That said, the CFB mode does have characteristics that make it interesting for some applications.

One property of the CFB mode is in the way errors are handled. The CFB mode effectively turns the block cipher into a *self-synchronising* stream cipher. An error in some CFB-encrypted ciphertext block c_i will be inherited by the corresponding plaintext block m_i that is recovered. And, since x_{i+1} will contain the incorrect c_i, the recovery of subsequent message blocks will be garbled until the source register x_j for some $j > i$ is free from the influence of c_i. This will happen when c_i has been shifted out of the register and so at most $\lceil \frac{b}{t} \rceil + 1$ plaintext blocks will be garbled by a single ciphertext error. So, provided sufficiently many ciphertext bits are received without error, correct decryption can be recovered.

Note that if bits of the ciphertext are dropped, then the CFB mode requires the realignment of the t-bit boundaries for resynchronisation to be successful. In the case of $t = 1$ this would happen as soon as b bits are recovered correctly and there would be no need to maintain synchronisation between the sender and the receiver by external means.

4.2.2 Output Feedback Mode

The *Output Feedback (OFB)* mode is a second stream cipher mode, but here encryption does not depend on previous ciphertexts. If we follow FIPS 81 [546] then the OFB mode has a parameter t that specifies the size of the feedback value used to update each x_i. However, for security reasons that are outlined below, we set $t = b$ and use a full feedback value. This is reflected in SP 800-38A [552] and this is the description that we give here. Following SP-800-38A the OFB mode is used to provide s bits of keystream at each iteration of the encryption function where $s = b$ for all but the last block of message when $s \leq b$ bits might be used depending on the size of the input block. Thus the OFB encryption of block m_i is given by

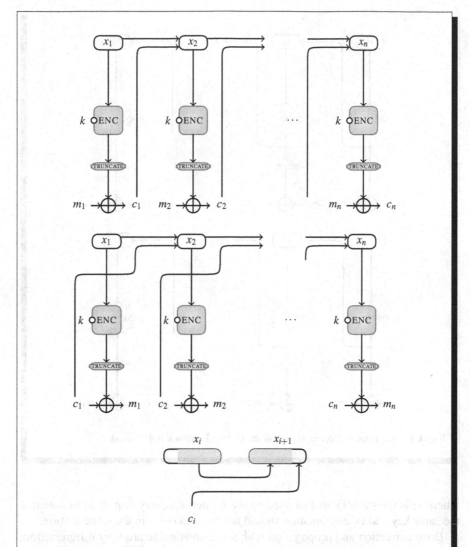

Fig. 4.3 Encryption and decryption with the Cipher Feedback (CFB) mode. The update of the state register from x_i to x_{i+1} is shown in greater detail (see text). First $b - s$ bits of x_i are shifted to the left and then c_i is used to replace the missing bits on the right.

$$c_i = m_i \oplus \text{MSB}_s(\text{ENC}_k(x_i)) \quad \text{and} \quad x_{i+1} = \text{ENC}_k(x_i) \quad \text{for} \quad 1 \leq i \leq n$$

where x_1 is some chosen initial value IV. Decryption is identical to encryption and the decryption of ciphertext block c_i is given by

$$m_i = c_i \oplus \text{MSB}_t(\text{ENC}_k(x_i)) \quad \text{and} \quad x_{i+1} = \text{ENC}_k(x_i) \quad \text{for} \quad 1 \leq i \leq n$$

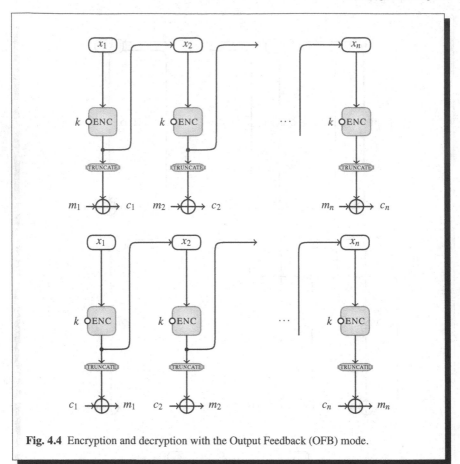

Fig. 4.4 Encryption and decryption with the Output Feedback (OFB) mode.

where x_1 is the same IV as that used by the sender. It is very important to note that the same key and IV combination should not be used twice in the same session.

Error correction and propagation within the OFB mode are very different from that experienced in the CFB mode. In the OFB mode, bits received in error lead directly to erroneous bits in the plaintext. But there is no error propagation. If ciphertext bits are dropped then alignment between the keystream generated by the receiver and that generated by the sender will be lost unless externally controlled resynchronisation takes place. The essential difference between the CFB and the OFB mode is that the CFB mode is a *self-synchronising* stream cipher (by virtue of the ciphertext feedback mechanism) while the OFB mode is an example of what is called a *synchronous* stream cipher.

In the OFB mode the feedback is independent of the message and ciphertext blocks. Note that while FIPS 81 allows some flexibility in this regard, we only allow full-length feedback in the keystream generator and we have that $x_{i+1} = \text{ENC}_k(x_i)$. While this has an obvious performance benefit, it is an important precaution for adequate security in the OFB mode. If we did not have full feedback then the feedback function would not be a permutation mapping but would be a random mapping. The consequence of this change would be relatively short cycles of length about $\sqrt{2^b}$ instead of an expected cycle length of 2^{b-1} for the permutation mapping [198]. Since secure use of a stream cipher depends on the keystream not repeating, full feedback is required in the OFB mode.

4.2.3 Counter Mode

The *Counter (CTR)* mode is a third stream cipher mode. Like the OFB mode, encryption at any instance does not depend on the values of previous plaintexts and ciphertexts. Following SP-800-38A [552] the CTR mode is used to provide s bits of keystream at each iteration of the encryption function where $s = b$ for all but the last block of message. In the last block $s \leq b$ bits might be used depending on the size of the input block. In its most general form, the encryption of block m_i is given by

$$c_i = m_i \oplus \text{MSB}_s(\text{ENC}_k(x_i)) \quad \text{and} \quad x_{i+1} = \text{INCREMENT}(x_i) \quad \text{for} \quad 1 \leq i \leq n$$

where x_1 is some chosen initial value IV and the operation INCREMENT provides the value of x_{i+1} as a function of x_i. This is easily done by means of a simple integer counter but a variety of options are available. Typically, and for obvious efficiency reasons, we set $s = b$ though one may choose a subset of $s < n$ bits if desired. Decryption is identical to encryption and the decryption of block m_i is given by

$$m_i = c_i \oplus \text{MSB}_t(\text{ENC}_k(x_i)) \quad \text{and} \quad x_{i+1} = \text{INCREMENT}(x_i) \quad \text{for} \quad 1 \leq i \leq n$$

where x_1 is the same IV as that chosen by the sender. It is very important to note that the same key and IV combination should not be used twice in the same session.

As with the OFB mode, error correction and propagation within the CTR mode are very different to that seen in the CFB mode. Since the CTR mode is a synchronous stream cipher (like the OFB mode) if bits are dropped during transmission then alignment between the keystream being generated by the receiver and that generated by the sender will be lost unless resynchronisation takes place. As in the OFB mode, the feedback in the CTR mode is independent of the message and ciphertext blocks. Transmission errors in a ciphertext block affect only the corresponding plaintext block. One important difference between the OFB and the CTR modes is the random access property offered by the CTR mode. At any point during the encryption process we can specify the value of the counter needed to decrypt a given block of ciphertext. With the OFB mode we have little alternative but to generate the entire intermediate keystream sequence.

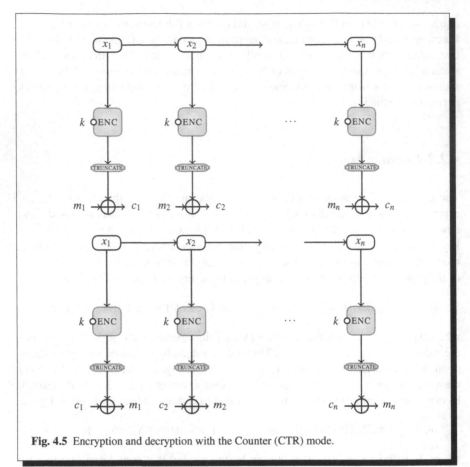

Fig. 4.5 Encryption and decryption with the Counter (CTR) mode.

4.3 Starting and Finishing

The modes of operation allow us to use a block cipher to encrypt messages that are longer than a single block. So far we haven't discussed how to deal with messages that are not exact multiples of the block size in length, *i.e.,* finishing. Here we provide some background for this issue but, since certain modes require the use of an *initial value (IV)*, we first consider how to start.

4.3.1 Choosing the IV

The CBC, CFB, OFB, and CTR modes all require the use of an initial value, alternatively called an initialisation value or vector, IV.

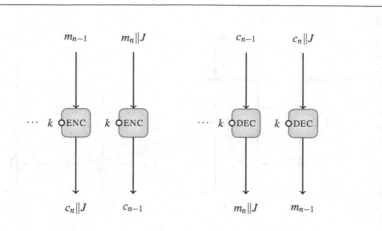

Fig. 4.6 Encryption and decryption with ciphertext stealing for the *Electronic Code Book* mode.

In the case of the OFB and CTR modes, reuse of the same key and IV combination will lead to the same keystream being generated. This would be catastrophic for security. So while the IV in the OFB and CTR modes can be known to some adversary, it should be used only once. Such an IV is sometimes referred to as a *nonce* and it can be implemented in a variety of ways, including as a simple counter.

By contrast, for the CBC and CFB modes we require the IV to be unpredictable. One way to achieve this would be to use a PRE-IV which can be a predictable nonce such as a counter. Then the IV that is used for the actual encryption operation can be generated as

$$IV = \text{ENC}_k(\text{PRE-IV}).$$

While SP-800-38A [552] allows the same encryption key to be used for both the IV generation and the encryption process, other theoretical work suggests that the IV be generated as

$$IV = \text{ENC}_{k'}(\text{PRE-IV})$$

using some key k' that is easily derived from the key k that is used for the encryption.

4.3.2 Padding

For OFB and CTR modes we don't need to think about padding. We just generate sufficient keystream to encrypt the message and throw away keystream we don't need. There is therefore no message expansion when using these modes.

Indeed, the same holds when we encrypt with 1-bit CFB mode. However, for s-bit CFB, CBC, and ECB modes we might need to pad some input block to a multiple of

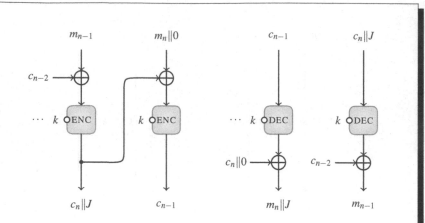

Fig. 4.7 Encryption and decryption with ciphertext stealing for the *Cipher Block Chaining* mode.

s bits in the case of CFB mode and a multiple of b bits in the cases of CBC and ECB. A variety of padding methods have been proposed in different standards. Work by several researchers [105, 600, 724] has shown the dangers of using inappropriate padding schemes under different attack models. The best padding scheme, and one that is championed by Black and Urtubia on theoretical grounds [105], is to append a 1 bit to the end of the message, starting a new message block if required, and then appending as many 0 bits as required to form an s- or b-bit block. This padding mechanism provides no ambiguity to the receiver in how to parse the recovered message after decryption.

Note that since we are considering the padding of incomplete blocks in ECB, CBC, and CFB modes we might be concerned that this causes a limited amount of message expansion. One way to provide secure encryption in ECB and CBC modes without padding a message, and therefore without any message expansion, is to use what is called *ciphertext stealing* [498].

4.3.3 Ciphertext Stealing

Ciphertext stealing is a very elegant way to securely encrypt messages whose length is not an exact multiple of the block size, and to do so when any message expansion cannot be tolerated. Ciphertext stealing does, however, require that the message be greater than b bits in length; that is, there must be at least two blocks in the encryption process.

Recall the encryption equations for ECB mode. For message block m_i and ciphertext block c_i encryption and decryption are given by

$$c_i = \text{ENC}_k(m_i) \quad \text{and} \quad m_i = \text{DEC}_k(c_i) \quad \text{for} \quad 1 \leq i \leq n.$$

Let us now suppose that $|m_n| < b$ and we do not wish to use padding. The last two encryption operations are illustrated in Fig. 4.6.

We compute $c_{i-1} = \text{ENC}_k(m_i)$ as normal, but consider this block as $c_n \| J$ where $|c_n| = |m_n|$ and J is $b - |m_n|$ bits of some temporary quantity. We then form a modified final block $m'_n = m_n \| J$ and we see that the modified m'_n is exactly b bits in length. We can then encrypt this block as normal to give a ciphertext block that we denote by c_{n-1}. The final two blocks of output are then given by c_{n-1} and c_n. For the receiver decryption is straightforward. The block c_{n-1} is received before c_n and so, on decryption, it provides $m_n \| J$. The receiver uses J to form an input ciphertext block $c_n \| J$ and on decryption this reveals m_{n-1} and the message can be easily reconstructed.

A very similar trick is used for CBC encryption and this is illustrated in Fig. 4.7. Recall that the encryption equations for c_i are given by

$$c_1 = \text{ENC}_k(m_1 \oplus \text{IV}) \quad \text{and} \quad c_i = \text{ENC}_k(m_i \oplus c_{i-1}) \quad \text{for} \quad 2 \leq i \leq n,$$

where IV is an initial value.

Now suppose that $|m_n| < b$. For the encryption of the last two blocks we proceed as follows. We compute $\text{ENC}_k(m_{n-1} \oplus c_{n-2})$ and denote the output as $c_n \| J$, where $|c_n| = |m_n|$ and J is an internal variable of length $b - |m_n|$ bits. We then generate a temporary last block $m'_n = m_n \| 0$, where the block ends with $b - |m_n|$ 0 bits. Computing $\text{ENC}_k(m'_n \oplus [c_n \| J])$ gives a b-bit output that we denote by c_{n-1}. The last two blocks of ciphertext that are sent to the receiver are c_{n-1} and c_n, which are b and $|m_n|$ bits in length respectively. The receiver can recover the correct plaintext as follows. On decrypting c_{n-1} we compute $\text{DEC}_k(c_{n-1}) \oplus [c_n \| 0]$, which yields $m_n \| J$. We then append the quantity J to c_n and compute $\text{DEC}_k(c_n \| J) \oplus c_{n-2}$, which yields m_{n-1}. Thus the last two blocks of the message m_{n-1} and m_n can be readily recovered.

Some mechanisms for ciphertext stealing are described by NIST [556].

4.4 Authentication

Block ciphers are used extensively in the provision of data authentication, and many *message authentication codes* (MACs) have been built around them. Different proposals might have only minor differences between them, and different strategies within standardisation groups has led to a confusing abundance of proposals. In this section we will try to concentrate on the ones that are most often deployed as well as looking at some newer, more efficient, proposals.

When using a message authentication code, the sender and receiver share a secret key k and both can compute an authentication tag $\mathcal{T}$ for the message that they exchange. The sender computes this tag, which is typically appended to the message. The receiver can then check that the received message has not been altered by computing a candidate tag $\mathscr{T}'$ under the action of the same, shared, secret key k and comparing $\mathscr{T}'$ with $\mathscr{T}$. A match authenticates the message. The security properties we demand of a message authentication code can be formalised [33] but we will present them informally as follows:

1. Given a set of message-tag pairs $(M_1, \mathscr{T}_1), \ldots, (M_n, \mathscr{T}_n)$ for a set of messages M_1, $\ldots, M_n$ that might be chosen adaptively by some attacker, it should be infeasible for an adversary to recover the secret key k that was used to create the tags.
2. Given a set of message-tag pairs $(M_1, \mathscr{T}_1), \ldots, (M_n, \mathscr{T}_n)$ for a set of messages M_1, $\ldots, M_n$ that might be chosen adaptively by some attacker, it should be infeasible for an adversary to provide a valid message-tag pair $(M', \mathscr{T}')$ for a message M' that has not previously been seen.

The first property encapsulates the idea of a *key recovery* attack. An adversary should not be able to recover the secret key that was used for the message authentication, and if the key k is κ bits in length, then our aim is to ensure that recovering the secret key requires a work effort of 2^{κ} operations.

The second property covers the idea of a *MAC forgery*. Clearly, key recovery leads to MAC forgery. However, in a weak scheme, it might be possible for an adversary to forge a tag for an unseen message without recovering or using any information about the secret key. Since a MAC can always be guessed, if the message authentication tag $\mathscr{T}$ is t bits in length then our aim is to ensure that a successful MAC forgery attack requires a work effort of 2^t operations.

4.4.1 CBC-MAC

When we look at the block diagram for the CBC mode, it is interesting to note that the encryption of the last block of plaintext is dependent, in a complex way, on the encryption of all the text that has preceded it. In some sense, the last block of ciphertext provides a cryptographic accumulation of the entire encryption process. We capture this intuition with the CBC-MAC and to compute the message authentication code we encrypt the plaintext using the CBC mode but we throw away the intermediate ciphertexts $c_1, \ldots, c_{n-1}$. So the processing of block m_i is given by

$$c_1 = \text{ENC}_k(m_1 \oplus \text{IV}) \quad \text{and} \quad c_i = \text{ENC}_k(m_i \oplus c_{i-1}) \quad \text{for } 1 \leq i \leq n$$

where, in its most general setting, IV has some chosen initial value. After the last block of message has been encrypted, there is an additional operation PROCESS that we will consider in more detail later. Thus we have that $\mathscr{T} = \text{PROCESS}(c_n)$. We note that some applications might optionally truncate this to arrive at a final message authentication tag that might be shorter than the block length. This process is illustrated in Fig. 4.8.

This method of forming a MAC when DES is the underlying block cipher is described in both FIPS 81 [546] and FIPS 113 [547]. In both cases the IV is set

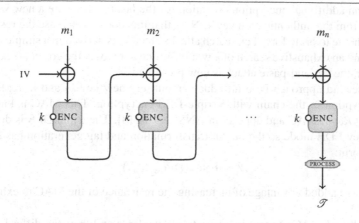

Fig. 4.8 Computing the CBC-MAC using the *cipher block chaining* mode. Different standards will specify a different final PROCESS as well as, potentially, truncation in the last step.

to zero, and there is no final PROCESS operation. However, the tag $\mathcal{T}$ might be optionally truncated by taking the necessary number of bits from the most significant part of c_n. In FIPS 113 the padding is defined as appending as many zeros as required to fill a full 64-bit block. When used with messages of variable length, however, this simple CBC-based MAC is not secure.

4.4.1.1 The Need for an Additional Process

The fact that we need some additional process to prevent a MAC forgery can be shown by a variety of *cut-and-paste* attacks. The following example is one of the simplest. Suppose that an adversary requests to see the message authentication tag for the single block message M_1 and further suppose that no additional process is used after the last encryption operation. Then we have that $\mathcal{T} = \text{ENC}_k(M_1 \oplus IV)$. This means that the two-block message $M_1 \| M_2$ where $M_2 = M_1 \oplus IV \oplus \mathcal{T}$ also has authentication tag $\mathcal{T}$. Thus an adversary is able to present a message-tag pair for a previously unseen message $M_1 \| M_2$. We need to prevent this.

4.4.1.2 The Choice of Additional Process

There are a variety of approaches that serve to mitigate the threat of cut-and-paste attacks. The first features in ISO 9797 [315]. Here the output transformation and tag generation has the form

$$\mathcal{T} = \text{ENC}_{k_2}(c_n).$$

We add an additional encryption operation for the last block where a new key k_2 is derived from the authentication key k. Note that this doesn't increase the resistance of the scheme to brute force key search attacks since k_2 is derived in a simple manner from k and an exhaustive search of k would suffice to recover the correct (k, k_2) pair. However, the cut-and-paste attack is now prevented.

The second approach is to introduce an entirely new key k_2 and to replace the last encryption in the chain with a triple-DES encryption. This MAC is known as the *ANSI Retail MAC* and appears in ANSI X9.19 [8]. The triple-DES is deployed in two-key EDE mode so the output transformation and tag generation has the following form:

$$\mathcal{T} = \text{ENC}_k(\text{DEC}_{k_2}(c_n)).$$

This has the added advantage of increasing the resistance of the MAC to exhaustive key search.

CBC-based MACs appear in a variety of standards and they can differ in subtle ways from one another. Particular block ciphers might be mandated in one standard and not another, different padding rules might be approved for use, and the option of truncating the output from the final transformation might also be supported in different ways. A full survey of the range of possibilities is outside our scope. However, the importance of the CBC-based MAC should be evident from the number of variants it has inspired.

4.4.2 OMAC

OMAC [320] is a simple MAC based on cipher block chaining. The culmination of a chain of increasingly efficient and elegant proposals such as EMAC [602], XCBC [102], and TMAC [423], OMAC deals with the problem of providing a provably secure CBC-based MAC algorithm for messages of any length, and does so in a manner that is particularly efficient in terms of the amount of key material required. After a flurry of activity, which included plans to adopt OMAC as a standardised MAC [544] and an ill-fated draft version of SP-800-38B [553] featuring a version of RMAC [330], NIST currently supports a version of CBC-MAC. More background on all these MAC efforts, along with the other modes of operation of a block cipher, can be found on the NIST website devoted to mode development [543].

The computations involved in OMAC are illustrated in Fig. 4.9. It is immediately clear that it is based on the CBC mode of operation with the IV set to zero. The construction is very simple, with two forms for the MAC computation, depending on whether the final block m_n fills an entire b-bit block or not. The final tag can be optionally truncated if a tag that is smaller than the block size is required. It is readily observed from Fig. 4.9 that two constants are introduced prior to the last invocation of the underlying block cipher. These are defined in the following way.

First we set $L = \text{ENC}_k(0)$ where k is the key used to compute the MAC and 0 is the all-zero input block. Then the b-bit output L is viewed as an element of $\text{GF}(2^b)$,

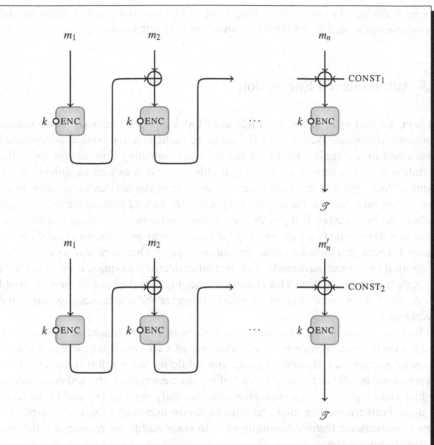

Fig. 4.9 Computing a message authentication code with OMAC. See the accompanying text for a description of CONST$_1$ and CONST$_2$. The upper chain of computation is used when $|m_n| = b$. The lower chain of computation is used when $|m_n| < b$ and we set $m'_n = m_n \| 1 \| 0 \cdots 0$ with sufficiently many zeros (possibly none) to fill a block.

see Sect. 3.1.1 for more background to field arithmetic. Here, the underlying field is defined using one of the polynomials suggested in [320] and the specific choice depends on the block size of the block cipher:

$$b = 64 \qquad f(X) = X^{64} + X^4 + X^3 + X + 1$$
$$b = 128 \qquad f(X) = X^{128} + X^7 + X^2 + X + 1$$
$$b = 256 \qquad f(X) = X^{256} + X^{10} + X^5 + X^2 + 1.$$

Multiplication of field elements in $GF(2^b)$ is then used to derive CONST$_1 = L \cdot X$ and CONST$_2 = L \cdot X^2$.

While a glance at Fig. 4.9 suggests that OMAC is a very elegant construction, its attractiveness will probably not be clear without digging into the surrounding research. Suffice it to say that the frugal use of key material and an accompanying proof of security make the OMAC construction very attractive.

4.5 Authenticated Encryption

In Sect. 4.4 we looked at how we might use a block cipher as the basis for a message authentication code. We saw that the message authentication code was computed as a result of a complete pass over the data, accumulating information about the plaintext along the way. If we were to provide encryption as well as authentication using a block cipher then it would seem reasonable to assume that we would require two passes over the data: the first to encrypt and the second to compute the message authentication code tag. Is it possible to combine these two operations to give some form of *authenticated encryption* (AE)? It will come as no surprise to find there are many different proposals for authenticated encryption. One particular style of mode is referred to as a *one-pass* mode. That is, authenticated encryption is provided with a single pass over the data. This equates to roughly half the computation as would be required for a more obvious approach, however these schemes are somewhat overlooked.

Instead, attention often focuses on a variety of *two-pass* schemes for authenticated encryption, including the obvious one of two completely separate operations of encryption and authentication. But while the work effort in using a two-pass scheme is little different from that of separate encryption and authentication, a tighter bundling of these two operations can simplify the way key and IV material is used. Furthermore, we might be able to obtain improved *proofs of security* for new constructions, thereby providing us with some additional reassurance that the constructions are sound.

From all the discussion of different authentication and encryption modes, some additional (and desirable) properties have been highlighted. For instance we might like to have some header information, say, that needs to be authenticated but not encrypted. In such a situation it would be possible to process the header without being required to decrypt the entire message first. Such schemes are referred to as *authenticated encryption with associated data* (AEAD) schemes.

With regards to performance and implementation, we might be concerned about whether the mode can be parallelised. We might also be interested in the property of being *online*, that is authentication and encryption can begin as soon as any part of the message is available, without the entire message (or even the length of the message) being known. As a consequence of all these tangential and orthogonal properties, the cryptography literature contains numerous proposals that differ in the number of passes, differ in their performance, and differ in their construction and promotion. Since this area is very active we will not go into too much detail. However, we will highlight two proposals that have gained some prominence; the first has already been widely adopted while the second is included because of its elegance.

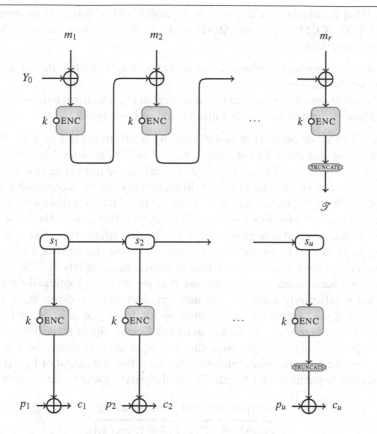

Fig. 4.10 Authenticated encryption with the CCM mode [554]. See the accompanying text for a description of the variables, but the upper chain of computation is used to derive an authentication tag $\mathscr{T}$ while the lower chain of computation is used to encrypt the payload using counter mode encryption. The final ciphertext sent to the receiver is given by $c_1 \| c_2 \| \cdots \| c_u \| (\mathscr{T} \oplus \text{ENC}_k(s_0))$.

4.5.1 CCM Mode

The *Counter with Cipher Block Chaining-Message Authentication Code (CCM)* Mode [554] bundles the counter mode for encryption with the cipher block chaining mode for message authentication. While combining these two operations into a single description, CCM remains a two-pass mode of use; the data is processed using the cipher block chaining mode to set the message-dependent value of a cryptographic accumulator, and then the data and accumulator are encrypted using the counter mode. This mode features in NIST Special Publication SP800-38C [554] as well as within the IEEE Standard 802.11 for wireless LANs [310]. The CCM mode [554] is defined for use with a block cipher with 128-bit blocks, essentially the AES. The CCM mode is an AEAD mode, so it operates on the secret key and the three quantities:

- *Nonce N*, a non-secret value that should not be repeated while the same encryption key is being used,
- *Associated data A*, which is to be authenticated but not encrypted, and
- *Payload P*, the data to be both authenticated and encrypted.

The CCM mode operates in two phases and is illustrated in Fig. 4.10. First we format all the data items into one string viewed as 128-bit blocks. So we have that $(N\|A\|P) = m_1\|m_2\|\cdots\|m_r$ for some r and clearly some formatting rule is required (and provided [554]) so that inputs of different sizes can be accommodated. Next we compute a cryptographic accumulation of the information in this formatted string using a variant of cipher block chaining. We set $Y_0 = \text{ENC}_k(m_0)$. The value $\mathcal{T}$ is a cryptographic accumulation of the material in the data string $m_1\|\cdots\|m_r$ and will be encrypted as part of the ciphertext. If desired, $\mathcal{T}$ can be shorter than a single block in which case only the necessary number of most significant bits of $\mathcal{T}$ are used.

Next we need to encrypt the payload P along with the (optionally truncated) $\mathcal{T}$. This is effectively accomplished using counter mode encryption. A rule for initialising s_0 and incrementing a counter s_i for $1 \leq i \leq u$ is required (and provided [554]), where u is the smallest number of block cipher invocations required to encrypt $P = p_1\|p_2\|\cdots\|p_u$. Note that p_u might not necessarily be a complete block, but since we are using counter mode encryption the output of $\text{ENC}_k(S_u)$ can be truncated to the necessary length. The final ciphertext sent to the receiver is

$$(\overbrace{c_1\|c_2\|\cdots\cdots\|c_u}^{\text{encrypted payload}} \| \overbrace{\mathcal{T} \oplus \text{ENC}_k(s_0)}^{\text{encrypted tag}}).$$

The actions of the receiver in decrypting and authenticating the message should be obvious and details can be found in [554]. While the CCM mode is accompanied by a *proof of security* [331], it is not without its critics and an alternative, the EAX mode [37], has been proposed.

4.5.2 EAX Mode

The EAX mode [37] is another mode for a block cipher that provides both authentication and encryption. Like CCM it is a two-pass mode; however, unlike CCM, it is *online* meaning that encryption and authentication can begin without having the entire message or even the length of the message at hand. As in the CCM mode we work with three quantities:

- *Nonce N*, a non-secret but non-repeating value,
- *Associated data A*, referred to as *Header* in [37], which is to be authenticated but not encrypted, and

The first person in the group has a birthday on a given day. To avoid a match, the second person is left with 364 options while the third has only 363 and so on, and so forth. The probability, therefore, that we don't have any matches among a group of t friends is

$$1 \times \frac{364}{365} \times \frac{363}{365} \times \cdots \times \frac{365-t}{365} = \frac{\prod_{i=0}^{i=t-1}(365-i)}{365^t}.$$

This means that the probability p that we have *at least* one match is given by

$$p = 1 - \left(1 \times \frac{364}{365} \times \frac{363}{365} \times \cdots \times \frac{365-t}{365}\right) = 1 - \frac{\prod_{i=0}^{i=t-1}(365-i)}{365^t}.$$

This can be evaluated for given values of t, and for $t = 23$ we have $p = 0.507$. For the more general case—where we have t people and n possible "birthdays"—it can be shown that

$$p \approx 1 - e^{-\frac{t^2}{2n}}.$$

So when there are n possibilities, after making approximately $t = \sqrt{n}$ choices it is more likely than not that we will see a collision.

Fig. 4.11 The *birthday paradox* refers to the following problem; how many people need to be in a room before it is more likely than not at least two people share a birthday?

- *Payload P*, the data to be both authenticated and encrypted.

At a high-level, the EAX mode is used as follows. First we compute two authentication tags $\mathcal{T}_N$ and $\mathcal{T}_A$, where these are the authentication tags for the nonce N and the associated data A respectively. These are computed using a message authentication code called OMAC (see Sect. 4.4) and are very closely related to the CBC mode of operation. These two instantiations of OMAC are differentiated from each other in a very simple way [37]. We then encrypt the payload P using the block cipher in counter mode where the counter is initialised with $\mathcal{T}_N$. This gives us the ciphertext $c_1 \| c_2 \| \cdots \| c_u$ and we compute a third authentication tag $\mathcal{T}_C$ on the ciphertext in much the same way as we computed $\mathcal{T}_N$ and $\mathcal{T}_A$. The authentication tag $\mathcal{T}$ for the entire message is then the (optionally truncated) quantity $\mathcal{T} = \mathcal{T}_N \oplus \mathcal{T}_C \oplus \mathcal{T}_A$. The final ciphertext is

$$\overbrace{(\ c_1 \| c_2 \| \cdots \cdots \| c_u}^{\text{encrypted payload}} \ \| \mathcal{T}).$$

The actions of the receiver and good background discussions on this mode can be found in [37].

4.6 Hashing

A hash function $h(\cdot)$ is a keyless algorithm that takes a variable length input x and returns an output y of fixed length n. It does so while satisfying the following properties:

- Given an output y it should be hard to find any input x such that $h(x) = y$. This is referred to as *preimage resistance*. In the absence of an analytical weakness, the effort required to compromise this property is expected to be 2^n operations.
- Given an associated input-output pair x_1 and y such that $h(x_1) = y$ it should be hard to find a second input x_2 such that $h(x_2) = y$. This is referred to as *second preimage resistance*. In the absence of an analytical weakness, the effort required to compromise this property is expected to be 2^n operations.
- If should be hard to find any pair of inputs x_1 and x_2 such that $h(x_1) = h(x_2)$. This is referred to as *collision resistance*. In the absence of an analytical weakness, the so-called birthday paradox, see Fig. 4.11 means that the effort required to compromise collision resistance is expected to be $2^{\frac{n}{2}}$ operations.

More details on hash functions, along with some dedicated designs, can be found in most cryptographic textbooks; see, for example, [493].

In this section, we are interested in using a good block cipher to build a good hash function. There are two reasons why we might like to do this. First, if we already have a deployed block cipher then it would be handy to build a hash function out of this component. In this way we might be able to leverage trust in the block cipher to make claims about the security of the hash function. Second, if we intend to implement both a block cipher and a hash function in some application, then we can save both space and implementation effort by reusing some of the components that we must anyway implement. However, the downside of building a hash function out of a block cipher is, typically, that hash functions of a dedicated design are faster than those based on a block cipher.

4.6.1 Three Important Constructions

Three hash function constructions are particularly important. The reader is likely to look at these constructions and consider alternative ways of connecting the basic operations. However, an exhaustive analysis of such simple constructions [104] has confirmed that these are the best ones to use. Before describing them in more detail, we need to establish a little bit of the theory behind hash function design.

4.6.1.1 The Iterated Construction

To accommodate an input of any length, many hash functions adopt an iterative design.[1] With this approach, an input $\mathcal{M}$ is divided into fixed-sized pieces $m_1 \| m_2 \| \cdots \| m_t$ which will be compressed one after the other. A *chaining variable* h_i, for $0 \le i \le t$, is used to carry the result of the computation forward at each intermediate step. The value of the chaining variable for the first iteration, h_0, is fixed and specified as a part of the hash function specifications. Then, at each iteration of what is termed the *compression function*, the current value of the chaining variable and the next block of input to be hashed are compressed together, with the output being the next value of the chaining variable. After the final part of the input (along with any specified padding) has been compressed, the hash function output $\mathcal{H}$ is typically the final value of the chaining variable though some further processing, such as truncation, is sometimes specified.

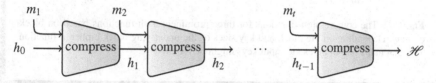

It is now very natural to ask how we might build a compression function from a block cipher.

4.6.1.2 The Compression Function

At any iteration of the compression function, we have essentially three quantities to consider:

- the current value of the chaining variable h_{i-1},
- the next value of the chaining variable h_i, and
- the word, or fixed quantity, of message m_i to be processed at this iteration.

A block cipher has two inputs, the message and the key, and in addition to these inputs we might imagine that we can use a simple operation such as exclusive-or to combine any of these quantities together. We might also like to incorporate a constant s within the computation. Thus, the output of the compression form can have the form $\mathrm{COMPRESS}(h_{i-1}, m_i) = \mathrm{ENC}_a(b) \oplus c$, where a, b, and c can each take one of the values $h_{i-1}, m_i, h_{i-1} \oplus m_i, s$, giving $4^3 = 64$ possibilities.

These were first studied by Preneel *et al.* [615] and proofs of security (and examples of insecurity) for all 64 variants were later systematically treated by Black *et al.* [104]. There are twelve arrangements that are deemed to be secure, of which three are particularly important. These are illustrated in Fig. 4.12.

---

[1] Since hash function design is a large and diverse field, we restrict our attention to the class of constructions exemplified by the *MD-family*. This is sufficient for our purposes.

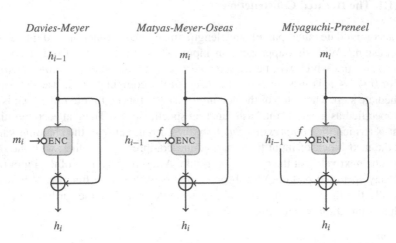

Fig. 4.12 The compression functions for three prominent hash functions based on block ciphers. Depending on the block and key sizes of the underlying block cipher, a function $f(\cdot)$ that maps output blocks to input keys might be required.

The three most prominent examples are known as Davies-Meyer, for which there is no easy citation (but see [493] for more background information), Matyas-Meyer-Oseas [484] and Miyaguchi-Preneel [613, 506]. More details about these constructions, and other variants, can be found in different sources [493, 613, 104], but here we mention some important points.

4.6.1.3 Structural Properties

While weak keys and complementation properties might not be immediately relevant to encryption when using a block cipher (though much depends on the exact form of the property in question) they can be vitally important for the security of a hash function built around a block cipher. The hash functions described here have been proven to be secure under the black-box model; *i.e.,* that we assume that the adversary cannot take advantage of any additional properties or weaknesses of the underlying block cipher. While this should be an appropriate assumption for most good block ciphers, DES is the classic example of a block cipher that must be used with care in some higher-level construction.

4.6.1.4 Block and Key Lengths

Both DES and two out of three instances of the AES do not use keys and blocks of the same length. Thus, if we are using a hash function construction such as Matyas-Meyer-Oseas or Miyaguchi-Preneel, we will need to specify exactly how an output block giving the next value of the chaining variable is to be mapped to a suitable key for the next iteration of the compression function. Note that for the Davies-Meyer construction this is not an issue since the message to be hashed is used as the key and the chaining variables are the input to and output of the compression function. Nevertheless, a fully specified hash function will need to address issues such as appending padding bits and other information to ensure that hashing is secure.

4.6.1.5 Performance

It is worth considering the performance of the three variants in Fig. 4.12 in terms of the underlying block cipher. To do this we define what is termed the *rate* of the construction. Suppose that the underlying block cipher has a b-bit block size and a κ-bit key size. The rate of the hash function is the proportion of a b-bit message block that is hashed in a single block cipher invocation. A rate of 1 means that the hashing performance is (essentially) similar to the encryption performance, a rate less than 1 means that the hashing is slower than the encryption, and a rate greater than 1 means that the hashing is potentially faster than the encryption. The rates of the different schemes for some block cipher choices are given below.

Cipher	Davies-Meyer	Matyas-Meyer-Oseas	Miyaguchi-Preneel
DES	$\frac{7}{8}$	1	1
AES-128	1	1	1
AES-192	$\frac{3}{2}$	1	1
AES-256	2	1	1

Note that there is some hesitation in saying that a rate greater than 1 necessarily means that the hashing is faster than the encryption. For each of the constructions in Fig. 4.12 a new "key" is used at each iteration of the compression function. Thus there is an additional computational overhead in repeatedly generating new subkeys when comparing hashing and encryption performance. Depending on the complexity of the key schedule this can be substantial.

4.6.1.6 Security

For the hash function constructions we have seen so far, the final output from the hash function is either the chaining variable or some truncation of that chaining variable. However the chaining variable is the output from the block cipher and so the hash output is limited in size to at most b bits. Looking back to the properties of a hash function, we see that the resistance of such a hash function to collision attacks will be no greater than $\sqrt{2^b}$. For a 64-bit or even a 128-bit block cipher such as DES and the AES, this is unlikely to provide sufficient security for future applications. The problem for DES, however, is particularly acute since a collision search that requires around 2^{32} operations is not exacting and hasn't been for some considerable time.

While we do have a 256-bit block version of Rijndael, the forerunner to the AES, this version of the cipher is not standardised [551] and will be rarely supported in software or hardware. So there has been, and continues to be, considerable interest in what are termed *double block length constructions*. Here we use the block cipher in a slightly more complicated arrangement but we hope the resultant hash function offers a resistance of 2^b operations against collision search attacks.

4.6.2 Double Block Length Constructions

To provide resistance to collision search attacks a single output block from commonly used block ciphers is not sufficient. Instead we require a hash output that consists of (at least) two blocks of output from the block cipher.

Two important constructions—MDC-2 and MDC-4—are intended to do this and were designed for use with DES [497, 698, 397]. They are built around the Matyas-Meyer-Oseas construction, with the MDC-4 compression function essentially consisting of two invocations of the MDC-2 compression function. In what follows we concentrate on a description of MDC-2, and while designed for DES, this general construction can be used with other good block ciphers.

The compression function for MDC-2 is illustrated in Fig. 4.13. As we see in the case of DES, the same 64-bit message block is used as an input to two parallel invocations of the Matyas-Meyer-Oseas hash function. For the first call of the compression function the chaining variables are set to the values

$$g_0 = \{ \texttt{0x5252525252525252} \} \quad \text{and} \quad h_0 = \{ \texttt{0x2525252525252525} \}.$$

The two functions $f_1(\cdot)$ and $f_2(\cdot)$ are chosen so that

- $f_1(x) \neq f_2(x)$ for any x,
- $f_1(g_0) \neq f_2(h_0)$, and
- $f_1(x)$ and $f_2(y)$ can never be a DES weak key for any inputs x and y.

The two parallel strands of hashing are mixed by swapping the rightmost 32-bit halves of the two strands after the Matyas-Meyer-Oseas hashing operation (see Fig. 4.13). The security of the MDC-2 and MDC-4 constructions has been studied closely [493, 613, 397], and by considering structural properties of DES, the work effort required to compromise MDC-2 and MDC-4 when based on DES is given below [613, 397].

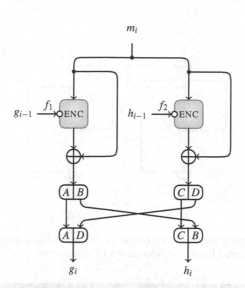

Fig. 4.13 The compression function for MDC-2. Since MDC-2 was designed for DES the functions $f_1(\cdot)$ and $f_2(\cdot)$ map 64-bit output blocks to 56-bit input keys by omitting every eighth bit and setting the second and third bits to $\{10\}$ and $\{01\}$ respectively. Thus different keys are used in each strand and they cannot be DES weak keys.

	MDC-2	MDC-4
Preimage attack	2^{55}	2^{109}
2nd preimage attack	2^{55}	2^{109}
Collision attack	$2^{51.5}$	2^{56}

If the underlying block cipher is a b-bit block cipher then to hash b bits of message we need two invocations of the block cipher. Thus the rate of MDC-2 is $\frac{1}{2}$. Since MDC-4 is built around two consecutive invocations of MDC-2 it has a rate of $\frac{1}{4}$. Other constructions with similar performance have been proposed [432] but it is natural to ask whether there are more efficient constructions. Unfortunately, it is not clear that there are. Knudsen and Lai [392, 383], later with Preneel [393], demonstrated generic attacks that apply against large classes of double block length hash function constructions of rate 1. In the right circumstances there are attacks that can provide a preimage with 2^b operations and 2^b words of storage. Since the hash output is $2b$ bits, these constructions are clearly compromised and currently there is no trusted double block length hash function construction with rate 1.

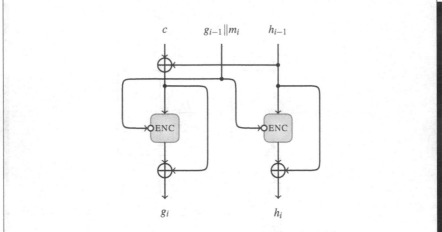

Fig. 4.14 The compression function for Hirose's construction based around, say, AES-256, where each of the named inputs corresponds to a 128-bit quantity.

4.6.2.1 Hirose's Construction

Described as a plausible construction for a double block length construction, the proposal of Hirose [297] is interesting for several reasons. Among them, the designer takes advantage of block ciphers that offer a key size that is twice the block size, for instance, AES-256. Further, while there are two invocations of the encryption function the design of Hirose requires only one single key expansion. This reuse of the key expansion gives a modest performance gain though the construction itself remains of rate $\frac{1}{2}$ when using AES-256. The construction of the compression function is illustrated in Fig. 4.14 where c denotes a nonzero constant and where some of the labels have been renamed for consistency with previous schemes.

4.6.3 The SHA-3 Competition

Previous sections considered the problem of building a hash function out of a good block cipher. Yet the most commonly deployed hash functions, MD5 [637] and SHA-1 [548], have a dedicated design. So does block cipher know-how tell us anything about such dedicated hash functions?

It would be too much of a digression to cover specific hash functions in great detail. Suffice it to say that if we look closely at MD5 and SHA-1, say, we notice that they use the Davies-Meyer construction that was illustrated in Fig. 4.12. The central computation in MD5 and SHA-1—that is, the compression function computation with the feedforward removed—is reversible and acts as a block cipher. The message input for these hash functions is used as the key for the block cipher and there is even a very simple (and we now know insecure [735, 734]) key schedule. So in some sense we have many more deployed block ciphers than we might ever have imagined, though they have very unusual features such as large block sizes (*e.g.*, 128 and 160 bits) and enormously long "keys" (512 bits). Some researchers have even taken these block ciphers and considered other constructions [349, 651] or the underlying block ciphers themselves, for instance, SHACAL-1 [277] and SHACAL-2 [278], of which the first has been studied extensively [66, 221, 305, 373].

Since these early efforts, much has changed in the field of hash functions. Nevertheless, there is strong interaction between hash functions and block ciphers, with many of the design and analysis techniques being common to both. Indeed, most of the high-profile failures of dedicated hash functions have been due to faults in the "key schedule", which confirms a commonly held view, see Sect. 8.5, that there is much to learn about this component in both hash functions and block ciphers.

The SHA-3 contest [541] received 64 submissions, of which 51 were accepted to the first round. Of these, 14 advanced to the second round with the most extreme distillation of block cipher know-how being evidenced by the four candidates that build on AES design techniques. Two of these—Grøstl [253] and Fugue [272]— use a combination of S-box/MDS primitives that we saw in the AES, see Chap. 3, though with different specific choices. Meanwhile ECHO [40] and SHAvite-3 [60] reuse the AES round function itself. Analysis of the SHA-3 candidates will undoubtedly provide many new advances in hash function design. It would be surprising if these didn't provide block cipher insights too.

4.7 Getting to the Source

In this chapter we have looked at how we might use a block cipher. First, we considered ways in which we could extend the basic primitive to encrypt messages of any length and we also saw how to achieve stream cipher encryption when using certain modes of use for a block cipher. Second, we surveyed some of the prominent ways in which a block cipher might be used as the basis for a message authentication code or a hash function. However, we were only able to give a brief taste of the work that has been happening in this area, work which is still very much ongoing. Naturally we focused our attention on constructions that are based around a block cipher, but the reader should be aware that this approach is complemented by many others. A very substantial body of work on both authentication and hashing remain outside the scope of this chapter.

Encryption modes   [316, 414, 543, 546, 552]
Authentication modes   [102, 103, 218, 247, 261, 320, 330, 344, 413, 423]
                          [540, 543, 544, 553, 554, 644, 743]
Mode analysis   [50, 51, 75, 76, 196, 197, 255, 329, 331, 332, 333]
                          [334, 387, 398, 439, 460, 489, 499, 503, 602, 617]
                          [728]
Hash functions   [30, 100, 101, 299, 325, 392, 393, 397, 399, 432]
                          [484, 495, 497, 615, 647, 698, 747]
Starting and finishing   [105, 300, 498, 600, 724, 754]

Chapter 5
Brute Force Attacks

The one attack that can always be mounted against any block cipher is a search for the key. No amount of clever design can prevent this attack and the designers' aim is to ensure that this is the best attack available to the adversary.

Over time the protection offered by a key of fixed length will be eroded. Computer hardware will improve and by appealing to *Moore's Law* [510] we can try and take this into account. One interpretation of Moore's Law is that the amount of computation power available for the same cost will double roughly every 18 months. Despite some discussion on the future validity of Moore's Law, it remains an essential aid in predicting future computer performance. There is, of course, an active industry in discussion papers [107, 224] and a range of models estimate what has been termed the *true cost* of exhaustive search. Taken together, these help us make the appropriate choice when deploying cryptography.

The most important aspect of exhaustive search is that it can be easily parallelised. So by throwing extra resources at the problem—where resources might be money used to buy additional hardware or extra computers offering spare computing cycles—we are able to solve the problem quicker. Such exhaustive search efforts can be either hardware- or software-based, with software-based efforts relying on the Internet to marshal sufficient computing power.

Since a hardware implementation of an algorithm will be faster than a software implementation, it seems inevitable that exhaustive search would best be done in hardware. However, the one-time development and fabrication costs for a dedicated machine can be enormous and only those with a suitable budget can overcome this initial barrier[1]. By contrast, notwithstanding the time required to get a project off the ground, a software-based effort using the Internet is very cheap (particularly so if people are freely contributing spare computing cycles). And while the hardware-based efforts appear to be very fast, the computing power available via the Internet can be significant. In the 1999 DES key search undertaken simultaneously across the Internet and with a dedicated hardware device, the hardware device continuously

---

[1] Actually there are opportunities for niche hardware projects at a reasonable budget as evidenced by the DES Cracker [226].

tested around 92×10^9 DES keys each second. Initially, this was by far the majority of the computing power being contributed to the search. But as more people joined via the Internet, the computing power of the software contribution exceeded that of hardware until, during peak testing, 62% of the computing power was coming from the Internet.

For an attack that demonstrates little sophistication, it is surprising that exhaustive search efforts manage unfailingly to capture the attention of the media. As Diffie writes [226];

> If it is a mystery why so many [...], myself included, have worked to refine and solidify their estimates, it is an even greater mystery that in the late 1990's, some people have actually begun to carry out key searches.

At the end of the day, perhaps it's just that we want to see it done for real.

5.1 Time–Memory Trade-offs

In considering exhaustive search we have not considered any role for memory. We take a key, we test it, and we move on. However, many tasks can be aided with the use of memory and exhaustive key search is no different.

The most obvious application of memory in the cryptanalysis of a block cipher is that of mounting what is termed a *dictionary attack*. Ahead of time, an adversary can compute all possible encryptions of a given chosen plaintext. The results can be stored in a table which is sorted on the value of the ciphertext. Then, when a ciphertext is intercepted, a match in the table gives a candidate value for the encryption key. The time required for the off-line precomputation is given by 2^κ operations and we require what we typically refer to as 2^κ *words* of storage where the size of the word will depend on the block cipher in question. At the time of cryptanalysis, that is, when the adversary has intercepted a ciphertext, it remains to complete a table lookup in a sorted table, which we assume can be accomplished easily.

We therefore have two extreme situations for the adversary: exhaustive search, which requires essentially no memory and 2^κ work, and the dictionary attack, which requires 2^κ words of memory and little time. This suggests a trade-off.

5.1.1 Hellman's Time–Memory Trade-off

The classic *time–memory trade-off* is described by Hellman [288]. This trade-off is embodied in a chosen plaintext attack and, like the dictionary attack, requires precomputation. However, the storage requirements when compared to the dictionary attack are greatly reduced.

We randomly make s initial choices for the encryption key. These are denoted by $SP_1, \ldots, SP_s$ and from each of these we derive a parallel chain of encryption keys. For chain r and at each step i for $1 \le i \le t$, we encrypt the chosen plaintext m under an encryption key $x_{(r,i)}$. The key $x_{(r,i)} = R(\text{ENC}_{x_{(r,i-1)}}(m))$, where R is a simple function (typically called a *reduction* function) that maps a b-bit block cipher output to a κ-bit key. Note that, depending on the key sizes, R might not be a reduction function at all, but since much of this early work was designed for use with DES we will keep this notation. After t iterations, we establish the end points of the chains $\text{EP}_1, \ldots, \text{EP}_s$ and we store the s pairs of values $(\text{SP}_1, \text{EP}_1), \ldots, (\text{SP}_s, \text{EP}_s)$ in a table. The table requires $2s\kappa$ bits of storage and one needs to do st encryption operations. This precomputation stage of the Hellman time–memory trade-off is illustrated in Fig. 5.1.

Now we assume that the adversary intercepts a ciphertext c that is the encryption of m under some key k. Immediately, the adversary can compute $k' = R(c)$ and check to see whether k' is equal to one of the end points $\text{EP}_1, \ldots, \text{EP}_s$:

1. If $k' = R(c) = \text{EP}_j$, say, then the adversary starts with the corresponding SP_j and computes through the chain until the value $x_{(j,s-1)}$ is obtained. This provides a candidate value $x_{(j,s-1)}$ for the encryption key k. Note that even though the table is computed in its entirety during the precomputation phase, only the end points are saved and the adversary needs to recover a candidate key from the starting point of a chain.
2. If $R(c)$ is not one of the end points, then the adversary checks to see if it might have appeared in the second to last column. The adversary computes $R(\text{ENC}_{R(c)}(m))$ and once again looks for a match among the end points. If there is no match then the key was not in the second to last column and the adversary searches back through the table in an iterative fashion.

The effectiveness of this attack will depend on two factors: (i) the number of distinct keys that are "covered" during the construction of the tables in Fig. 5.1 and (ii) the chance of analysis giving a *false alarm*, that is, an incorrect key value.

First we consider the issue of coverage. If all the values computed during the construction of the table in Fig. 5.1 are distinct, then the precomputation phase would cover $\frac{st}{2^\kappa}$ of all key values. However, when we generate the key chains there is a chance that some key values will appear several times. Analysis [288, 426] reveals that for a given table with s and t large and with $st^2 \approx 2^\kappa$, the probability that the st keys covered by a single table are distinct, which we denote by $\text{DISTINCT}(1)$, can be approximated by

$$\text{DISTINCT}(1) \approx \frac{0.8st}{2^\kappa}.$$

So there is a slight degradation in the expected coverage of $\frac{st}{2^\kappa}$, but not too much. To ensure that we cover more keys during our attack it might be tempting to increase the size of s or t. However, it has been shown [288, 426] that there is little reason to increase these values beyond the point at which $st^2 = 2^\kappa$. This is because once a table is sufficiently large, the addition of new rows or columns is unlikely to greatly increase the number of key candidates covered by the table. It would instead lead to more duplication. Instead, a better way to increase the coverage of the keyspace is to construct an independent table of the same size.

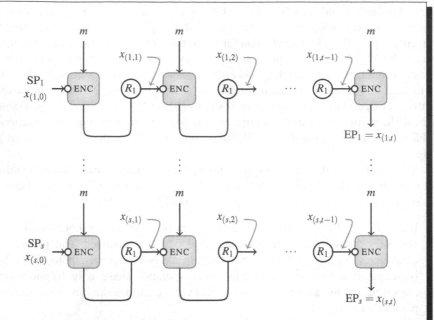

Fig. 5.1 Hellman's time–memory trade-off, illustrated with one table. While one table is constructed using one particular reduction function R_1, we typically use n different tables, each built using n different reduction functions $R_1, \ldots, R_n$. The tables consist of s rows and t columns of which only the first and last columns are stored. The lookup tables are sorted on the values of the last column.

So we construct n different tables where the reduction function R is replaced by a set of distinct functions $R_1, \ldots, R_n$, one for each table. We are effectively working with n independent tables and the probability that the stn keys covered by the n tables are distinct, DISTINCT(n), can be lower bounded by

$$\text{DISTINCT}(n) \geq 1 - (1 - p(1))^n \approx 1 - e^{-\frac{0.8stn}{2^\kappa}}.$$

To obtain good coverage of the keyspace, we set $stn = 2^\kappa$ and using $n = t$ tables we expect a success rate of at least

$$\text{DISTINCT}(n) \geq 1 - e^{-0.8} \approx 55\%.$$

To compute the online time required for cryptanalysis we observe that if a key lies in a particular row r of the table then it can be recovered with t operations; if the key lies in column $t - i$ for $1 \leq i \leq t$ then we require i operations to arrive at the end point EP$_r$ and $t - i$ operations to recover the encryption key from the corresponding SP$_r$. Note that we can simultaneously search across all s rows of a table since the lookup table is sorted by end points EP$_1, \ldots,$ EP$_s$; finding a match with some end point

		Precomputation	Memory	Online analysis
		st^2	st	t^2
$s = 2^{20}$	$t = 2^{18}$	2^{56}	2^{38}	2^{36}
$s = 2^{16}$	$t = 2^{20}$	2^{56}	2^{36}	2^{40}
$s = 2^{12}$	$t = 2^{22}$	2^{56}	2^{34}	2^{44}
$s = 2^{8}$	$t = 2^{24}$	2^{56}	2^{32}	2^{48}
Dictionary attack		2^{56}	2^{56}	1
Exhaustive search		–	1	2^{56}

Fig. 5.2 Different parameter choices for Hellman's time–memory trade-off when applied to DES.

EP_j immediately identifies the correct starting point from among all rows. Since we check $n = t$ tables sequentially this yields an online work effort of t^2 operations.

We need to address the issue of false alarms. Depending on the reduction function R_i it is possible that for some ciphertext c there is a match in a table, but with an incorrect key value. Hellman [288] provides an upper bound of $\frac{st(t+1)}{2^{\kappa+1}}$ for the probability of a false alarm occurring in this way. Given that we set $st^2 = 2^\kappa$ the probability of a false alarm is roughly $\frac{1}{2}$. So the key initially recovered using Hellman's time–memory trade-off is correct with probability 50% and rejecting such a false alarm requires at most t operations (that is, by checking all the way along a row of a table). So by considering false alarms, our online work effort would increase from t^2 operations to $\frac{3t^2}{2}$ operations. However, given the typical size of t, we often ignore the impact of false alarms on online processing time.

So for a good chance of success we choose $n = t$ and we construct t tables, each requiring $2m\kappa$ bits of storage and st^2 operations to precompute. We have the restriction that $st^2 = 2^\kappa$ and so a variety of trade-offs are possible, some of which are illustrated in Fig. 5.2 for DES. To put some perspective on these figures we might consider the row identified by $s = 2^8$ and $t = 2^{24}$, for which we require $2^{32} \times 2^7$ bits or 64 Gbytes of storage and an encryption time equivalent to 2^{48} DES operations. Under the assumption that it took nine days for the 1998 DES cracker to exhaust the DES keyspace, we might expect to exhaust the DES keyspace in around 51 minutes using comparable technology and this specific time–memory trade-off. The original paper of Hellman [288] included a costing for a device that would implement a key search attack on DES using a time–memory trade-off. It was provisionally estimated that such a machine would cost around $4 million and give a time-to-solution of around one day. A more recent survey using modern performance improvements [426] suggests that the cost of attacking DES with success probability 50% and a time to solution of one hour is around $260,000.

5.1.2 More Advanced Considerations

Since the basic time–memory trade-off was published there have been enhancements and further proposals. Most important among them is the use of *distinguished points*, which is attributed to Rivest [640]. Here we adopt some simple criteria (such as the ten leftmost bits being all 0) by which we identify a *distinguished point*. Such distinguished points can be used as end point to the key chains computed in Hellman's attack. So when computing the tables in Fig. 5.1, instead of letting every row consist of t steps along the chain, we generate key values until we encounter a distinguished point. We then store the distinguished point as the end point to the chain, along with the starting point. In the work of [697] it is proposed also that we store the number of iterations required to travel from the start to the end of a key chain.

To see why this might be advantageous consider the online actions of the adversary. With Hellman tables the cryptanalyst computes a potential value for the end point and checks for its appearance in the precomputed tables. This requires a memory lookup, which is a rather costly operation in practice. By contrast, when using distinguished points the adversary computes forward from the intercepted ciphertext until a distinguished point is reached. Only then is the lookup table accessed to see whether or not the distinguished point is present.

Some other important structural advantages to using distinguished points have also been highlighted in the literature [122, 697]. With Hellman tables there is the risk that two chains of computation merge and it is unlikely that this will be recognised from the value of the end points. We would thus waste valuable storage space covering the same key values twice. With distinguished points, if two chain computations merge then the end point in both cases would be the same distinguished point. We can thus detect the merge, store only the longer chain since we wish to cover as many key values as possible with our computation, and use the space saved to store new chains. We might implement a maximum number of iterations when generating our chains with distinguished points. Then we can decide to abandon a chain computation that is taking too long to reach a distinguished point and assume that the chain computation is probably stuck in a loop. So at the cost of some additional precomputation, we can store a set of tables that are free from some of the structural problems that might complicate analysis with Hellman tables.

As remarked in [122, 578, 697], the use of distinguished points has become commonly accepted. There are some problems though. The rows in the tables we generate are of different lengths so the estimation of the key coverage provided by these tables is more complex than for the case of Hellman tables. This specific issue is covered in [697], where a careful analysis of the success rate when using distinguished points is provided. More importantly in practice, it is observed in [578] that the number of false alarms when using distinguished points can become a practical problem. If we have detected a merge when manufacturing the table then we will only save the longer chain. However, long chains have a greater probability of being involved in a merge and they offer a greater opportunity for false alarms. Further, eliminating a false alarm requires computing to the end of the chain which will

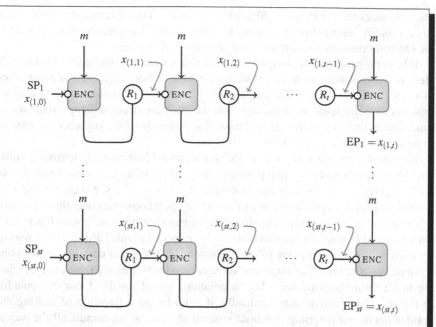

Fig. 5.3 The table construction for Oechslin's time–memory trade-off using *rainbow tables*. We build one table using t difference functions $R_1 \ldots R_t$, where the table consists of st rows and t columns of which only the first and last are stored. The lookup table is sorted on the values of the last column. This contrasts with Hellman's approach using t independent tables (see Fig. 5.1).

be more computationally expensive on longer chains. With these thoughts in mind Oechslin introduced so-called *rainbow tables* [578].

5.1.2.1 Oechslin's Time–Memory Trade-off

While they are similar to Hellman tables, Oechslin's tables have the advantages of distinguished points but fewer drawbacks. They are constructed in the following way.

We randomly make st initial choices for the encryption key. These are denoted by $SP_1, \ldots, SP_{st}$ and from each of these we derive a parallel chain of encryption keys. In contrast to the Hellman time–memory trade-off, we use a different mapping function at each step of the chain. More particularly, for chain r and at each step i for $1 \le i \le t$, we encrypt the chosen message m under an encryption key $x_{(r,i)}$ and use a reduction function R_i. The key $x_{(r,i)}$ was given by $R_i(\text{ENC}_{x_{(r,i-1)}}(m))$, where R_i is a simple function that maps a b-bit block cipher output to a κ-bit key. After t iterations, we establish the endpoints of the chains $EP_1, \ldots, EP_{st}$ and we store the st

pairs of values $(SP_1, EP_1), \ldots, (SP_{st}, EP_{st})$ in a table. The table requires $2st\kappa$ bits of storage and st^2 encryption operations to compute and this precomputation stage of the Oechslin time–memory trade-off is illustrated in Fig. 5.3.

It is worth pausing and comparing our situation with that devised by Hellman. In Hellman's time–memory trade-off we have t tables where each table is constructed using one redundancy function and the table has s rows and t columns. In Oechslin's time–memory trade-off we have one table which is constructed using t redundancy functions and it has st rows and t columns. The coverage of the keyspace is expected to be similar in both cases [578].

How do we recover a key using Oechslin's tables? Just as with Hellman's tables we observe that if a key lies in a particular row r of the table then it can be recovered with t operations; if the key lies in column $t - i$ for $1 \leq i \leq t$ then we require i operations to arrive at the end point EP_r and $t - i$ operations to recover the encryption key from the corresponding SP_r. Note that we can simultaneously search across all s rows of a table since the lookup table is sorted by end points $EP_1, \ldots, EP_{st}$; finding a match with some end point EP_j immediately identifies the correct starting point from among all rows. The important difference with Hellman's tables is that when we move from checking for a key in column i to column $i - 1$ our computation to the end of the row changes radically. It is no longer a question of making one application of our mapping function; instead, we have to recompute all the way to the end point. In the case of Hellman's tables we were applying the same reduction function at each step of the chain and so moving back one column in the table required only one iteration of our mapping function. It remains the case, however, that we can simultaneously check all rows once we have reached a candidate end point and so the number of online operations to find a candidate key is given by

$$\sum_{i=1}^{t-1} i = \frac{t(t-1)}{2}.$$

This is half the online work effort of Hellman's approach and yet has the same key coverage, precomputation, and memory requirements.

5.1.2.2 Summary

We have briefly described some attacks that exploit a time–memory trade-off. However, more sophisticated analysis is possible. Amirazizi and Hellman [9] consider a *time–memory–processor* trade-off where the cost of memory and the cost of processing units are separated. This creates the possibility of a trade-off between the two quantities. Schroeppel and Shamir [670] also consider time–memory trade-offs when applied to certain problems. However, for our purposes, the time–memory trade-off serves to highlight the importance of memory in a cryptanalytic attack.

It might be interesting to close this section by considering trends in the cost of memory. While there is no reason to change the often-used adage that "memory costs more than time" [288], the cost of memory has fallen considerably. The cost of storage media is dropping quickly, as evidenced by trends in recent consumer devices, and these reduced costs are being matched by faster access. So while memory costs more than time, it is perhaps not as expensive as it once was.

5.2 Meet-in-the-Middle Attacks

The term *meet-in-the-middle attack* is pretty clear about its intentions. In fact, the term features in many different types of analysis and finds application in many cryptographic attacks. The most prominent examples are in assessing the security of proposals for multiple encryption.

5.2.1 Double Encryption

Recall from Sect. 2.4.1 that we might define a block cipher EE_2 by means of double encryption using two different keys, each of length κ, where

$$EE_2(m) = ENC_{k_2}(ENC_{k_1}(m)).$$

A naïve exhaustive key search to find k_1 and k_2 would require $2^{2\kappa}$ steps. However, by using memory an attacker can reduce the time for an attack to about 2^{κ} steps. This meet-in-the-middle attack on double encryption is illustrated in Fig. 5.4.

The adversary chooses a message m and then computes a table of all encryptions $ENC_{k_1}(m)$ for all possible values of the key k_1. This requires a one-time off-line work effort of 2^{κ} operations. The table is then sorted by the value of $ENC_{k_1}(m)$ and requires $b2^{\kappa}$ bits of storage. We casually refer to this as 2^{κ} *words* of memory. After intercepting the ciphertext c the adversary undoes the second encryption operation with all values of k_2 and looks for a match in the precomputed table. Such a match gives candidate values for the pair of keys (k_1, k_2). For typical values of b and κ we expect false alarms but candidate values can be tested for correctness against additional message-ciphertext pairs. In the case of double-DES we require a precomputation of 2^{56} operations and 2^{56} words of memory for the table. However, once the table has been computed the online work effort, at the time of decryption, would be 2^{56} operations.

There is a variant of this attack for the known message scenario. Here the adversary computes the table once he knows the message m. The rest of the attack is the same as before and, in the worst case, the adversary needs to do $2^{\kappa+1}$ encryptions.

It should be no surprise that there is a more gradual trade-off between time and memory. Suppose that we guess the s leftmost bits of k_1. Then with these single bits fixed, we can perform a computation of $2^{\kappa-s}$ operations to build a table of $2^{\kappa-s}$ words of memory. We might then exhaustively try all 2^{56} values to k_2 and look for a match. If we don't find one, we assume that our guess for the s leftmost bits was

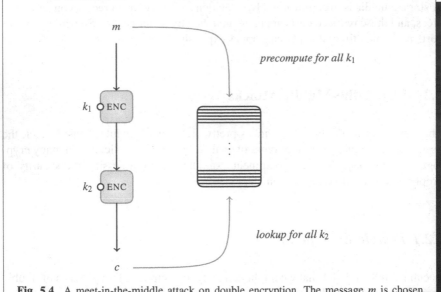

Fig. 5.4 A meet-in-the-middle attack on double encryption. The message m is chosen ahead of time and the set of partial encryptions under all possible values for k_1 is stored. When the ciphertext c is intercepted, the adversary searches over all k_2 until a match is found in the table.

incorrect. We then try another guess, compile another table of $2^{\kappa-s}$ words writing over the one we have already tried, and continue the attack. The total storage requirements remain $2^{\kappa-s}$ words but the time requirements will be given by $2^s(2^{\kappa-s}+2^\kappa)$ operations. If $s = 0$ we have the standard meet-in-the-middle attack with one set of 2^κ operations forming the precomputation and the second set of 2^κ operations forming the online computation. If $s = \kappa$ the attack coincides with naïve exhaustive key search and we have a work effort of $2^{2\kappa}$ operations.

More detailed analysis by Van Oorschot and Wiener [583] permits some performance improvements for meet-in-the-middle attacks. Using w words of memory, a meet-in-the-middle attack on double-DES requires $\frac{7 \times 2^{(3 \times 28)}}{\sqrt{w}}$ operations. Roughly speaking this amounts to $2^{56+\alpha+3}$ operations using $2^{56-2\alpha}$ words of memory for moderate integer values α.

We should note that the requirements for all attacks on double encryption are still greater than they were for attacks on DES. However, they can be much less than those required for naïve exhaustive search. Thus we typically avoid double-DES when attempting to strengthen DES and we move straight to triple-DES.

5.2.2 Triple Encryption

When looking at Fig. 5.2.1 it should come as no surprise that a generic meet-in-the-middle attack can be applied to triple-DES as well as double-DES. While there is a naïve application of the meet-in-the-middle attack, there are also more sophisticated variants that apply to the two-key variants of triple encryption.

5.2.2.1 Two-Key Triple Encryption

Merkle and Hellman provided the first attack on two-key triple encryption [494]. It is a chosen message attack that, for DES, requires a precomputation of 2^{56} single-DES encryptions, 2^{56} words of storage, and 2^{56} encryption during online computation. Consider the following representation of two-key triple encryption where A and B denote two points internal to the computation.

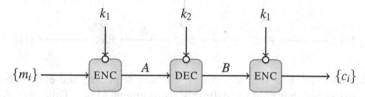

During a stage of precomputation we assume that $A = 0$ and we use all values of the key k_1 to compute a set of messages $\{m_0, \ldots, m_{2^\kappa - 1}\}$ where $m_i = \text{DEC}_i(0)$. We store this information in a table T of triples denoted by $(m_i, i, 0)$. Then for all these values of m_i we mount a chosen plaintext attack and we request the corresponding ciphertext c_i. We therefore have

$$c_i = \text{ENC}_{k_1}(\text{DEC}_{k_2}(\text{ENC}_{k_1}(m_i))).$$

Now consider the fact that $A = 0$. As we trace through the encryption process we see that $B = \text{DEC}_{k_2}(A)$, but since we have that $A = 0$, $B = \text{DEC}_{k_2}(0)$ and so the candidate value B must also lie in the set $\{m_0, \ldots, m_{2^\kappa - 1}\}$. For a given c_i, for $0 \leq i \leq 2^\kappa - 1$, the value of B is given by $B = b_i = \text{DEC}_i(c_i)$ and so we can add the set of triples $(b_i, i, 1)$ to the precomputed table T. The table doubles in size (which is not too onerous) and the process requires 2^{56} operations.

We then sort the table T according to the first entry in all triples and we look for a match in the values of m_i and b_j. This gives us values of i and j. Note that we are not expecting $i = j$ (which would happen when $k_1 = k_2$) but rather we are looking for any $m_i = b_j$ from which we would conclude that $k_1 = i$ and $k_2 = j$ are candidates for the secret keys. Depending on the respective block and key sizes the suggested values for (k_1, k_2) may need to be checked against additional message-ciphertext pairs.

Clearly, the demands on the cryptanalyst are prohibitive in terms of at least one of precomputation, storage, data, or online processing. This is why Merkle and Hellman refer to their attack as being a *certificational attack* [494]. The phrase is typically used to indicate that an attack is not practical, but it illustrates a surprising property of the encryption mechanism or potentially points the way to further cryptanalytic improvement. And indeed, in this particular case there is a known plaintext improvement to the Merkle-Hellman attack that is due to Van Oorschot and Wiener [582], and which confirms this intuition.

The requirements for the attack by Van Oorschot and Wiener [582] can be expressed in the following way. For a block cipher with b-bit blocks and κ-bit key, if 2^t known message-ciphertext pairs are available then an attack on two-key triple encryption (EDE$_2$) requires storage proportional to 2^t and work effort of $2^{b+\kappa-t}$ online operations. To follow the outline of the attack, consider Merkle and Hellman's attack on two-key triple encryption where A and B again denote the two points that are internal to the computation.

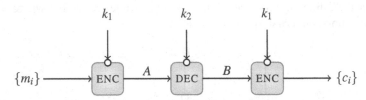

Merkle and Hellman set the value $A = 0$ and requested the encryption of all possible messages that could satisfy this requirement [494]. This gave a chosen message attack. In contrast, Van Oorschot and Wiener [582] consider the case of a set of 2^t known message-ciphertext pairs which are first organised into a table of pairs (m, c) and sorted by the values of the message m. This requires 2^t words of memory.

Next we guess that A takes some value a. This is correct with probability 2^{-b}. Under the assumption that $A = a$, we run through every value of k_1 and compute the message $m_i = \text{DEC}_i(a)$ for $0 \le i \le 2^\kappa - 1$. If the value m_i is among the set of known message-ciphertext pairs (m, c) then we know the corresponding value of the ciphertext c_i and we can compute the anticipated value for B as $b_i = \text{DEC}_{k_1}(c_i)$ for $k_1 = i$. We keep the pairs of values (b_i, i) in a table sorted on the values of b_i and this requires a work effort of around 2^κ operations and $2^{t+\kappa-b}$ words of memory. Next we search over all possible values of k_2 and compute $b_j = \text{ENC}_j(a)$ for $0 \le j \le 2^\kappa - 1$. If we find some value of b_j in the table (b_i, i) then we take (i, j) as a candidate for the values of k_1 and k_2 during encryption. This can be checked against other known message-ciphertext pairs.

The probability that we are correct in our guess for some a from among all 2^t pairs is $\frac{2^t}{2^b} = 2^{t-b}$ and the expected number of values of a that need to be tried [582] is given by 2^{b-t}. Therefore the total running time is $2^{b-t} \times 2^\kappa = 2^{b+\kappa-t}$ and different trade-offs for this attack are shown in Table 5.1.

We note that all the attacks in this section can be thwarted by changing the middle decryption to an encryption (though we would lose backward compatibility with single encryption) or by moving to three instead of two independent keys. Despite

Variant	Time	Precomputation	Memory	Data
EE_2	$2^{2\kappa}$	–	–	$\lceil \frac{2\kappa}{b} \rceil$ known
EE_2	2^{κ}	2^{κ}	2^{κ}	$\lceil \frac{2\kappa}{b} \rceil$ chosen
EE_2	$2^{\kappa+1}$	–	2^{κ}	$\lceil \frac{2\kappa}{b} \rceil$ known
EE_2	$2^{\kappa+s}$	–	$2^{\kappa-s}$	$\lceil \frac{2\kappa}{b} \rceil$ known
EE_2	$2^{\kappa+\alpha+3}$	–	$2^{\kappa-2\alpha}$	$\lceil \frac{2\kappa}{b} \rceil$ known
EDE_2	$2^{2\kappa}$	–	–	$\lceil \frac{2\kappa}{b} \rceil$ known
EDE_2	2^{κ}	2^{κ}	2^{κ}	2^{κ} chosen
EDE_2	$2^{b+\kappa-t}$	2^{t}	2^{t}	2^{t} known
EEE_2	$2^{2\kappa}$	–	–	$\lceil \frac{2\kappa}{b} \rceil$ known
EDE_3	$2^{3\kappa}$	–	–	$\lceil \frac{3\kappa}{b} \rceil$ known
EDE_3	$2^{2\kappa}$	2^{κ}	2^{κ}	$\lceil \frac{3\kappa}{b} \rceil$ chosen
EDE_3	2^{κ}	$2^{2\kappa}$	$2^{2\kappa}$	$\lceil \frac{3\kappa}{b} \rceil$ chosen
EDE_3	$2^{\kappa+s}$	$2^{2\kappa-s}$	$2^{2\kappa-s}$	$\lceil \frac{3\kappa}{b} \rceil$ chosen
EEE_3	$2^{3\kappa}$	–	–	$\lceil \frac{3\kappa}{b} \rceil$ known
EEE_3	$2^{2\kappa}$	2^{κ}	2^{κ}	$\lceil \frac{3\kappa}{b} \rceil$ chosen
EEE_3	2^{κ}	$2^{2\kappa}$	$2^{2\kappa}$	$\lceil \frac{3\kappa}{b} \rceil$ chosen
EEE_3	$2^{\kappa+s}$	$2^{2\kappa-s}$	$2^{2\kappa-s}$	$\lceil \frac{3\kappa}{b} \rceil$ chosen

Table 5.1 Basic attacks on double and triple encryption where the basic encrypting unit is a b-bit block cipher with a κ-bit key. This is a generalised version of the DES-specific table given in Table 2.10, along with the obvious trade-offs when trading 2^s words of memory for time in a more finely tuned attack.

this, the two-key variant of triple encryption (EDE_2) features in standards [7, 314] and is widely deployed.

5.2.2.2 Three-Key Triple Encryption

Despite the added difficulty of working with additional key material, three-key triple encryption (either EEE_3 or EDE_3) typically offers the best security options. In general, the EDE_3 configuration allows backward compatibility with single encryption by setting $k_1 = k_2 = k_3$. However, in the specific case of DES we can achieve backward compatibility in the EEE_3 configuration by setting (say) $k_1 = k_2 = w$ where w is one of the four DES weak keys, see Sect. 2.3. Triple encryption, in its most basic form, is illustrated below and we mark the intermediate points A and B:

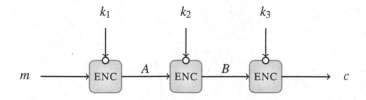

The time and memory requirements for exhaustive search and table lookup will be obvious. For the meet-in-the-middle attacks we might consider building a table for the values at point A. This would require 2^κ operations for precomputation and 2^κ words of storage. After receiving the ciphertext we would then decrypt backwards through two steps of the triple encryption to find a match in the table. This requires around $2^{2\kappa}$ operations. Of course, we could compile a table for the values at point B but this would increase the storage requirements massively (though reduce online work effort). These trade-offs are included in Table 5.1. Detailed analysis by Lucks [455] allows some improvement, but the practical impact is slight.

5.3 Getting to the Source

Here we give a selection of references to many of the topics covered in this chapter. The reader might note that work has also been done on the appropriate way to define modes of operation for multiple encryption and much of this work is referenced in Chap. 4.

Key length discussions	[107, 224]
DES exhaustive search	[160, 215, 226, 419, 426, 624, 744, 745]
Double and triple encryption	[36, 191, 279, 301, 455, 494, 549, 582]
Time–memory trade-offs	[9, 122, 279, 288, 455, 494, 578, 583, 640]
	[697]

Chapter 6
Differential Cryptanalysis: The Idea

Differential cryptanalysis is one of the most powerful attacks available to the block cipher cryptanalyst. Invented by Biham and Shamir [81] but identified more than ten years earlier by the designers of DES [167], this attack has been used widely on many different cryptographic primitives.

Our treatment of differential cryptanalysis comes in two parts. In this chapter, we introduce the basic ideas behind differential cryptanalysis and we do this by example. While we start with painfully simple block ciphers, we will gradually work our way towards more complex ciphers, introducing the main features of differential cryptanalysis along the way. This approach will, inevitably, gloss over some of the sophisticated problems that we might encounter in real life. So we set aside Chap. 8 to consider some of the more detailed aspects of differential cryptanalysis.

To motivate the whole idea of differential cryptanalysis, however, consider a very simple cryptosystem. Let m, c, and k be b-bit strings where m represents the message (or plaintext), c the ciphertext, and k the secret key. We can encrypt the message m in the following way:

$$c = m \oplus k$$

where $\oplus$ is the (bitwise) exclusive-or operation. If the key k is chosen at random and used only once, then the cryptanalyst can gain no information about m from observing the ciphertext c. In Shannon's framework (see Sect. 1.3) this cryptosystem is unconditionally secure under a ciphertext-only attack.

What happens if we use the key twice? Assume we use key k to encrypt m_0 and m_1 to give c_0 and c_1. Then an attacker intercepting both ciphertexts can compute

$$c_0 \oplus c_1 = (m_0 \oplus k) \oplus (m_1 \oplus k) = m_0 \oplus m_1.$$

In other words, the cryptanalyst recovers the exclusive-or of two plaintext messages directly from the intercepted ciphertexts. If the messages contain redundancy, for example, if they represent text from a natural (spoken) language, then the attacker can deduce information about the plaintext from the exclusive-or of the ciphertexts.

This is not particularly sophisticated, but it does highlight the key (no pun intended) idea behind differential cryptanalysis. While we might not get much infor-

$$c = S[m \oplus k_0] \oplus k_1$$

x	0 1 2 3 4 5 6 7 8 9 a b c d e f
$S[x]$	6 4 c 5 0 7 2 e 1 f 3 d 8 a 9 b

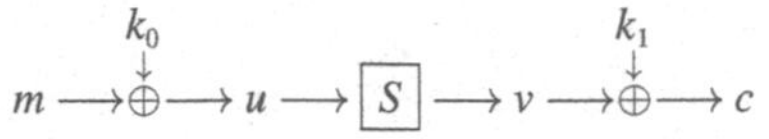

$$m \longrightarrow \oplus \xrightarrow{\ \ k_0\ \downarrow\ \ } u \longrightarrow \boxed{S} \longrightarrow v \longrightarrow \oplus \xrightarrow{\ \ k_1\ \downarrow\ \ } c$$

Fig. 6.1 Encryption with CIPHERONE.

mation from considering a single message and ciphertext, we might gain much more by considering pairs of messages and ciphertext. In our simple example the action of the secret key k could be entirely removed by simply manipulating the ciphertexts. In reality things will be far more complex, but looking at the difference between pairs of plaintext as they are encrypted—and choosing a notion of difference that allows us to ignore the action of the key for at least part of the analysis—forms the main idea behind differential cryptanalysis.

6.1 Getting Started

To start properly we need to consider a slightly more complex encryption system. The ciphertext will be computed using a block cipher called CIPHERONE. This is a four-bit block cipher with an eight-bit key and it uses a four-bit look-up box $S[\cdot]$. Encryption with CIPHERONE is described in Fig. 6.1 where the four-bit message m is encrypted using an eight-bit key $k_0\|k_1$ where both k_0 and k_1 are four-bit randomly chosen round keys. $S[\cdot]$ is a four-bit permutation and we assume that the details of $S[\cdot]$ are public (Kerckhoffs' assumption).

Consider the *differential behaviour* of CIPHERONE, *i.e.*, let us trace a difference between two plaintexts that are encrypted with CIPHERONE. Assume that a cryptanalyst knows two message inputs m_0 and m_1 along with the corresponding ciphertexts c_0 and c_1. The computation of these ciphertexts can be written in a stepwise fashion in the following way:

CIPHERONE$(m_0, k_0\|k_1)$	CIPHERONE$(m_1, k_0\|k_1)$
$u_0 = m_0 \oplus k_0$	$u_1 = m_1 \oplus k_0$
$v_0 = S[u_0]$	$v_1 = S[u_1]$
$c_0 = v_0 \oplus k_1$	$c_1 = v_1 \oplus k_1$

Since the value of k_0 is secret the cryptanalyst doesn't know the values of u_0 or u_1, which are internal to the cipher. However, the cryptanalyst *does* know the value of the *difference* between these two internal values since

$$u_0 \oplus u_1 = (m_0 \oplus k_0) \oplus (m_1 \oplus k_0) = m_0 \oplus m_1.$$

This can be used to find the secret value of k_1 in the following way. Given two message-ciphertext pairs (m_0, c_0) and (m_1, c_1) the cryptanalyst can compute the value of $u_0 \oplus u_1$. Next the cryptanalyst can guess the value of k_1 and compute the values of v_0 and v_1 directly from c_0 and c_1. Furthermore, since $S[\cdot]$ is publicly known and invertible, the cryptanalyst can compute $S^{-1}[v_0]$ and $S^{-1}[v_1]$. Now, the cryptanalyst cannot compare these values to u_0 and u_1 directly since these values are unknown. However, for the correct value of k_1, the cryptanalyst does know that

$$u_0 \oplus u_1 = S^{-1}[v_0] \oplus S^{-1}[v_1].$$

The value $u_0 \oplus u_1$ is known since it is equal to $m_0 \oplus m_1$. So the cryptanalyst can simply try all values t of k_1 one by one and whenever the expected value of $S^{-1}[t \oplus c_0] \oplus S^{-1}[t \oplus c_1]$ is obtained the value t is noted as a candidate for the secret key k_1. If more than one candidate for the key k_1 remains, the attack can be repeated on other messages and ciphertexts. An example of the attack is given in Fig. 6.2.

There are of course certain conditions that ensured our attack worked for our demonstration and it might not be clear why these particular message-ciphertext pairs worked so well. Nevertheless, this simple attack on a simple cipher allows us to highlight two important points:

1. Even though we might not know the value of internal variables during the encryption process, we were able to state the difference between variables at certain points during encryption.
2. We were able to recover key information by guessing values to part of the secret key and testing whether some differential condition held. This is the most common method of recovering key material when performing differential cryptanalysis.

We will now extend the first of these ideas. In our toy example CIPHERONE we were able to categorically state the value of a difference between internal variables. In more complex ciphers we might not be so sure.

Suppose we have two message-ciphertext pairs $\{(\mathtt{a} \to 9), (5 \to 6)\}$ and $\{(9 \to 7), (8 \to 0)\}$. For the first pair we have

$$
\begin{array}{ccccccccc}
& k_0 & & & & & & k_1 & \\
& \downarrow & & & & & & \downarrow & \\
\mathtt{a} \longrightarrow & \oplus & \longrightarrow u_0 \longrightarrow & \boxed{S} & \longrightarrow v_0 \longrightarrow & \oplus & \longrightarrow 9
\end{array}
$$

$$
\begin{array}{ccccccccc}
& k_0 & & & & & & k_1 & \\
& \downarrow & & & & & & \downarrow & \\
5 \longrightarrow & \oplus & \longrightarrow u_1 \longrightarrow & \boxed{S} & \longrightarrow v_1 \longrightarrow & \oplus & \longrightarrow 6
\end{array}
$$

so we know that $u_0 \oplus u_1 = \mathtt{a} \oplus 5 = \mathtt{f}$.

Step 1. Compute the inverse permutation $R[\cdot]$ from $S[\cdot]$.

x	0 1 2 3 4 5 6 7 8 9 a b c d e f
$R[x]$	4 8 6 a 1 3 0 5 c e d f 2 b 7 9

Step 2. For each value of t compute $u' = R[t \oplus 9] \oplus R[t \oplus 6]$.

t	0 1 2 3 4 5 6 7 8 9 a b c d e f
u'	e b e e d 8 d f f d 8 d e e b e

Step 3. Values of t giving $u' = \mathtt{f}$ are guesses for k_1 so $k_1 \in \{7,8\}$.

For the second pair we have

$$
\begin{array}{ccccccccc}
& k_0 & & & & & & k_1 & \\
& \downarrow & & & & & & \downarrow & \\
9 \longrightarrow & \oplus & \longrightarrow u_0 \longrightarrow & \boxed{S} & \longrightarrow v_0 \longrightarrow & \oplus & \longrightarrow 7
\end{array}
$$

$$
\begin{array}{ccccccccc}
& k_0 & & & & & & k_1 & \\
& \downarrow & & & & & & \downarrow & \\
8 \longrightarrow & \oplus & \longrightarrow u_1 \longrightarrow & \boxed{S} & \longrightarrow v_1 \longrightarrow & \oplus & \longrightarrow 0
\end{array}
$$

and so we know that $u_0 \oplus u_1 = 9 \oplus 8 = 1$.

Step 4. For each value of t compute $u' = R[t \oplus 7] \oplus R[t \oplus 0]$.

t	0 1 2 3 4 5 6 7 8 9 a b c d e f
u'	1 8 5 b b 5 8 1 5 9 2 b b 2 9 5

Step 5. Values of t giving $u' = 1$ are guesses for k_1 so $k_1 \in \{0,7\}$.
Step 6. Combining results for both message-ciphertext pairs we have $k_1 \in \{7,8\}$ and $k_1 \in \{0,7\}$. We deduce that $k_1 = 7$.
Rewriting encryption equations for $(\mathtt{a} \to 9)$ gives $k_0 = \mathtt{a} \oplus S^{-1}[9 \oplus k_1]$ and so we use $k_1 = 7$ to give $k_0 = \mathtt{d}$.

Fig. 6.2 Recovering the key $k_0 \| k_1$ from CipherOne.

$$c = S[S[m \oplus k_0] \oplus k_1] \oplus k_2$$

x	0 1 2 3 4 5 6 7 8 9 a b c d e f
$S[x]$	6 4 c 5 0 7 2 e 1 f 3 d 8 a 9 b

$$m \to \oplus \to u \to \boxed{S} \to v \to \oplus \to w \to \boxed{S} \to x \to \oplus \to c$$

with k_0 above the first $\oplus$, k_1 above the second $\oplus$, and k_2 above the third $\oplus$.

Fig. 6.3 Encryption with CIPHERTWO.

6.1.1 Working with Less Certainty

Next we consider the slightly more complex CIPHERTWO given in Fig. 6.3. We assume that a cryptanalyst knows two message inputs m_0 and m_1 along with the ciphertexts c_0 and c_1. The computation of these ciphertexts can be written in a step-wise fashion in the following way:

CIPHERTWO$(m_0, k_0 \| k_1 \| k_2)$	CIPHERTWO$(m_1, k_0 \| k_1 \| k_2)$
$u_0 = m_0 \oplus k_0$	$u_1 = m_1 \oplus k_0$
$v_0 = S[u_0]$	$v_1 = S[u_1]$
$w_0 = v_0 \oplus k_1$	$w_1 = v_1 \oplus k_1$
$x_0 = S[w_0]$	$x_1 = S[w_1]$
$c_0 = x_0 \oplus k_2$	$c_1 = x_1 \oplus k_2$

An attacker might try to attack CIPHERTWO in much the same way as CIPHERONE though he will soon encounter a problem. As before, we can work backwards and guess the value of k_2 to compute x_0 and x_1. This allows us to get w_0 and w_1. We can't compute the values for v_0 and v_1 but, since we know the values of w_0 and w_1, we know $w_0 \oplus w_1$ and hence we know $v_0 \oplus v_1$. Once again, note the power of being able to work with differences!

So we have worked a long way back but are still left with unknown variables internal to the encryption process. Starting with the plaintext and working forwards we know m_0 and m_1 and we also know $u_0 \oplus u_1$ since this is equal to $m_0 \oplus m_1$. But this is of little help since the S-box is non-linear with respect to exclusive-or and this prevents us from deducing the value of the difference $v_0 \oplus v_1$. Thus an attacker cannot immediately verify a guess for k_2. However, there is a way around this problem, and this provides the next piece of the differential puzzle.

To move forwards we need to understand the behaviour of $S[\cdot]$. Consider two inputs to $S[\cdot]$, i and j, where j is the bitwise complement of i. Another way of saying this is to write $j = i \oplus \mathtt{f}$, where the latter is hexadecimal notation for the all-1 word.

Table 6.1 Inputs and output relations for i and $j = i \oplus \mathtt{f}$ across $S[\cdot]$.

i	j	$S[i]$	$S[j]$	$S[i] \oplus S[j]$
0	f	6	b	d
1	e	4	9	d
2	d	c	a	6
3	c	5	8	d
4	b	0	d	d
5	a	7	3	4
6	9	2	f	d
7	8	e	1	f
8	7	1	e	f
9	6	f	2	d
a	5	3	7	4
b	4	d	0	d
c	3	8	5	d
d	2	a	c	6
e	1	9	4	d
f	0	b	6	d

For all values of i we can compute j and we can compute the difference between the outputs $S[i] \oplus S[j]$. The results are given in Table 6.1.

It is worth considering Table 6.1 in some detail. We have tabulated all possible inputs (i, j) that satisfy $(i \oplus j) = \mathtt{f}$ and we have evaluated the difference in the outputs that result. It is important to note the fifth column has an uneven distribution of values. Not all 16 possible values occur and some values occur more frequently than others. In this particular case the output difference $\mathtt{d}$ occurs in ten out of 16 cases.

Let us return to attacking CIPHERTWO. We have already noted that if we guess k_2 we can compute back from the ciphertexts to get the value $w_0 \oplus w_1$. However, we are unable to verify our guess for k_2 since we don't know what value of $w_0 \oplus w_1$ to expect.

As attackers we have the luxury of a chosen plaintext attack; that is, we get to choose the messages m_0 and m_1 and request the values of the ciphertext that result. To attack CIPHERTWO we won't need to choose the exact values of m_0 and m_1. Instead we pick m_0 at random and set $m_1 = m_0 \oplus \mathtt{f}$.

Working forwards through CIPHERTWO we can immediately see that $u_0 \oplus u_1 = m_0 \oplus m_1 = \mathtt{f}$ though, of course, we don't know the actual values of u_0 or u_1. But we do know that if u_0 takes all values with equal likelihood, and if $u_0 \oplus u_1 = \mathtt{f}$, then the probability that $S[u_0] \oplus S[u_1] = \mathtt{d}$ is $\frac{10}{16}$. This follows directly from Table 6.1. In our attack a guess for k_2 allows us to compute $w_0 \oplus w_1 = v_0 \oplus v_1$ and in our chosen plaintext attack we know that $v_0 \oplus v_1 = \mathtt{d}$ with a probability of $\frac{10}{16}$. If we assume

$$c = S[S[S[m \oplus k_0] \oplus k_1] \oplus k_2] \oplus k_3$$

x	0 1 2 3 4 5 6 7 8 9 a b c d e f
$S[x]$	6 4 c 5 0 7 2 e 1 f 3 d 8 a 9 b

$$m \to \oplus \to \boxed{S} \to \oplus \to \boxed{S} \to x \to \oplus \to y \to \boxed{S} \to z \to \oplus \to c$$
$$\quad k_0 \qquad\quad k_1 \qquad\qquad k_2 \qquad\qquad k_3$$

Fig. 6.4 Encryption with CIPHERTHREE.

that an incorrect guess for k_2 leads to garbage with the result that $w_0 \oplus w_1$ "looks random", then the attacker can gradually work out the correct value of k_2.

One way to do this in practice is to keep a series of counters $T_0, \ldots, T_{|k_2|-1}$, one for each possible value of k_2, which are initialised to 0 at the start of the attack. For any given message-ciphertext pair, we run through all possible values i to k_2 and if we get $v_0 \oplus v_1 = $ d then we increment the counter T_i by 1. Note that the correct value of k_2 gives $v_0 \oplus v_1 = $ d with probability $\frac{10}{16}$ while any incorrect value of the key will give $v_0 \oplus v_1 = $ d with (an average) probability of $\frac{1}{16}$. If we use N message-ciphertext pairs, then the counter for the correct value of k_2 (which we might refer to as the *signal*) will be much larger than the counters for the other key values (the *noise*)[1]. For instance, if we use 16 message-ciphertext pairs then the counter for the correct key value will have a value close to 10 ($16 \times \frac{10}{16}$) while counters for the incorrect key values are likely to have a value close to 1 ($16 \times \frac{1}{16}$). It is worth observing that in attacks on a real cipher, the memory cost of keeping a large number of counters for a large number of message-ciphertext pairs can be substantial.

The key to extending our basic attack on CIPHERONE was given in Table 6.1. This mapped out the possible evolution of the difference f across $S[\cdot]$. In fact, we can do this for all possible input differences and this gives us a *difference distribution table* for $S[\cdot]$. This is presented in Table 6.2. With such a table to hand, we can now mount far more powerful attacks.

---

[1] Intuitively, every time a pair of texts has the predicted differences the signal becomes stronger. So the amount of data we might expect to need in a successful differential attack will be proportional to p^{-1}. In practice things are a little more complex and this is discussed more in Chap. 8.

Table 6.2 The difference distribution table for $S[\cdot]$. There is a row for each input difference d_{in} and the frequency with which a given output difference d_{out} occurs is given across the row. The entry (d_{in}, d_{out}) divided by 16 gives the probability that a difference d_{in} gives difference d_{out} when taken over all possible pairs with difference d_{in}.

	0	1	2	3	4	5	6	7	8	9	a	b	c	d	e	f
0	16	-	-	-	-	-	-	-	-	-	-	-	-	-	-	-
1	-	-	6	-	-	-	-	2	-	2	-	-	2	-	4	-
2	-	6	6	-	-	-	-	-	-	2	2	-	-	-	-	-
3	-	-	-	6	-	2	-	-	2	-	-	-	4	-	2	-
4	-	-	-	2	-	2	4	-	-	2	2	2	-	-	2	-
5	-	2	2	-	4	-	-	4	2	-	-	2	-	-	-	-
6	-	-	2	-	4	-	-	2	2	-	2	2	2	-	-	-
7	-	-	-	-	-	4	4	-	2	2	2	2	-	-	-	-
8	-	-	-	-	-	2	-	2	4	-	-	4	-	2	-	2
9	-	2	-	-	-	2	2	2	-	4	2	-	-	-	-	2
a	-	-	-	-	2	2	-	-	-	4	4	-	2	2	-	-
b	-	-	-	2	2	-	2	2	2	-	-	4	-	-	2	-
c	-	4	-	2	-	2	-	-	2	-	-	-	-	6	-	-
d	-	-	-	-	-	2	2	-	-	-	-	-	6	2	-	4
e	-	2	-	4	2	-	-	-	-	2	-	-	-	-	-	6
f	-	-	-	-	2	-	2	-	-	-	-	-	-	10	-	2

6.2 Introducing Characteristics

Table 6.2 gave the total number of input pairs satisfying a given input difference (indexed by row) which leads to a pair of outputs with a given difference (indexed by column). As examples,

- two pairs of inputs to $S[\cdot]$ differing in all four bits will lead to outputs that differ in all bits,
- ten pairs of inputs that differ in all bit positions will yield output pairs that differ in all but the second least significant bit position,
- a pair with input difference 1 can never give an output pair with difference 1,

and so on and so forth.

More abstractly, a pair (α, β) for which two inputs with difference α lead to two outputs with difference β is called a (differential) *characteristic* across the operation $S[\cdot]$. It will be denoted by $\alpha \xrightarrow{S} \beta$. A characteristic has a probability associated with it and, as an example, we have already seen that the characteristic $\mathtt{f} \xrightarrow{S} \mathtt{d}$ holds with probability $\frac{10}{16}$. To illustrate the power of this abstraction, we introduce the latest version of our cipher, CIPHERTHREE, in Fig. 6.4.

An attacker trying to mount the attack we used on CIPHERTWO will be able to compute the values y_0 and y_1 for a correct guess for the value of k_3. They can readily go on to compute $x_0 \oplus x_1$. To verify whether such a guess for the value of k_3 is correct, the attacker will try to push information about the messages (which he can choose) through two applications of the S-box. Looking at Table 6.2 we saw that two messages satisfying the difference f will yield two outputs from $S[\cdot]$ with a difference d with a probability of $\frac{10}{16}$.

If k_1 is a randomly chosen four-bit key, then the difference of d in the inputs to the second application of $S[\cdot]$ (a difference directly inherited as the output from the first $S[\cdot]$) will lead to $x_0 \oplus x_1 = $ c with probability $\frac{6}{16}$. This is obtained by looking at Table 6.2. We can therefore combine the actions of the two S-boxes by using the characteristic f $\xrightarrow{S}$ d over the first $S[\cdot]$ and the characteristic d $\xrightarrow{S}$ c over the second $S[\cdot]$. If we assume that these characteristics are independent, then the probability of the two-round characteristic f $\xrightarrow{S}$ d $\xrightarrow{S}$ c is $\frac{10}{16} \times \frac{6}{16}$. Thus an attacker who chooses pairs of messages related by the difference f can expect the difference $y_0 \oplus y_1$ to take the value c with probability $\frac{15}{64}$. Given sufficient message-ciphertext pairs, it should be possible to find the key k_3 just as we did when attacking CIPHERTWO.

To highlight the last major elements of a differential cryptanalytic attack we need a more sophisticated cipher. This cipher will be called CIPHERFOUR and is now beginning to resemble a real-life cipher unlike our previous toy examples.

6.2.1 Joining Characteristics

CIPHERFOUR is an iterated cipher with r rounds that operates on 16-bit blocks. Both the message and the ciphertext blocks will be 16 bits long, as will the $r + 1$ round keys which are independent and chosen uniformly at random. Encryption with CIPHERFOUR is outlined in Fig. 6.5 and illustrated in Fig. 6.6. It can be shown that each bit of the ciphertext depends on each bit of the plaintext after four rounds of encryption and CIPHERFOUR appears to have many of the features of a real-life block cipher. Note that in CIPHERFOUR there is no permutation $P[\cdot]$ in the rth round; in addition to having no cryptographic importance it would serve as a distraction to our analysis.

Let us consider a differential attack on CIPHERFOUR. First note that all components are linear with respect to the exclusive-or operation except the S-box. The S-box is the same as that used in CIPHERTHREE and so the distribution of differences across $S[\cdot]$ can be found in Table 6.2.

If we are trying to mount an attack similar in style to that used on CIPHERTHREE, then the first thing we need to do is to find a characteristic which predicts the difference between two partially encrypted messages after $r - 1$ rounds. If such a characteristic can be identified that holds with a sufficiently high probability, then, in principle, we should be able to find the round key k_r.

To build characteristics over several rounds, one normally tries to find one-round characteristics (of high probability) that can be joined together. So, with CIPHER-FOUR, we will need to identify input differences to four S-boxes for which high-

x	0	1	2	3	4	5	6	7	8	9	a	b	c	d	e	f
$S[x]$	6	4	c	5	0	7	2	e	1	f	3	d	8	a	9	b

i	0	1	2	3	4	5	6	7
$P[i]$	0	4	8	12	1	5	9	13
i	8	9	10	11	12	13	14	15
$P[i]$	2	6	10	14	3	7	11	15

1. Set $u_0 = m$.
2. For $i := 1$ to $r - 1$ do:

   a. Combine the round key k_{i-1} with u_{i-1} so that $a_i = u_{i-1} \oplus k_{i-1}$.
   b. Divide a_i into four nibbles $a_i = A_0 || \ldots || A_3$.
   c. Compute $S[A_0] || \ldots || S[A_3]$ where $S[\cdot]$ is defined above.
   d. Write $S[A_0] || \ldots || S[A_3]$ as $y_{15} \ldots y_0 = S[A_0] || \ldots || S[A_3]$.
   e. Permute bit y_i to position j according to the permutation $j = P[i]$ defined above.
   f. Set $u_i = y_{15} || y_{11} || \ldots || y_4 || y_0$.

3. *(Last round)*

   a. Combine the round key k_{r-1} with u_{r-1} so that $a_r = u_{r-1} \oplus k_{r-1}$.
   b. Divide a_r into four nibbles $a_r = A_0 || \ldots || A_3$.
   c. Compute $S[A_0] || \ldots || S[A_3]$ where $S[\cdot]$ is defined above.
   d. Write $S[A_0] || \ldots || S[A_3]$ as $y = y_{15} \ldots y_0 = S[A_0] || \ldots || S[A_3]$.
   e. Output $y \oplus k_r$ as ciphertext.

Fig. 6.5 Encryption with r rounds of CIPHERFOUR.

probability output differences exist. Then, since $P[\cdot]$ is linear, one can easily compute the output difference from one round. Let us denote by

$$(\alpha_1, \alpha_2, \alpha_3, \alpha_4) \xrightarrow{S} (\beta_1, \beta_2, \beta_3, \beta_4)$$

the input and output differences to four S-boxes, where α_i is the difference between the values A_i input to the relevant S-box and β_i is the difference between the output values $S[A_i]$. Since the action of $P[\cdot]$ is fixed, we can predict with certainty how the difference will evolve across the permutation

$$(\beta_1, \beta_2, \beta_3, \beta_4) \xrightarrow{P} (\gamma_1, \gamma_2, \gamma_3, \gamma_4)$$

where $\beta_1 || \beta_2 || \beta_3 || \beta_4$ is the 16-bit input difference[2] to $P[\cdot]$ and $\gamma_1 || \gamma_2 || \gamma_3 || \gamma_4$ is the 16-bit output difference from $P[\cdot]$. As a result, over the full round $\mathscr{R}$,

$$(\alpha_1, \alpha_2, \alpha_3, \alpha_4) \xrightarrow{\mathscr{R}} (\gamma_1, \gamma_2, \gamma_3, \gamma_4),$$

---

[2] Note the interchange between the bit- and nibble-based representation.

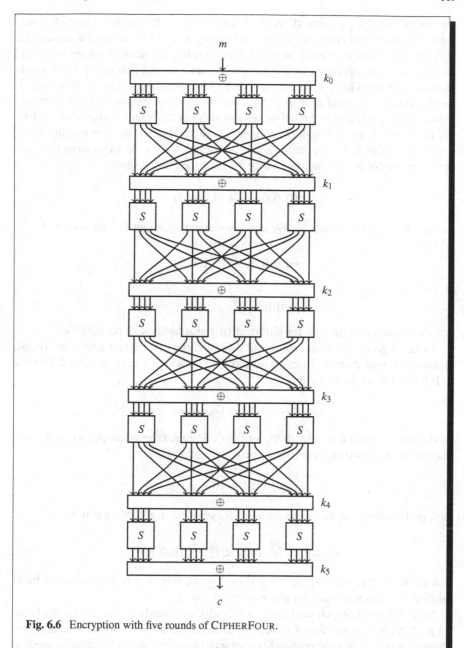

Fig. 6.6 Encryption with five rounds of CIPHERFOUR.

which denotes a nibble-based, one-round characteristic for CIPHERFOUR.

Referring to Table 6.2 we note that an input difference of 0 gives an output difference of 0 with probability 1. Clearly, equal inputs to the S-box $S[\cdot]$ lead to equal outputs! Certainly, choosing all input differences to all four S-boxes to be 0 would lead to a one-round characteristic of probability 1. But this would be useless for cryptanalysis since we would, in effect, be encrypting the same message twice and getting the same ciphertext as a result. So we deduce that a *nontrivial* one-round characteristic must have a maximum of three S-boxes taking a 0 input difference. It would seem to be a good strategy to look for a one-round characteristic for CIPHER-FOUR that has a 0 input to three S-boxes and an input difference f to one S-box, let's say the fourth. Why f? Recall from Table 6.2 that the biggest entry in the difference distribution table is for the input-output difference $f \to d$. So expressing this as a characteristic for the whole input block we have the characteristic

$$(0,0,0,f) \xrightarrow{S} (0,0,0,d)$$

that holds with probability $\frac{10}{16}$. We can immediately account for the action of $P[\cdot]$, which gives

$$(0,0,0,d) \xrightarrow{P} (1,1,0,1)$$

and so

$$(0,0,0,f) \xrightarrow{\mathscr{R}} (1,1,0,1)$$

is a one-round characteristic for CIPHERFOUR that holds with probability $\frac{10}{16}$.

So far so good. Now we can extend this characteristic by one additional round, and we must find characteristics across three S-boxes with a starting input difference 1. It follows from Table 6.2 that

$$(1,1,0,1) \xrightarrow{S} (2,2,0,2)$$

holds with a (maximum) probability of $\left(\frac{6}{16}\right)^3$. So under the assumption that the two one-round characteristics are independent and noting that

$$(2,2,0,2) \xrightarrow{P} (0,0,d,0)$$

with probability 1, we have a two-round characteristic for CIPHERFOUR,

$$(0,0,0,f) \xrightarrow{\mathscr{R}} (1,1,0,1) \xrightarrow{\mathscr{R}} (0,0,d,0),$$

that holds with probability $\frac{10}{16} \times \left(\frac{6}{16}\right)^3 = \frac{135}{4096}$. In this way it is possible to build multi-round characteristics for any number of rounds.

Note that our approach has been to add additional rounds by looking for the highest probability characteristic for each S-box from Table 6.2. However, this does not always lead to the highest probability characteristics over several rounds. A specific choice of difference might be optimal over one round but other cipher components (in this case $P[\cdot]$) might mean that much less optimal differences will need to be used later.

As an example consider inputs to CIPHERFOUR with the difference $(0,0,2,0)$. It follows from Table 6.2 and the permutation P that

$$(0,0,2,0) \xrightarrow{S} (0,0,2,0) \text{ and } (0,0,2,0) \xrightarrow{P} (0,0,2,0)$$

hold with probability $\frac{6}{16}$ and 1 respectively. The one-round characteristic for CI-PHERFOUR

$$(0,0,2,0) \xrightarrow{\mathcal{R}} (0,0,2,0)$$

holds with probability $\frac{6}{16}$. Since the input difference equals the output difference for one round, the characteristic can be concatenated with itself to produce a characteristic over as many rounds as we would like. Such a characteristic is called an *iterative* characteristic and we have that

$$(0,0,2,0) \xrightarrow{\mathcal{R}} (0,0,2,0) \xrightarrow{\mathcal{R}} (0,0,2,0)$$

is a two-round characteristic of probability $(\frac{6}{16})^2$,

$$(0,0,2,0) \xrightarrow{\mathcal{R}} (0,0,2,0) \xrightarrow{\mathcal{R}} (0,0,2,0) \xrightarrow{\mathcal{R}} (0,0,2,0)$$

is a three-round characteristic of probability $(\frac{6}{16})^3$, and

$$(0,0,2,0) \xrightarrow{\mathcal{R}} (0,0,2,0) \xrightarrow{\mathcal{R}} (0,0,2,0) \xrightarrow{\mathcal{R}} (0,0,2,0) \xrightarrow{\mathcal{R}} (0,0,2,0)$$

is a four-round characteristic that holds with probability $(\frac{6}{16})^4$ and so on. Note that the probability of the two-round iterative characteristic $(\frac{6}{16})^2$ is bigger than the probability of the previous two-round characteristic $\frac{10}{16} \times (\frac{6}{16})^3$.

To estimate the probability of our characteristic we have assumed that its evolution over successive rounds is independent. In this way the cumulative probability can be computed as the product of probabilities of one-round characteristics. In a real-life cipher this may or may not be a reasonable assumption and to determine its validity we often need some detailed analysis and experimentation.

Now we can consider attacking five rounds of CIPHERFOUR. To do this we will use the four-round characteristic

$$(0,0,2,0) \xrightarrow{\mathcal{R}} (0,0,2,0) \xrightarrow{\mathcal{R}} (0,0,2,0) \xrightarrow{\mathcal{R}} (0,0,2,0) \xrightarrow{\mathcal{R}} (0,0,2,0)$$

which has a (conjectured) probability of $(\frac{6}{16})^4$. Experimental verification of this is described in Fig. 6.7.

In an attack the cryptanalyst chooses message pairs with difference $(0,0,2,0)$ as specified in the characteristic. Using the expected difference in the partially encrypted ciphertexts after four rounds, the attacker will try to identify (some bits of) the last subkey k_5. This is done by computing backwards from the ciphertexts, as was done in earlier attacks on this chapter. However, there could be a problem. The probability of our specific difference occurring is 0.02 and yet the probability of any given difference occurring at random is $\frac{1}{16} \simeq 0.06 > 0.02$. The probability of the characteristic we have identified could be too small to effectively distinguish the correct key information.

Consider CIPHERFOUR with five rounds using randomly chosen subkeys $k_0 = $ 5b92, $k_1 = $ 064b, $k_2 = $ 1e03, $k_3 = $ a55f, $k_4 = $ ecbd, and $k_5 = $ 7ca5. An exhaustive search over all 2^{16} pairs of messages with difference $(0,0,2,0)$ reveals that $1,300$ pairs follow the four-round characteristic and the difference in the ciphertexts after each round of encryption is $(0,0,2,0)$. For this specific key, this yields a probability of approximately 0.02. For five other randomly chosen subkey sets, the number of such pairs were $1,312$, $1,290$, $1,328$, $1,318$, and $1,228$ providing an average of $1,296$. Thus, for all six sets of keys, the probability of the four-round characteristic was approximately 0.02. Our expected value was $\left(\frac{6}{16}\right)^4 \simeq 0.02$.

Fig. 6.7 Experimental confirmation of the four-round characteristic for CIPHERFOUR.

Perhaps there are other characteristics that hold with a better probability? If so, the cryptanalyst might use those instead. However, for a block cipher with a large block size, it might be hard to search for such improved characteristics. Instead, there are two more tricks that help us to exploit the full power of differential cryptanalysis.

6.2.2 Introducing Differentials

In trying to attack a five-round version of CIPHERFOUR we identified a good four-round characteristic that we would like to use. However, the probability of the characteristic holding appears to be too small to use directly.

Note that an attacker cannot verify whether a specific message pair starting with a given difference actually gives the desired difference after every round (as specified by the characteristic). We say that any pair that follows the characteristic is a *right pair*. A pair which deviates from the characteristic at some point is called a *wrong pair*.

Let us take a closer look at CIPHERFOUR and the four-round characteristic. The attacker can use the ciphertexts that are output after five rounds to compute candidate differences in the partially encrypted text after four rounds. This is for all values of part, or all, of k_5. Guesses which give rise to the expected difference specified by the characteristic give candidate values for the secret key bits.

It is important to note that the attacker is only concerned with the difference in the partially encrypted inputs after four rounds. The differences in the first, second, and third rounds are not used. Indeed, since the attacker does not (yet) know the values of the subkeys in these rounds, he has no way of determining these differences. So in our example, instead of using the four-round characteristic in full detail the attacker is really only interested in the probability that the following happens:

$$(0,0,2,0) \xrightarrow{\mathscr{R}} ? \xrightarrow{\mathscr{R}} ? \xrightarrow{\mathscr{R}} ? \xrightarrow{\mathscr{R}} (0,0,2,0)$$

where ? indicates an unknown (and irrelevant) value. Such a structure is called a *differential* and a differential may contain many characteristics, all starting and ending with the same difference. In our particular case we know that this differential contains at least one characteristic, namely,

$$(0,0,2,0) \xrightarrow{\mathscr{R}} (0,0,2,0) \xrightarrow{\mathscr{R}} (0,0,2,0) \xrightarrow{\mathscr{R}} (0,0,2,0) \xrightarrow{\mathscr{R}} (0,0,2,0).$$

But it also contains at least three other possible characteristics. They are

$$(0,0,2,0) \xrightarrow{\mathscr{R}} (0,0,0,2) \xrightarrow{\mathscr{R}} (0,0,0,1) \xrightarrow{\mathscr{R}} (0,0,1,0) \xrightarrow{\mathscr{R}} (0,0,2,0),$$

$$(0,0,2,0) \xrightarrow{\mathscr{R}} (0,0,0,2) \xrightarrow{\mathscr{R}} (0,0,1,0) \xrightarrow{\mathscr{R}} (0,0,2,0) \xrightarrow{\mathscr{R}} (0,0,2,0), \text{ and}$$

$$(0,0,2,0) \xrightarrow{\mathscr{R}} (0,0,2,0) \xrightarrow{\mathscr{R}} (0,0,0,2) \xrightarrow{\mathscr{R}} (0,0,1,0) \xrightarrow{\mathscr{R}} (0,0,2,0).$$

For each round, for all these characteristics, the inputs to three S-boxes are equal, giving rise to a 0 difference, and the combinations through the remaining S-box are $1 \xrightarrow{S} 2$, $2 \xrightarrow{S} 1$, or $2 \xrightarrow{S} 2$. Thus, according to Table 6.2, all such one-round characteristics have a probability of $\frac{6}{16}$ and the four-round characteristics shown above all have a probability of $(\frac{6}{16})^4$ if we assume the probability of a four-round characteristic can be calculated as the product of the four, individual, one-round characteristics.

What does this mean in practice? Analysis of a single characteristic suggests we are using the characteristic

$$(0,0,2,0) \xrightarrow{\mathscr{R}} (0,0,2,0) \xrightarrow{\mathscr{R}} (0,0,2,0) \xrightarrow{\mathscr{R}} (0,0,2,0) \xrightarrow{\mathscr{R}} (0,0,2,0)$$

which held with probability $(\frac{6}{16})^4$. But a little extra consideration leads us to deduce that the attacker is only really interested in a structure of the form

$$(0,0,2,0) \xrightarrow{\mathscr{R}} ? \xrightarrow{\mathscr{R}} ? \xrightarrow{\mathscr{R}} ? \xrightarrow{\mathscr{R}} (0,0,2,0).$$

We do not know the exact probability that this differential holds. But we do know that there are four characteristics contained within this differential and that they *each* hold with probability $(\frac{6}{16})^4$. This suggests that the probability that the differential holds will be at least as large as the sum of the probabilities of the four characteristics. Depending on a variety of factors, the probability of the differential should therefore be around, or greater than, $4 \times (\frac{6}{16})^4 = \frac{81}{1024}$. Some experimental confirmation of this is given in Fig. 6.8. Our cryptanalyst is in a much better position than we might have expected!

Consider five rounds of CIPHERFOUR using the subkeys of Example 6.7. Consider all 2^{16} message pairs with difference $(0,0,2,0)$. An exhaustive search reveals that $5,080$ pairs give a difference $(0,0,2,0)$ after four rounds of encryption. For this specific key, this yields a probability for the differential $(0,0,2,0) \to ? \to ? \to ? \to (0,0,2,0)$ of approximately 0.078. For five other randomly chosen sets of subkeys (the same as those used in Example 6.7) the number of pairs were $5,760, 4,640, 5,060, 5,542$, and $5,776$ respectively. This provides an average of $5,310$ corresponding to an average probability for the four-round differential of 0.081.

Fig. 6.8 Experimental confirmation of the four-round differential for CIPHERFOUR.

Consider five rounds of CIPHERFOUR using the same subkeys as Fig. 6.7. Consider all 2^{16} pairs of messages with difference $(0,0,2,0)$. An exhaustive search reveals that $7,216$ of these pairs would not be discarded in a filtering process. For the five other sets of subkeys from Example 6.7, the number of pairs not filtered were $7,842, 6,638, 7,292, 7,478$, and $7,856$ providing an overall average of $7,387$. It may be helpful to compare this to the figure derived in Fig. 6.8. There we had that, on average, $5,310$ pairs satisfied our target differential. Thus we might expect $\frac{5310}{7387} \simeq 70\%$ of filtered pairs to be useful in our attack whereas only $\frac{5310}{65536} \simeq 8\%$ would have been useful without filtering.

Fig. 6.9 Experimental confirmation of the process of filtering for CIPHERFOUR.

6.3 Filtering

The basic aim of differential cryptanalysis is to identify a statistically unusual distribution in the differences that occur. This is the signal we are trying to detect, and it can be masked by many other pairs that are not following the anticipated characteristic or differential. It can therefore be very desirable to eliminate such wrong pairs from consideration as soon as possible. This makes searching for the signal that much easier.

It is often possible to identify wrong pairs—pairs that have not followed a characteristic—by looking at the ciphertext alone. When such wrong pairs have been identified, they should be discarded from consideration and not used any further in the attack. This is advantageous because pairs that have not behaved as hoped cannot be expected to contribute useful information when recovering key information.

To illustrate the technique of *filtering* we continue with CIPHERFOUR. We now examine closely how differences propagate through the fifth (and final) round of the cipher when the plaintext pair is a right pair. In the fifth round we have

$$(0,0,2,0) \xrightarrow{S} (0,0,h,0),$$

where h can take any of the values in $\{1, 2, 9, a\}$. This can be deduced by considering the row with input difference 2 in Table 6.2. Since $P[\cdot]$ is not used in the last round, these four difference values must become part of the ciphertext after five rounds. Therefore, for each message pair, the attacker can inspect the corresponding ciphertext pair and immediately determine whether a pair is a wrong pair or, potentially, a right pair. In our example, 12 bits of the difference in the ciphertexts must be 0 bits and the remaining four bits can take only four specific values. The process of discarding wrong pairs is called *filtering* and a good filtering technique is essential for the success of many differential attacks. Fig. 6.9 gives an example of filtering for CIPHERFOUR.

6.3.1 Recovering Key Information

We can now put all the pieces together and construct a key recovery attack on CI-PHERFOUR. In our attack we will only recover four bits of the key k_5 used in the last round. These bits correspond to the four key bits that affect the S-box in the fifth encryption round, for which the output difference is nonzero. We call these key bits the *target* key bits.

From the results given in Fig. 6.8, we believe that the four-round differential

$$(0, 0, 2, 0) \xrightarrow{\mathscr{R}} ? \xrightarrow{\mathscr{R}} ? \xrightarrow{\mathscr{R}} ? \xrightarrow{\mathscr{R}} (0, 0, 2, 0)$$

holds with a probability of $\frac{5310}{65536} \simeq 0.08$. From Fig. 6.9 we assume that a pair survives the filtering process with probability $\frac{7387}{65536} \simeq 0.11$.

Suppose that the cryptanalyst receives the encryption of t message pairs which satisfy the starting difference $(0, 0, 2, 0)$. From the ciphertext difference alone, the attacker can filter out pairs that are certainly wrong. If the plaintexts are randomly chosen one expects $t \times \frac{7387}{65536} \simeq t \times 0.11$ pairs to survive filtering. Among these we will certainly have the pairs that satisfy the differential and there will be $t \times 0.08$ such pairs.

Now we know that as we try out each value to the target bits, the correct value will yield the expected difference for pairs that satisfy the differential. This means that pairs satisfying the differential will always suggest the right value of the target bits. Thus, with t pairs of texts in the attack, the correct value of the target bits will be suggested $t \times 0.08$ times.

We assume that target bits with an incorrect value give the expected difference (as specified by the differential) with a probability of $\frac{1}{16}$. Then we expect a pair not satisfying the differential but still passing the filter to suggest one (incorrect) value of the target bits on average. Thus, over t chosen message pairs, we would expect roughly $t \times (0.11 - 0.08) = t \times 0.03$ incorrect values for the target bits to be suggested. By using sufficiently many messages we expect the correct value of the target bits to be the most suggested value. For example, if $t = 500$ then we would

Consider five rounds of CIPHERFOUR using the same subkeys as in Fig. 6.7 along with the differential attack outlined in Sect. 6.3.1. With eight pairs of chosen messages, that is, 16 chosen messages, the target key bits of k_5 (the third nibble a) were suggested four times while the other 15 values of the target bits were each suggested less than four times. The attack was then implemented for a total of 100 different sets of keys. With 32 chosen messages the target key bits were the most suggested values in 64% of the cases; with 64, 128, and 256 chosen messages, respectively, the target key bits were the most suggested values in 76%, 85%, and 96% of the cases respectively.

Fig. 6.10 Experimental confirmation of the differential attack of Sect. 6.3.1.

expect the correct target value to be suggested 40 times while an incorrect target value should be suggested 15 times.

Our attack finds four key bits from a five-round version of CIPHERFOUR, see Fig. 6.10. Of course, there are many simple enhancements to this attack to derive more key information. To find other key bits one can implement a similar differential attack using other differentials. It might also be possible to determine some key bits in the first round of the cipher[3]. Once all 16 bits of the first subkey or the last subkey are determined, the attacker can peel off one round of the cipher and continue his attack on a weaker version. In a differential attack the most difficult part is to find *any* secret key bits. Once some key bits have been found it is rarely hard to recover the rest.

6.4 Getting to the Source

There are a great number of publications on differential cryptanalysis. Here we provide just a few references to some sources for additional study.

Classical papers	[81, 433]
Tutorial descriptions	[294, 700]
Other references	[12, 15, 39, 63, 65, 73, 79, 80, 142, 202, 219, 246, 252]
	[284, 285, 287, 295, 303, 318, 353, 354, 355, 389, 427]
	[507, 509, 514, 515, 517, 574, 575, 627, 634, 675, 676]
	[687, 701, 713, 749, 753]

---

[3] In CIPHERFOUR we can find values to the key bits in the first round that influence the S-box with nonzero differential input.

Chapter 7
Linear Cryptanalysis: The Idea

After differential cryptanalysis, linear cryptanalysis provides the most important general technique for analysing a block cipher. Generally speaking it tends to be far less successful than differential cryptanalysis, the most prominent exception to this rule of thumb being DES; see Chap. 2. It is an important technique and has one very specific advantage over differential cryptanalysis, namely that the cryptanalysis requires only known plaintext rather than chosen plaintext.

The development of linear cryptanalysis and its application to DES is due to Matsui [476, 475]. However, linear cryptanalytic techniques had been used earlier in an attack on FEAL-4 by Tardy-Corfdir and Gilbert [705]. Interestingly, apart from a few enhancements (see Chap. 8) linear cryptanalysis has evolved little since the early systematic treatment of Matsui. As an attack it seems to be intrinsically less versatile than differential cryptanalysis.

Just as with differential cryptanalysis we will illustrate some of the basic principles of linear cryptanalysis with a very simple cryptosystem. Let m, c, and k be b-bit strings, where m represents the message, c the ciphertext, and k the key, where

$$c = m \oplus k.$$

We already know that this system is weak when the key is reused. However, let us observe the following. If we write the quantities m, c, and k in terms of bits m_{b-1}, ..., m_0, c_{b-1}, ..., c_0, and k_{b-1}, ..., k_0 then the following b equations always hold;

$$c_0 = m_0 \oplus k_0$$
$$\vdots$$
$$c_{b-1} = m_{b-1} \oplus k_{b-1}.$$

In a known plaintext attack we know the values of m_i and c_i for $0 \leq i \leq b - 1$ and we can rearrange the equations and compute the values of k_0, ..., k_{b-1}.

Consider a slightly more complex example but with a restricted encryption scheme that operates on four-bit inputs and outputs. We might specify a (bad) encryption method where the bits of ciphertext are given by the following equations:

$$c_3 = m_3 \oplus m_1 \oplus m_0 \oplus k_3 \oplus k_1 \oplus k_0,$$
$$c_2 = m_2 \oplus m_0 \oplus k_2 \oplus k_0,$$
$$c_1 = m_3 \oplus m_2 \oplus k_3 \oplus k_2, \text{ and}$$
$$c_0 = m_1 \oplus m_0 \oplus k_1 \oplus k_0.$$

Yet we can easily rearrange these equations to give expressions for the key bits:

$$k_0 = m_0 \oplus c_0 \oplus c_3,$$
$$k_1 = m_1 \oplus c_0 \oplus c_2 \oplus c_3,$$
$$k_2 = m_2 \oplus c_1 \oplus c_1 \oplus c_2, \text{ and}$$
$$k_3 = m_3 \oplus c_0 \oplus c_1 \oplus c_2 \oplus c_3.$$

The encryption key can easily be deduced as a set of equations involving the message and the ciphertext.

The more advanced reader might be a little uncomfortable with these motivational examples. After all, we can express the weakness of these toy encryption methods in many different and more elegant ways. Nevertheless, these examples do help to motivate the basic aim of linear cryptanalysis, that is, of constructing equations that express bits of the key in terms of bits of the message and ciphertext.

7.1 Getting Started

Things have so far been easy because we have been dealing with linear encryption systems. Let us look at a slightly more complex encryption system which we have named CIPHERA. This is a four-bit block cipher with an eight-bit key and the cipher uses a four-bit look-up box $S[\cdot]$. Encryption with CIPHERA is identical to encryption with CIPHERONE from Chap. 6 except that the S-box has been changed. The four-bit message m is encrypted using an eight-bit key $k_0 \| k_1$ where both k_0 and k_1 are four-bit randomly chosen round keys, and $S[\cdot]$ is a four-bit permutation.

The approach of our first examples—of writing the key bits directly as linear equations of the message and ciphertext bits—will not be directly applicable here since $S[\cdot]$ is not linear. So, instead, our approach will be to "approximate" this non-linear component.

To illustrate, assume that an attacker knows a message m and the corresponding ciphertext c. The computation of the ciphertext can be split into the following operations:

$$u = m \oplus k_0, v = S[u], \text{ and } c = v \oplus k_1.$$

We now introduce our first set of notation for linear cryptanalysis. We will view our blocks of input, output, and key as column vectors of bits (though this will rarely be reflected in our notation). So if we wish to identify specific bits of a b-bit vector x we can do so by pre-multiplying our column vector by a row vector which acts as a

$$c = S[m \oplus k_0] \oplus k_1$$

x	0	1	2	3	4	5	6	7	8	9	a	b	c	d	e	f
$S[x]$	f	e	b	c	6	d	7	8	0	3	9	a	4	2	1	5

$$m \longrightarrow \oplus \longrightarrow u \longrightarrow \boxed{S} \longrightarrow v \longrightarrow \oplus \longrightarrow c$$

Fig. 7.1 Encryption with CIPHERA.

mask. When identifying a single bit, the masking vector consists of a single 1 with all remaining coefficients 0. To illustrate,

$$(1,0,0,0) \times \begin{pmatrix} m_3 \\ m_2 \\ m_1 \\ m_0 \end{pmatrix} = m_3, \text{ and } (0,0,1,0) \times \begin{pmatrix} m_3 \\ m_2 \\ m_1 \\ m_0 \end{pmatrix} = m_1.$$

Even better, if we wish to refer to the sum of two bits from the same vector then we apply the appropriate mask. For example, referring to our second example in the introduction to this chapter,

$$(1,0,1,1) \times \begin{pmatrix} m_3 \\ m_2 \\ m_1 \\ m_0 \end{pmatrix} \oplus (1,0,1,1) \times \begin{pmatrix} k_3 \\ k_2 \\ k_1 \\ k_0 \end{pmatrix} = m_3 \oplus m_1 \oplus m_0 \oplus k_3 \oplus k_1 \oplus k_0.$$

This will be familiar to many as the *scalar product* of vectors in linear algebra.

By introducing the concept of a *linear mask* we can formulate our linear cryptanalytic attacks in more general and elegant terms. Dropping the $\times$ notation, the first equation from our preamble has the form

$$c_3 = m_3 \oplus m_1 \oplus m_0 \oplus k_3 \oplus k_1 \oplus k_0,$$

which can be written as $\alpha \cdot c = \beta \cdot m \oplus \beta \cdot k$, where α and β are masks and

$$\alpha = (1,0,0,0) \text{ and } \beta = (1,0,1,1).$$

Returning to the task at hand, that of *approximating* the function $S[\cdot]$ in CI-PHERA, let us assume that we can find some masks (α, β) so that $(\alpha \cdot x) = (\beta \cdot S[x])$ with a probability $p \neq \frac{1}{2}$. This means that the (modulo 2) sum of certain bits of the input to S equals the (modulo 2) sum of certain bits in the output of S with some probability. Then, considering CIPHERA it follows that

$$(\alpha \cdot m) = (\alpha \cdot k_0) \oplus (\alpha \cdot u) \quad \text{with probability 1}$$
$$(\alpha \cdot u) = (\beta \cdot v) \qquad\qquad \text{with probability } p$$
$$(\beta \cdot v) = (\beta \cdot k_1) \oplus (\beta \cdot c) \quad \text{with probability 1.}$$

We can just add these equations together to get

$$(\alpha \cdot m) \oplus (\alpha \cdot u) \oplus (\beta \cdot v) = (\alpha \cdot k_0) \oplus (\alpha \cdot u) \oplus (\beta \cdot v) \oplus (\beta \cdot k_1) \oplus (\beta \cdot c),$$

which, by rearranging and allowing terms to cancel, simplifies to

$$(\alpha \cdot m) \oplus (\beta \cdot c) = (\alpha \cdot k_0) \oplus (\beta \cdot k_1) \quad \text{with probability } p.$$

It is important to observe the form of this equation. On the left we have an expression involving the sum of bits of message m and ciphertext c. On the right we have an expression involving the sum of bits of the round keys k_0 and k_1. Given a message and the corresponding ciphertext we can evaluate the left-hand side of this equation. If our equation always held, *i.e.*, $p = 1$, then we would immediately deduce one bit of information about the key material. For the cryptanalyst this would be an ideal situation. However, the cryptanalyst would be equally happy if the equation never held, that is, if $p = 0$. Why? Well, if our equation holds with probability $p = 0$ then we know that

$$(\alpha \cdot m) \oplus (\beta \cdot c) = (\alpha \cdot k_0) \oplus (\beta \cdot k_1) \quad \text{with probability 0}$$
$$\text{and so } (\alpha \cdot m) \oplus (\beta \cdot c) \oplus 1 = (\alpha \cdot k_0) \oplus (\beta \cdot k_1) \quad \text{with probability 1.}$$

Once we have a known message and ciphertext we can compute the left-hand side of the equation and this gives one bit of information about the key.

So the cryptanalyst is happy if our equation always holds or never holds. The worst case for the cryptanalyst is when the probability $p = \frac{1}{2}$. In this case the left-hand side of the equation equals the right-hand side exactly half the time. When averaged over all possible inputs this gives no information about the bit of key material: it might be 0 or 1 with equal probability. This is, of course, the ideal case for the cipher designer! So the art of linear cryptanalysis is to choose masks α and β so that equations in our linear approximation hold with probability $p = \frac{1}{2} + \varepsilon$ where ε, which is known as the *bias*, is nonzero. The bias can be positive or negative, but we are aiming to have $0 < |\varepsilon| \leq \frac{1}{2}$, and the larger $|\varepsilon|$ is, the better things will be for the attacker.

For concreteness consider the S-box $S[\cdot]$ given as follows:

x	0	1	2	3	4	5	6	7	8	9	a	b	c	d	e	f
S[x]	f	e	b	c	6	d	7	8	0	3	9	a	4	2	1	5

Now consider the masks $\alpha = (1,0,0,1)$ and $\beta = (0,0,1,0)$ and add another two rows to our table.

x	0	1	2	3	4	5	6	7	8	9	a	b	c	d	e	f
S[x]	f	e	b	c	6	d	7	8	0	3	9	a	4	2	1	5
$\alpha \cdot x$	0	1	0	1	0	1	0	1	1	0	1	0	1	0	1	0
$\beta \cdot S[x]$	1	1	1	0	1	0	1	0	0	1	0	1	0	1	0	0

If we count the number of times that $\alpha \cdot x = \beta \cdot S[x]$ then we find two matches and so $\alpha \cdot x = \beta \cdot S[x]$ with probability $\frac{2}{16}$. Equivalently we might say that $(\alpha \cdot x) \oplus 1 = \beta \cdot S[x]$ with probability $\frac{14}{16}$. So returning to the cryptanalysis of CIPHERA we have that $\alpha \cdot u \oplus 1 = \beta \cdot v$ with probability $\frac{14}{16}$ and so

$$(\alpha \cdot m) \oplus (\beta \cdot c) \oplus 1 = (\alpha \cdot k_0) \oplus (\beta \cdot k_1) \text{ with probability } \frac{14}{16}.$$

We might now be able to envisage the foundation to an attack. We use two counters T_0 and T_1 which are initialised to $T_0 = T_1 = 0$ and we request the encryptions of N known plaintexts. For each plaintext-ciphertext pair, we compute the left-hand side of our equation, which is either 0 or 1. This will be an estimate for the value of $(\alpha \cdot k_0) \oplus (\beta \cdot k_1)$. We increment the counter T_0 by 1 if we evaluate the left-hand side of our equation to 0 and we increment T_1 by 1 if we evaluate the left-hand side of our equation to 1.

Consider the situation after processing N plaintexts. While we do not know the value of $(\alpha \cdot k_0) \oplus (\beta \cdot k_1)$, we do know that it is either 0 or 1. We also know that the estimate we make for the value of this key bit is correct with probability $\frac{14}{16}$. If in reality $(\alpha \cdot k_0) \oplus (\beta \cdot k_1) = 1$ then our counter T_0 should have the value $\frac{2N}{16}$ while our counter T_1 should have the value $\frac{14N}{16}$. Similarly, if $(\alpha \cdot k_0) \oplus (\beta \cdot k_1) = 0$ then our counter T_0 should have the value $\frac{14N}{16}$ while our counter T_1 should have the value $\frac{2N}{16}$. Therefore, by looking at the value of counter T_0, say, and observing whether it has the value $\frac{2N}{16}$ or $\frac{14N}{16}$, we can determine the value of one bit of key material, namely, $(\alpha \cdot k_0) \oplus (\beta \cdot k_1)$. The likelihood of being right in this attack can be increased by repeating the procedure with more texts.

More concisely, if we let s denote the value of the right-hand side of our target equation for one (secret) value of the key, then using N texts one would expect the counter T_s to have the value pN and the counter $T_{s\oplus 1}$ to have the value $(1 - p)N$. Thus, if $p \neq \frac{1}{2}$, then with a sufficiently high N it will be possible to determine the value of s and recover one bit of key information.

7.2 Joining Approximations

Let us now look at the slightly more complex cipher CIPHERB; see Fig. 7.2.

We can assume that an attacker knows a message m and the corresponding ciphertext c. To start we can establish the following equations that hold with certainty for any masks α, β, and γ:

$$c = S[S[m \oplus k_0] \oplus k_1] \oplus k_2$$

x	0	1	2	3	4	5	6	7	8	9	a	b	c	d	e	f
$S[x]$	f	e	b	c	6	d	7	8	0	3	9	a	4	2	1	5

$$m \to \oplus \to u \to \boxed{S} \to v \to \oplus \to w \to \boxed{S} \to x \to \oplus \to c$$

Fig. 7.2 Encryption with CIPHERB.

$$(\alpha \cdot m) = (\alpha \cdot k_0) \oplus (\alpha \cdot u),$$
$$(\beta \cdot v) = (\beta \cdot k_1) \oplus (\beta \cdot w), \text{ and}$$
$$(\gamma \cdot x) = (\gamma \cdot k_2) \oplus (\gamma \cdot c).$$

Compared with our earlier analysis a complication arises from having two occurrences of $S[\cdot]$ during encryption. However, we assume that we can find specific values to the masks α, β, and γ so that

$$\alpha \cdot u = \beta \cdot S[u] = \beta \cdot v \quad \text{with probability } p_1 \neq \tfrac{1}{2} \text{ and}$$
$$\beta \cdot w = \gamma \cdot S[w] = \gamma \cdot x \quad \text{with probability } p_2 \neq \tfrac{1}{2}.$$

This gives us five equations which link together to cover the entire encryption process. We can see this by adding the first three equations,

$$(\alpha \cdot m) \oplus (\beta \cdot v) \oplus (\gamma \cdot x) = (\alpha \cdot u) \oplus (\beta \cdot w) \oplus (\gamma \cdot c) \oplus (\alpha \cdot k_0) \oplus (\beta \cdot k_1) \oplus (\gamma \cdot k_2)$$

and rearranging to give

$$(\alpha \cdot m) \oplus (\gamma \cdot c) \oplus (\alpha \cdot u) \oplus (\beta \cdot v) \oplus (\beta \cdot w) \oplus (\gamma \cdot x) = (\alpha \cdot k_0) \oplus (\beta \cdot k_1) \oplus (\gamma \cdot k_2),$$

where the right-hand side is constant. We also know the following relations hold:

$$(\alpha \cdot u) = (\beta \cdot v) \quad \text{with probability } p_1,$$
$$(\beta \cdot w) = (\gamma \cdot x) \quad \text{with probability } p_2,$$

where the probabilities are taken over all possible values of u and w. These will allow us to remove the internal variables $\alpha \cdot u$, $\beta \cdot v$, $\beta \cdot w$, and $\gamma \cdot x$ from our linear approximation to CIPHERB. The important question to resolve is, what is the probability that the approximation holds when all the internal variables have canceled out? Assuming that the two events $(\alpha \cdot u) = (\beta \cdot v)$ and $(\beta \cdot w) = (\gamma \cdot x)$ are independent, this probability is easy to calculate.

1. In the case $\alpha \cdot u = \beta \cdot v$ and $\beta \cdot w = \gamma \cdot x$, then we have that

$$(\alpha \cdot m) \oplus (\gamma \cdot c) = (\alpha \cdot k_0) \oplus (\beta \cdot k_1) \oplus (\gamma \cdot k_2)$$

2. A similar equation results if $\alpha \cdot u = (\beta \cdot v) \oplus 1$ and $\beta \cdot w = (\gamma \cdot x) \oplus 1$.
3. If $\alpha \cdot u = \beta \cdot v$ while $\beta \cdot w = \gamma \cdot x \oplus 1$, or *vice versa*, then we have that

$$(\alpha \cdot m) \oplus (\gamma \cdot c) \oplus 1 = (\alpha \cdot k_0) \oplus (\beta \cdot k_1) \oplus (\gamma \cdot k_2).$$

The probability of the first event is $p_1 \times p_2$ and the probability of the second event is $(1 - p_1) \times (1 - p_2)$. Taken together, this means that the equation

$$(\alpha \cdot m) \oplus (\gamma \cdot c) = (\alpha \cdot k_0) \oplus (\beta \cdot k_1) \oplus (\gamma \cdot k_2)$$

holds with probability $p_1 p_2 + (1 - p_1)(1 - p_2)$. If this probability is sufficiently far away from $\frac{1}{2}$ one can determine the value of the right side of the equation with sufficiently many pairs of plaintexts and ciphertexts. The method is exactly the same as for our attack on CIPHERA.

How do we evaluate how far away $p_1 p_2 + (1 - p_1)(1 - p_2)$ is from $\frac{1}{2}$? One approach is to work with biases and to set $p_1 = \frac{1}{2} + \varepsilon_1$ and $p_2 = \frac{1}{2} + \varepsilon_2$. Then, evaluating $p_1 p_2 + (1 - p_1)(1 - p_2)$ gives

$$
\begin{aligned}
&p_1 p_2 + 1 - p_1 - p_2 + p_1 p_2 \\
&= 1 - p_1 - p_2 + 2 p_1 p_2 \\
&= 1 - \frac{1}{2} - \varepsilon_1 - \frac{1}{2} - \varepsilon_2 + 2(\frac{1}{4} + \frac{\varepsilon_1}{2} + \frac{\varepsilon_2}{2} + \varepsilon_1 \varepsilon_2) \\
&= \frac{1}{2} + 2\varepsilon_1 \varepsilon_2.
\end{aligned}
$$

In Chap. 8 this result is generalised and it is often referred to as the *piling-up lemma*. It's not new, but it is a handy result that says that the sum of m (independent) Boolean expressions with probabilities p_i for $i = 1, \ldots, m$ is

$$\frac{1}{2} + 2^{m-1} \prod_{i=1}^{m} \left(p_i - \frac{1}{2} \right).$$

Returning to CIPHERB and working with $S[\cdot]$, we need to find masks α, β, and γ so that

$$(\alpha \cdot m) \oplus (\gamma \cdot c) = (\alpha \cdot k_0) \oplus (\beta \cdot k_1) \oplus (\gamma \cdot k_2)$$

has the highest possible bias. To do this, we typically construct a table of linear approximations as shown in Table 7.1. This table lists the probabilities that the sum of certain input bits of a equals the sum of certain output bits of $S[a]$.

When looking at Table 7.1 we notice that there are three entries with the value -6. The corresponding linear relations for these entries will give the highest (nontrivial) bias through $S[\cdot]$. Consider the entry $(9, 2)$ with the value -6. This is the combination that was used in Sec. 7.1. Such a pair (α, β), where α is the input mask and β the

Table 7.1 Linear approximation table for $S[\cdot]$. The input mask values run down the first column, while the output mask values run across the first row. We have omitted the 0 values to the input and output mask since these will result in trivial approximations. If we divide entry (i, j) by 16 and add $\frac{1}{2}$ then this gives the probability that an input masked by i equals the output masked by j, where the probability is taken over all possible inputs. For clarity '.' is used to denote a zero value.

	1	2	3	4	5	6	7	8	9	a	b	c	d	e	f
1	-2	.	2	.	-2	4	-2	2	4	2	.	-2	.	2	.
2	2	-2	.	-2	.	.	2	2	4	.	2	4	-2	-2	.
3	4	2	2	-2	2	.	.	.	.	2	-2	-2	-2	.	4
4	.	-2	2	2	-2	.	.	-4	.	2	2	2	2	.	4
5	-2	2	.	2	4	.	2	-2	4	.	-2	.	2	-2	.
6	-2	.	2	.	2	4	2	2	-4	2	.	2	.	-2	.
7	.	.	.	4	.	-4	.	.	.	.	4	.	4	.	.
8	.	-2	2	-4	.	2	2	-4	.	-2	-2	.	.	2	-2
9	-2	-6	.	.	2	-2	.	2	.	.	-2	-2	.	.	2
a	-2	.	-6	-2	.	2	.	-2	.	2	.	.	-2	.	2
b	.	.	.	2	-2	2	-2	.	.	-4	-4	2	-2	-2	2
c	.	.	.	-2	-2	-2	-2	.	.	4	-4	2	2	-2	-2
d	-2	.	2	2	.	-2	.	-2	.	2	.	.	-6	.	-2
e	2	-2	.	.	2	2	-4	-2	.	.	2	-2	.	-4	-2
f	-4	2	2	-4	.	-2	-2	.	.	-2	2	.	.	-2	2

output mask, is called a (linear) *characteristic*, and we will denote this by $\alpha \xrightarrow{S} \beta$. This explicitly denotes the function across which the approximation holds. Just as in the case of a differential characteristic, a linear characteristic has a probability, and hence a bias, associated with it.

Now, returning to our approximation it follows, by inspection of Table 7.1, that choosing $\alpha = \beta = \gamma = \mathrm{d}$, gives the highest bias for approximations across CIPHERB. One therefore gets for CIPHERB that

$$(m \cdot \mathrm{d}) \oplus (c \cdot \mathrm{d}) = (k_0 \cdot \mathrm{d}) \oplus (k_1 \cdot \mathrm{d}) \oplus (k_2 \cdot \mathrm{d})$$

with probability

$$\frac{1}{8} \times \frac{1}{8} + \frac{7}{8} \times \frac{7}{8} = \frac{25}{32} = \frac{1}{2} + \frac{9}{32}.$$

This is the probability of the characteristic ($\mathrm{d} \xrightarrow{S} \mathrm{d} \xrightarrow{S} \mathrm{d}$). In an attack on CIPHERB we would calculate $(m \cdot \mathrm{d}) \oplus (c \cdot \mathrm{d})$ for N known messages. If the result were 0 more often than 1 then we would guess that $(k_0 \oplus k_1 \oplus k_2) \cdot \mathrm{d}$ is 0. If the result were 1 more often than 0 then we would guess that $(k_0 \oplus k_1 \oplus k_2) \cdot \mathrm{d}$ is 1.

While we will look at this in more detail in the next chapter, it turns out that to get a good probability of success one needs about $N = |p - 1/2|^{-2}$ known messages.

$$c = S[S[S[m \oplus k_0] \oplus k_1] \oplus k_2] \oplus k_3$$

x	0	1	2	3	4	5	6	7	8	9	a	b	c	d	e	f
$S[x]$	f	e	b	c	6	d	7	8	0	3	9	a	4	2	1	5

$$m \to \oplus \overset{k_0}{\underset{}{\to}} \boxed{S} \to \oplus \overset{k_1}{\underset{}{\to}} \boxed{S} \to x \to \oplus \overset{k_2}{\underset{}{\to}} y \to \boxed{S} \to z \to \oplus \overset{k_3}{\underset{}{\to}} c$$

Fig. 7.3 Encryption with CIPHERC.

In this particular case we would have that $N = \left(\frac{9}{32}\right)^{-2} \simeq 12$ and so perhaps a dozen known messages should suffice.

7.2.1 Deducing More Key

In our attacks so far we have only recovered a single bit of key information, namely the exclusive-or of certain bits of the round keys. If we were attacking a cipher in this way then we'd recover a single bit of key and we'd be left trying to recover the rest of the key by some other means. One bit of key material (while regrettable) probably won't help the attacker too much if the final goal is recovery of the complete key.

The first suggestion might be to use another linear approximation. This would likely involve different bits of key material and so we would recover an additional bit of key in this way. This is certainly reasonable, though it is not always possible to find alternative approximations with a sufficiently good bias. Instead, it would be better if we could use the one good approximation that we have already identified and recover more than one bit of key information. To illustrate this, we consider a third cipher which has been named CIPHERC. This is a three-round cipher and we assume that the attacker knows a message m and the corresponding ciphertext c.

CIPHERC is similar to CIPHERB but with an extra round of encryption. From our work on CIPHERB we have already identified the following linear approximation (written here in terms of the variables for CIPHERC)

$$(m \cdot \mathsf{d}) \oplus (y \cdot \mathsf{d}) = (k_0 \cdot \mathsf{d}) \oplus (k_1 \cdot \mathsf{d}) \oplus (k_2 \cdot \mathsf{d}).$$

This holds with probability $\frac{1}{2} + \frac{9}{32}$. In other words, the sum of three (specified) bits of a plaintext equals the sum of three (specified) bits of the intermediate value y plus the sum of some (specified) bits of the round keys with probability $\frac{1}{2} + \frac{9}{32}$.

Taking inspiration from previous attacks, it is natural for the attacker to guess the value of k_3 and to unwind the last round of encryption. When the guess for k_3

is correct, the attacker correctly computes the value of the intermediate value y. He can therefore correctly evaluate the left-hand side of

$$(m \cdot \mathrm{d}) \oplus (y \cdot \mathrm{d}) = (k_0 \cdot \mathrm{d}) \oplus (k_1 \cdot \mathrm{d}) \oplus (k_2 \cdot \mathrm{d}).$$

If the attacker is given N messages and their corresponding ciphertexts, then for a possible value i of k_3 he can compute the value $y' = S^{-1}[c \oplus i]$ from a given ciphertext c. The attacker then uses the corresponding message to compute

$$(m \cdot \mathrm{d}) \oplus (y' \cdot \mathrm{d}).$$

For each possible guess i we will keep two counters $T_0^{(i)}$ and $T_1^{(i)}$. We will increment $T_0^{(i)}$ if $(m \cdot \mathrm{d}) \oplus (y' \cdot \mathrm{d})$ gives a 0 and increment $T_1^{(i)}$ otherwise.

Note that while the attacker doesn't know the value of $(k_0 \cdot \mathrm{d}) \oplus (k_1 \cdot \mathrm{d}) \oplus (k_2 \cdot \mathrm{d})$ he does know that it is constant provided the key doesn't change. So if the attacker has guessed the correct value v, say, for k_3 then when he examines the counters $T_0^{(v)}$ and $T_1^{(v)}$, one counter should show a value close to $\frac{N}{2} + \frac{9N}{32}$ and the other should show a value close to $\frac{N}{2} - \frac{9N}{32}$. It is assumed that all the other pairs of counters, which correspond to incorrect guesses for k_3, will show roughly the same value for both counters in the pair, namely about $\frac{N}{2}$.

Thus, when looking at the pairs of counters, the attacker could learn (a) the correct value v to the key k_3 by finding the twin counters that show the expected imbalance, and (b) the value of $(k_0 \oplus k_1 \oplus k_2) \cdot \mathrm{d}$ since within the twin counters with imbalance, the largest counter value will (in this case) indicate the value of the key bit $(k_0 \oplus k_1 \oplus k_2) \cdot \mathrm{d}$.

This is a simple trick to recover information from the last round of encryption. We can do something similar with the first round. In this way, therefore, it is possible to determine more than one bit of key information in a linear cryptanalytic attack.

The main consideration in practice is the management of the counters. In general, the attacker needs to have the computing power to search exhaustively through all values of the round key k_3; there are $2^{|k_3|}$ possible keys and this means that the attacker sets up $2^{|k_3|}$ pairs of counters $(T_0^{(i)}, T_1^{(i)})$ for $0 \leq i \leq 2^{|k_3|} - 1$. These may well have to handle large quantities of data.

As in the case of recovering a single bit with linear cryptanalysis, it turns out that to get a good probability of success in an attack of this sort the attacker needs about $N = c|p - 1/2|^{-2}$ (or $N = c|\varepsilon|^{-2}$ where the bias is ε) known plaintexts. The constant $c \geq 2$ will vary depending on the attack and the block cipher, but it will be larger than that required in the recovery of a single key bit. This is quite intuitive: in the case of recovering a single bit we had to choose between two counters; here, when recovering $|k_3|$ bits we need to choose between $2^{|k_3|}$ pairs of counters. To attain the same level of confidence we will need more data.

x	0	1	2	3	4	5	6	7
$S[x]$	f	e	b	c	6	d	7	8

x	8	9	a	b	c	d	e	f
$S[x]$	0	3	9	a	4	2	1	5

i	0	1	2	3	4	5	6	7
$P[i]$	0	4	8	12	1	5	9	13

i	8	9	10	11	12	13	14	15
$P[i]$	2	6	10	14	3	7	11	15

1. Set $u_0 = m$.
2. For $i := 1$ to $r - 1$ do:

   a. Combine the round key k_{i-1} with u_{i-1} so that $a_i = u_{i-1} \oplus k_{i-1}$.
   b. Divide a_i into four nibbles $a_i = A_0 \| \ldots \| A_3$.
   c. Compute $S[A_0] \| \ldots \| S[A_3]$ where $S[\cdot]$ is defined above.
   d. Write $S[A_0] \| \ldots \| S[A_3]$ as $y_{15} \ldots y_0 = S[A_0] \| \ldots \| S[A_3]$.
   e. Permute bit y_i to position j according to the permutation $j = P[i]$ defined above.
   f. Set $a_i = y_{15} \| y_{11} \| \ldots \| y_4 \| y_0$.

3. (Last round)

   a. Combine the round key k_{r-1} with u_{r-1} so that $a_r = u_{r-1} \oplus k_{r-1}$.
   b. Divide a_r into four nibbles $a_r = A_0 \| \ldots \| A_3$.
   c. Compute $S[A_0] \| \ldots \| S[A_3]$ where $S[\cdot]$ is defined above.
   d. Write $S[A_0] \| \ldots \| S[A_3]$ as $y_{15} \ldots y_0 = S[A_0] \| \ldots \| S[A_3]$.
   e. Output $y \oplus k_r$ as ciphertext.

Fig. 7.4 Encryption with r rounds of CIPHERD.

7.3 Putting Things Together

Finally we arrive at our most sophisticated example, CIPHERD. Apart from the small block size this cipher has many attributes that we find in real-life ciphers; it is identical to CIPHERFOUR from Chap. 6 but uses a different S-box. CIPHERD operates on 16-bit blocks and it is an iterated cipher with r rounds. The $r + 1$ 16-bit round keys are independent and chosen uniformly at random. Encryption with CIPHERD is described in Fig. 7.4 and illustrated in Fig. 7.5. Recall from the description of CIPHERFOUR in Chap. 6 that there is no diffusion permutation in the final round.

Let us consider a linear cryptanalytic attack on CIPHERD. First note that all components except for the S-box are linear with respect to the exclusive-or operation. These will give linear equations that always hold. The S-box of CIPHERD is the same as in the previous example, so the linear relations through $S[\cdot]$ are given in Table 7.1. The first thing to do in attacking CIPHERD is to find a good linear characteristic over as many rounds as possible. In this example we shall specify a

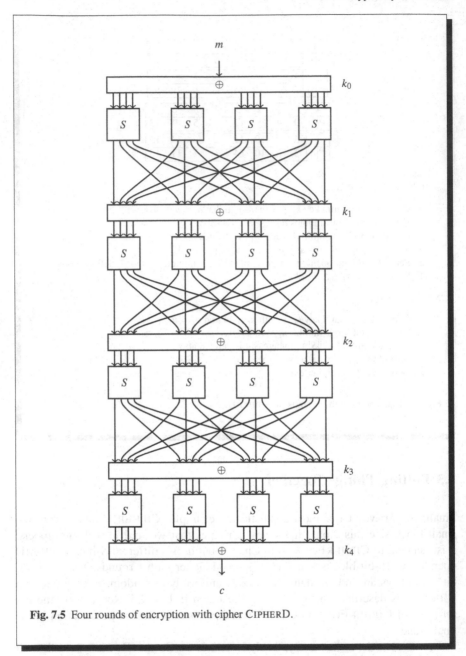

Fig. 7.5 Four rounds of encryption with cipher CIPHERD.

characteristic over the first $r-1$ rounds of an r-round version of CIPHERD. If this can be achieved using an approximation with a sufficiently good bias, then we can use our techniques from Sec. 7.2.1 to find the round key k_r.

To build a characteristic over several rounds, one normally tries to find good one-round characteristics that can be combined. It follows that we must find input and output masks to the S-boxes which will lead to linear characteristics with a large bias. Then since the permutation P is linear one can compute the masks of one round. Let

$$(\alpha_0\,\alpha_1\,\alpha_2\,\alpha_3) \xrightarrow{S} (\beta_0\,\beta_1\,\beta_2\,\beta_3)$$

denote the input and output masks through each of the four S-boxes, and let

$$(\beta_0\,\beta_1\,\beta_2\,\beta_3) \xrightarrow{P} (\gamma_0\,\gamma_1\,\gamma_2\,\gamma_3)$$

denote the input and output masks through the linear function P. Then

$$(\alpha_0\,\alpha_1\,\alpha_2\,\alpha_3) \xrightarrow{\mathscr{R}} (\gamma_0\,\gamma_1\,\gamma_2\,\gamma_3)$$

will denote a one-round characteristic for CIPHERD.

Referring to Table 7.1 we recall that input and output masks of 0 will give a trivial approximation (involving no bits) holding with probability 1. Clearly choosing 0 input masks to all four S-boxes is useless for cryptanalysis since this would mean that we were not tracing any bits in the linear approximation. However, in a nontrivial one-round approximation to CIPHERD, we can have up to three S-boxes with a 0 input mask. Therefore the most promising one-round approximation for CIPHERD will have a 0 input to three S-boxes and an input difference such as d to the fourth S-box. As an example,

$$(000\mathrm{d}) \xrightarrow{S} (000\mathrm{d})$$

is a perfectly good linear characteristic that holds with probability $\frac{1}{2} - \frac{6}{16}$. We can trace the action of P on the linear characteristic and obtain

$$(000\mathrm{d}) \xrightarrow{\mathscr{R}} (1101)$$

for one round of CIPHERD that holds with probability $\frac{1}{2} - \frac{6}{16}$.

Now consider the next round. If this characteristic were extended by one additional round, then the cryptanalyst would have to find linear characteristics through three instances of $S[\cdot]$ which would all require input mask 1. It follows from Table 7.1 that a good choice could be

$$(1101) \xrightarrow{S} (6606) \xrightarrow{P} (0\mathrm{d}\mathrm{d}0),$$

which holds with a probability of $\frac{1}{2} + 2^2\left(\frac{4}{16}\right)^3 = \frac{1}{2} + \frac{1}{16}$, where we have used the piling-up lemma; see page 133.

Thus, under the assumption that the two one-round linear characteristics are independent, we have that

$$(000\mathrm{d}) \xrightarrow{\mathscr{R}} (1101) \xrightarrow{\mathscr{R}} (0\mathrm{d}\mathrm{d}0)$$

is a two-round linear characteristic for CIPHERD that holds with probability

$$\frac{1}{8} \times \frac{9}{16} + \frac{7}{8} \times \frac{7}{16} = \frac{29}{64} = \frac{1}{2} - \frac{3}{64}.$$

It is possible to continue this style of argument and to build s-round linear characteristics for any number of rounds. For each additional round we try to find the most biased linear approximation across each S-box using Table 7.1 and concatenate the new one-round linear characteristic onto the existing characteristic.

However, this does not always lead to linear characteristics with the greatest biases. Consider using the input mask (8000) across the ensemble of four S-boxes. From Table 7.1 we see that

$$(8000) \xrightarrow{S} (8000)$$

holds with probability $\frac{1}{2} - \frac{4}{16}$. When we incorporate the action of P we find that

$$(8000) \xrightarrow{\mathscr{R}} (8000)$$

is a one-round linear characteristic for CIPHERD that holds with probability $\frac{1}{2} - \frac{4}{16}$. Over one round this is worse than the linear characteristic we used previously,

$$(000d) \xrightarrow{\mathscr{R}} (1101).$$

However, the input and output masks for the new one-round linear characteristic are equal. This means that the characteristic can be concatenated with itself over as many rounds as we need. Such a characteristic is called an *iterative* characteristic and as a consequence we have that

$$(8000) \xrightarrow{\mathscr{R}} (8000) \xrightarrow{\mathscr{R}} (8000)$$

is a two-round linear characteristic that holds with probability

$$\left(\frac{1}{4}\right)^2 + \left(\frac{3}{4}\right)^2 = \frac{5}{8} = \frac{1}{2} + \frac{1}{8}.$$

We have assumed that the constituent one-round linear characteristics act independently, but if this is the case then the resultant bias $\frac{1}{8}$ will be higher than the bias $\frac{1}{16}$ that we would have used in the previous case. So now considering CIPHERD with (say) five rounds, we can build the following four-round characteristic:

$$(8000) \xrightarrow{\mathscr{R}} (8000) \xrightarrow{\mathscr{R}} (8000) \xrightarrow{\mathscr{R}} (8000) \xrightarrow{\mathscr{R}} (8000).$$

Using the piling-up lemma (see page 133) and the assumption of independent rounds, this four-round linear characteristic holds with a probability of

Consider CIPHERD with four rounds using the randomly chosen round keys $k_0 = $ 5b92, $k_1 = $ 064b, $k_2 = $ 1e03, $k_3 = $ a55f, $k_4 = $ ecbd. Consider all 2^{16} plaintexts and set both input and output masks to $(8, 0, 0, 0)$. According to previous analysis this linear characteristic will have an average bias of $\frac{1}{32}$. An exhaustive search reveals that in this case the four-round characteristic has a bias of about $\frac{1}{9}$. For 99 other randomly chosen sets of round keys the biases were recorded yielding an average bias of $\frac{1}{18.2}$ in the 100 tests. It is noted that the recorded biases varied from $\frac{1}{6}$ to $\frac{1}{2048}$. Thus for some keys the biases are too low to be useful.

Fig. 7.6 Experimental analysis of the linear characteristic $(8000) \xrightarrow{\mathscr{R}} (8000) \xrightarrow{\mathscr{R}} (8000)$ $\xrightarrow{\mathscr{R}} (8000) \xrightarrow{\mathscr{R}} (8000)$.

$$\frac{1}{2} + 2^3 \left(\frac{1}{4}\right)^4 = \frac{17}{32} = \frac{1}{2} + \frac{1}{32}.$$

We can use this to attack a five-round version of CIPHERD.

7.3.1 Introducing Linear Hulls

It is always instructive to implement the attacks we discuss and to check that experiments confirm our intuition. So let's examine the four-round linear characteristic with which we aim to attack a five-round version of CIPHERD. By construction we deduced that

$$(8000) \xrightarrow{\mathscr{R}} (8000) \xrightarrow{\mathscr{R}} (8000) \xrightarrow{\mathscr{R}} (8000) \xrightarrow{\mathscr{R}} (8000)$$

held with probability $\frac{1}{2} + \frac{1}{32}$. The results of some limited experiments are given in Fig. 7.6.

The results are interesting in (at least) two aspects. First, there is a strong key dependence in the bias recorded for a linear characteristic. For some keys the bias experienced in practice is much greater than that for other keys. We will examine this issue more in Chap. 8. Second, the average bias in practice was around $\frac{1}{18.2}$ instead of the $\frac{1}{32}$ predicted theoretically. There are clearly additional factors at work.

Closer analysis of our linear characteristic shows some interesting features. Recall that our linear characteristic has the form

$$(8000) \xrightarrow{\mathscr{R}} (8000) \xrightarrow{\mathscr{R}} (8000) \xrightarrow{\mathscr{R}} (8000) \xrightarrow{\mathscr{R}} (8000)$$

and connects the leftmost bit of the message to the leftmost bit of the ciphertext after four rounds of encryption. Yet, as in the case of differentials highlighted in Chap. 6, there may well be other linear characteristics that have the same starting and ending mask. Indeed, the same input and output bits can be related in other ways:

Consider a linear attack on CIPHERD with five rounds using the linear hull $(8,0,0,0) \xrightarrow{4\mathscr{R}}$ $(8,0,0,0)$ of average bias $\frac{1}{18.2}$ and a randomly chosen key. With $c = 3$ one gets $c|\varepsilon|^{-2} \simeq$ 994. Generate 1,000 random known messages. For all values of four bits to the last round key, decrypt the ciphertexts one round and compute the resulting bias. The key which produces the highest bias is taken as the guess for the secret bits. In 100 such tests, the correct value of the four key bits resulted in the highest bias 45 times and the second-highest bias nine times.

Fig. 7.7 Experimental attacks on five-round CIPHERD.

$$(8000) \xrightarrow{\mathscr{R}} (0800) \xrightarrow{\mathscr{R}} (4000) \xrightarrow{\mathscr{R}} (8000) \xrightarrow{\mathscr{R}} (8000)$$
$$(8000) \xrightarrow{\mathscr{R}} (8000) \xrightarrow{\mathscr{R}} (0800) \xrightarrow{\mathscr{R}} (4000) \xrightarrow{\mathscr{R}} (8000)$$

It can be established that the bias for these new linear characteristics is $\frac{1}{32}$, and all three characteristics are included in what is called a *linear hull* [565]:

$$(8,0,0,0) \xrightarrow{\mathscr{R}} (*,*,*,*) \xrightarrow{\mathscr{R}} (*,*,*,*) \xrightarrow{\mathscr{R}} (*,*,*,*) \xrightarrow{\mathscr{R}} (8,0,0,0).$$

The asterisks indicate that the mask can take any value and we really don't care too much about it. However, these internal masks may well introduce different key bits to the different approximations. The net result is that the bias we see in reality will be a complex key-dependent interaction between several linear characteristics, something that is commonly described in the literature as a *linear hull effect* and which is illustrated in Fig. 7.6 for CIPHERD.

However, we can continue with our attack on five-round CIPHERD using a four-round linear hull. As in previous attacks, the idea is to take the ciphertext and to compute backwards through the last round of the cipher for each guess for the key in the last round. With a correct guess the attacker should see the bias of the four-round linear hull reflected in the associated counter values. For an incorrect key guess we would expect the counters to have roughly equal values.

While the last round key has 16 bits in it, we observe that only four play any role in determining the value of the linear hull. The ciphertext mask for the linear hull only involves bits from one S-box. Out of the 16 key bits 12 have no effect. These bits will not be (immediately) recoverable without other techniques. However, we can recover the remaining four bits of key from the last round.

Using our results from before, we expect to require around $c|\varepsilon|^{-2}$ plaintexts for c a small constant where ε is the bias of the linear hull. The results of experiments are presented in Fig. 7.7. As expected, most of the time the correct value to the four key bits was recovered. Sometimes the correct value was missed since another pair of counters had a more biased appearance. Taking more text—that is, increasing the value of c—would help to lift the correct key guesses away from the incorrect guesses.

More difficult to deal with is the fluctuating bias of the linear hull when different round keys are used. As observed in Fig. 7.6, the bias that occurs in practice can differ significantly from the theoretical value when the key changes. Thus we see in Fig. 7.7 that there were cases in our experiments where the four key bits from the last round of encryption could not be determined. In such cases the values to the round keys were such that the bias of the linear hull was much lower than expected and insufficient data was available to reliably recover any key information.

7.3.2 A Unified Measure

Before closing this introduction to linear cryptanalysis, we highlight a nice way of providing a duality between the resistance of a cipher to differential and to linear cryptanalysis. For this we need to introduce the concept of the *correlation* between two Boolean functions.

Imagine that we have a set of s connecting linear approximations A_i and that the approximation $A_i = d_{i-1} \oplus d_i$ holds with probability $p_i = \frac{1}{2} + \varepsilon_i$ for $i = 1, \ldots, s$. Then if the s events are independent, one can use the piling-up lemma and say that the combined approximation A for $d_0 \oplus d_s$ holds with probability p where

$$p = \frac{1}{2} + 2^{s-1} \prod_{i=1}^{s} \left(p_i - \frac{1}{2} \right).$$

To provide an alternative view, consider two Boolean functions f and g which take as input n bits and return a single bit. Clearly, $f(b_0, \ldots, b_{n-1}) = g(b_0, \ldots, b_{n-1})$ with some probability p that depends on the form of f and g. Then the *correlation* [631, 140] between f and g is given by

$$corr(f, g) = 2p - 1.$$

Returning to our case of the piling-up lemma, we observe that

$$corr(d_0, d_s) = 2 \left(\frac{1}{2} + 2^{s-1} \prod_{i=1}^{s} \left(p_i - \frac{1}{2} \right) \right) - 1$$

$$= 2^s \prod_{i=1}^{s} \left(p_i - \frac{1}{2} \right)$$

$$= \prod_{i=1}^{s} (2p_i - 1) = \prod_{i=1}^{s} corr(d_{i-1}, d_i).$$

This means that to compute the resultant correlation when combining several approximations, we can just multiply the constituent correlations.

We know that the data requirements for a linear cryptanalytic attack depend on the inverse square of the bias, and so it will be useful to define the *squared correla-*

tion as $q_i = (2p_i - 1)^2$. The squared correlation for the overall approximation will be given by $q = (2p - 1)^2$, where $q = \prod_{i=1}^{s} q_i$.

Now we see that the data complexities for differential and linear attacks, which are classically given as p^{-1} and ε^{-2}, can in fact be replaced by p^{-1} and q^{-1} where p is the probability of the differential and q is the squared correlation of the linear hull. This unified approach to measuring the strength of a cipher with respect to differential and linear cryptanalysis has turned out to be very helpful in the literature.

7.4 Getting to the Source

As for differential cryptanalysis, there are a great number of publications on linear cryptanalysis. We provide references to a selection.

Classical papers	[475, 476]
Tutorial descriptions	[294, 476, 700]
Other references	[23, 24, 61, 123, 144, 164, 165, 250, 256, 286, 293]
	[335, 336, 356, 360, 396, 537, 576, 581, 654, 683, 708]
	[732]

Chapter 8
Advanced Topics

In this chapter we will be looking at several issues in depth. First we will return to differential and linear cryptanalysis. In Chaps. 6 and 7 our descriptions were driven by example but there are more formal foundations to these topics. So in this chapter we will provide pointers to these foundations and other work in the area. We will also use this chapter to emphasize links between analysis and design and we illustrate how block cipher design has advanced over the years, often in response to increasingly sophisticated attacks.

8.1 Differential Cryptanalysis Revisited

Previously we introduced increasingly complex ciphers to illustrate different features of a differential attack. However, we could have been much more formal and with this in mind we will set up some notation. We let $\mathcal{M}$ denote the set of all possible messages or plaintexts, and we let $\mathcal{K}$ denote the set of keys. For completeness we let $\mathcal{C}$ denote the set of ciphertexts.

Now consider the *difference* between two elements x and y of $\mathcal{M}$, and sometimes a restriction of $\mathcal{M}$ that we might denote G. We can denote this difference by

$$\Delta(x,y) = x \otimes y^{-1}$$

where y^{-1} is the inverse of y with respect to some operation $\otimes$ on the set G. For most ciphers G will be a set of fixed-length bit strings and $\otimes$ will be the exclusive-or operation. In such cases the inverse of y with respect to the exclusive-or operation is y itself. However, we can extend the notion of difference to other operations. For example, the inverse of y with respect to addition modulo 2^n would be $2^n - y$.

The important observation is that the appropriate choice of difference will depend on the block cipher. Typically the cryptanalyst chooses a notion of difference that coincides with the operation used to introduce key material. In such circumstances the difference between x and y before combining the key will be preserved:

$$(x \otimes k) \otimes (y \otimes k)^{-1} = x \otimes k \otimes k^{-1} \otimes y^{-1} = x \otimes y^{-1}.$$

The difficulties in differential cryptanalysis stem from predicting how the difference between intermediate texts will evolve through other components of the cipher. If h is a non-linear cipher component with respect to $\otimes$, then it can be difficult to predict with certainty the difference between outputs $h(x)$ and $h(x')$ when given the difference between inputs x and x'. Often, however, performance and implementation considerations in the design of the cipher have led to cipher components with a relatively small set of inputs and outputs, for instance, S-boxes. The *domain*, often abbreviated to Dom, is the set of possible inputs while the *range* is the set of possible outputs. With a small domain and range it is often possible to compute full differential tables, so-called *difference distribution tables*, for different components of the cipher. For DES and the AES the non-linear components are the S-boxes, and for each possible difference between two inputs we can list the frequency with which each possible output difference occurs. Such a table for a function h can be represented by a matrix $A = \{A_{i,j}\}$ where

$$A_{i,j} = |\{x \in \text{Dom}(h) : h(x) \otimes (h(x \otimes i^{-1}))^{-1} = j\}|.$$

In the specific case where we use exclusive-or as the difference we have that

$$A_{i,j} = |\{x \in \text{Dom}(h) : h(x) \oplus (h(x \oplus i)) = j\}|$$

and we say that a difference i leads to difference j through $h(\cdot)$ with probability $\frac{A_{i,j}}{|\text{Dom}(h)|}$, where the probability is taken over all possible inputs to $h(\cdot)$. The idea is to find good combinations of input and output differences for different components and to combine them together.

8.1.1 Joining Components

We saw in the simple ciphers of Chap. 6 that a differential attack required us to predict the evolution of a difference across multiple components. The primary tool was the notion of the differential characteristic, referred to simply as a characteristic when context allows.

An s-round **characteristic** *is a series of differences, notated as an $(s+1)$-tuple $(\alpha_0, \ldots, \alpha_s)$, where α_i is the anticipated value of Δc_i for the characteristic, i.e., the difference between the values of the partially encrypted messages where α_0 is the chosen value of the message difference $\Delta m = \Delta c_0$.*

The probability of an s-round characteristic can be computed from the conditional probability that $\Delta c_i = \alpha_i$ is the observed difference after i rounds given that $\Delta c_{i-1} = \alpha_{i-1}$ is the difference after $i-1$ rounds for $i = 1, \ldots, s$. More formally, the probability of a characteristic is given by

$$\mathrm{Pr}_{\mathcal{M},\mathcal{K}}(\Delta c_i = \alpha_i, \Delta c_{i-1} = \alpha_{i-1}, \dots, \Delta c_1 = \alpha_1 \mid \Delta m = \alpha_0),$$

where the probability is taken over all choices of the plaintext and the key. This probability can be hard to calculate. However, for certain ciphers it can be calculated from the probabilities of single-round characteristics as follows.

A sequence of stochastic variables $v_0, v_1, \dots, v_r$ is a **Markov chain** *if* $\mathrm{Pr}(v_{i+1} = \beta_{i+1} \mid v_i = \beta_i, \dots, v_0 = \beta_0) = \mathrm{Pr}(v_{i+1} = \beta_{i+1} \mid v_i = \beta_i)$ *for* $0 \le i < r$. *If* $\mathrm{Pr}(v_{i+1} = \beta \mid v_i = \alpha)$ *is independent of i for all α and β then a Markov chain is called* **homogeneous**.

Thus we speak of a Markov chain if the probabilities of the individual variables are independent of each other and of a *homogeneous chain* if the probability distribution is the same for all variables. This leads to the following.

An iterated cipher is called a **Markov cipher** *if there is an operation $\otimes$ defining Δ such that* $\mathrm{Pr}(\Delta c_1 = \beta \mid \Delta c_0 = \alpha, c_0 = \gamma)$ *is independent of γ for all α and β when the round key k is chosen uniformly at random.*

In a Markov cipher, therefore, the probability of a one-round characteristic taken over all keys is the same as the probability of the one-round characteristic taken over all keys and all plaintexts.

Our next concern is the accumulated behaviour of a chain of such one-round characteristics.

If an r-round iterated cipher is a Markov cipher and the r round subkeys are independent and generated uniformly at random, then the sequence of differences $\Delta m = \Delta c_0, \Delta c_1, \dots, \Delta c_r$ is a homogeneous Markov chain.

Thus for a Markov cipher with independent round keys, the probability of an s-round characteristic is the product of the probabilities of the s one-round characteristics or, more formally,

$$\mathrm{Pr}(\Delta c_s = \alpha_s, \Delta c_{s-1} = \alpha_{s-1}, \dots, \Delta c_1 = \alpha_1 \mid \Delta m = \alpha_0) =$$
$$\prod_{i=1}^{s} \mathrm{Pr}(\Delta c_1 = \alpha_i \mid \Delta m = \alpha_{i-1}).$$

DES and the AES with independent subkeys are Markov ciphers when the notion of difference is the exclusive-or operation. Experimental results on other ciphers [81] have tended to show that, provided round keys are not derived from a catastrophically simple key schedule algorithm, calculating the probability of a characteristic by multiplying single-round probabilities is a reasonable approximation that works for most practical purposes. To simplify statistical arguments it is often assumed that the round keys are independent and uniformly random.

As we saw in Chap. 6 there is more to an attack than a single characteristic. This leads us to revisit the broader definition of *differentials*.

An s-round **differential** *is a pair of differences (α_0, α_s), where α_s is the expected value of Δc_s and α_0 the chosen value of Δm.*

The probability of an s-round differential (α_0, α_s) is the conditional probability that given a difference $\Delta m = \alpha_0$ in the messages, the difference after s rounds of encryption is α_s. More formally, the probability of an s-round differential when $\Delta c_0 = \Delta m$ is given as

$$\Pr(\Delta c_s = \alpha_s \mid \Delta m = \alpha_0) = \sum_{\alpha_1} \sum_{\alpha_2} \cdots \sum_{\alpha_{s-1}} \prod_{i=1}^{s} \Pr(\Delta c_i = \alpha_i \mid \Delta c_{i-1} = \alpha_{i-1}).$$

Note that for any fixed values of $\alpha_1, \ldots, \alpha_{s-1}$ the probability of the differential is identical to the probability of a characteristic. Thus the probability of a differential is the sum of the probabilities of a set of characteristics. In a Markov cipher the probability of an s-round characteristic can be calculated easily as the product of the probabilities of s one-round characteristics. However, the probabilities of s-round differentials for large s are often hard to calculate. So while the differential is the real focus of the cryptanalyst, off-line analysis typically considers one or, occasionally, a few characteristics. Such analysis only delivers a lower bound to the probability of the differential.

8.1.2 Key Equivalence

In planning a differential attack the attacker usually has no information about the key. Therefore, when finding a good differential, the attacker is forced to compute the probability of a differential or characteristic over many keys, preferably over all possible keys. However, the pairs of message and ciphertext that the attacker receives are likely to be encrypted with the same *fixed* key. The probability of a differential over all choices of the plaintext and key might be different from the probability of the differential over all messages for a fixed key.

One way to deal with this problem is to assume that the vast majority of keys behave in a similar manner, and this leads us to what is termed the *hypothesis of stochastic equivalence*.

> The **hypothesis of stochastic equivalence** *states that for virtually all high probability s-round differentials (α, β) we have that $Pr_{\mathcal{M}}(\Delta c_s = \beta \mid \Delta m = \alpha, k = \ell) \approx Pr_{\mathcal{M}, \mathcal{K}}(\Delta c_s = \beta \mid \Delta m = \alpha,)$ holds for a substantial fraction of the key values ℓ.*

While this hypothesis seems to hold for most good ciphers, there are ciphers for which this might not be the case.

8.1.3 Key Recovery and Data Complexity

The starting point for a differential attack is the identification of a good characteristic and subsequent differential. Let us suppose that we have identified a good

s-round differential $(\Delta c_0, \Delta c_s)$ that holds with probability p. In what follows we will simplify this notation to (Δ_0, Δ_s).

A plaintext pair m and m' with difference Δ_0 is called a *right pair* with respect to a key k and an s-round differential if, when the pair is encrypted using k, the output difference Δ_s occurs as predicted by the differential. In a block cipher with b-bit blocks one can form 2^b pairs of plaintexts of any given difference Δ_0. Therefore there are about $p \cdot 2^b$ right pairs, where p is the probability of the differential.

If m and m' do not form a right pair, then they are said to form a *wrong pair* with respect to the key k and the differential. Generally speaking we'd prefer to reduce the number of wrong pairs that we deal with in our analysis. Note that it is possible for a pair to be wrong for one key but right for other keys.

More advanced differential attacks attempt to recover key information from the outer rounds of a cipher. Here we will focus on attacks that attempt to determine key material from the last round only. We will reuse earlier notation so that c_i represents the partial encryption of some starting text m after i rounds and we will use the exclusive-or notation of difference for simplicity. We will use g to denote the action of one round of encryption and so for round r and subkey k_r we have

$$c_r = g_{k_r}(c_{r-1}) \quad \text{or} \quad c_{r-1} = g_{k_r}^{-1}(c_r).$$

To recover information from the last round of a cipher, the cryptanalyst first identifies (by analysis) a differential for which the difference between the partially encrypted texts after $r-1$ rounds of encryptions, Δ_{r-1}, is known (either completely or partially) with some probability p. So when starting with two plaintexts m and m' the cryptanalyst predicts that $c_{r-1} \oplus c'_{r-1} = \Delta_{r-1}$ with probability p.

To mount an attack the cryptanalyst requests the ciphertexts c_r and c'_r for a pair of inputs m and m' with the right starting difference. The cryptanalyst then looks at the ciphertext pair. In some situations it is possible to immediately determine that a pair could not possibly be a right pair just by looking at the ciphertext. This process is called *filtering* and it was introduced in Sect. 6.3. If a pair cannot be a right pair then it can be discarded. This is advantageous to the attacker since wrong pairs only serve to cloud the subsequent analysis.

Since the cryptanalyst does not know the last-round key the inputs to the final round, c_{r-1} and c'_{r-1}, are unknown. However, the cryptanalyst can always guess the key (or part of the key) k'_r used in the last round and can then compute the quantity

$$g_{k'_r}^{-1}(c_r) \oplus g_{k'_r}^{-1}(c'_r).$$

For the attack we will need a set of intercepted text pairs and we will need a set of counters T_i, one corresponding to each of the possible values of k_r. We can assume there are τ possible values for k_r and so the τ counters can be numbered $T_0, \ldots, T_{\tau-1}$ where the subscript t is interpreted as the value of k_r, and T_t its associated counter.

The Generic Differential 1R-Attack with Differential of Probability p

1. *For all N intercepted text pairs (m,c) and (m',c') with $m \oplus m' = \Delta_0$:*

   - *If inspection of c and c' reveals the pair to be a wrong pair, reject it.*
   - *Otherwise, for all values t of k_r*
     - *Compute $g_t^{-1}(c_r) \oplus g_t^{-1}(c_r')$.*
     - *If $g_t^{-1}(c_r) \oplus g_t^{-1}(c_r') = \Delta_{r-1}$ increment counter T_t.*

2. *Identify the counter T_i, for $0 \le i \le \tau - 1$, with the expected value Np. Suppose this counter is T_s.*
3. *Guess that $k_r = s$.*

The step of trying all τ possible values of k_r is an interesting one. It is possible that several values for k_r give the predicted difference for some plaintext pair. These are referred to as *candidates* for k_r. We can denote this set of candidates as $\{\ell_1, \ell_2, \ldots, \ell_j\}$. Now if m and m' form a right pair then the correct value of k_r must be among the set of candidates, *i.e.*, $k_r \in \{\ell_1, \ell_2, \ldots, \ell_j\}$. However, if m and m' do not form a right pair, then what does the set of candidates look like? Certainly, the correct value of k_r cannot lie among the set of candidates. Otherwise the difference after $(r-1)$ rounds of encryption would have been the one specified in the differential, which means the pair must have been a right pair all along [631]. So for a wrong pair we typically assume that the set of key candidates is independent of k_r and effectively generated at random, though it cannot include the correct value to k_r. This has been referred to as *wrong-key randomisation* [281]. To illustrate that this is plausible, suppose that a cryptanalyst has the ciphertext c_r to hand and consider the guess t for the value of k_r. Consider the following situation where, as before, we casually denote the round function as g.

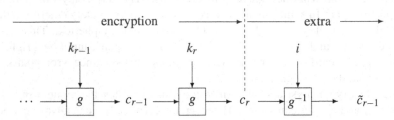

If i really does equal k_r then we have that $\tilde{c}_{r-1} = c_{r-1}$. However, if $i \ne k_r$ then $\tilde{c}_{r-1} = g_i^{-1}(g_{k_r}(c_{r-1}))$. Provided the function $g(\cdot)$ is not catastrophically weak we are doing additional processing beyond a full encryption. If the correlation in c_{r-1} is high enough it seems that for a good block cipher it is likely that $\tilde{c}_{r-1}$ would be "closer to random" than c_{r-1} for most incorrect values of i.

The net result is that if many different pairs m and m' are examined with the number of times a target difference appears being recorded for different candidate values of k_r, then we would expect the correct value of k_r to be suggested at a different rate than incorrect key values. In this way the correct key guess should be identifiable with sufficiently many texts.

It is obvious, but still worth observing, that while key recovery was expressed in terms of a 1R-attack, *i.e.*, recovering information from a single outer round, exactly the same techniques can be used to guess key information from more rounds. The typical limit on these techniques is due to the number of key candidates τ that we would need to search. Too many candidates, and the work effort becomes too large for the attack to be useful. Much therefore depends on the design features of the block cipher in question.

To estimate the data complexity of a differential attack we will consider the generic attack outlined above. For a successful differential attack the following must hold for a given pool of data:

1. There must be a reasonable chance that the identified differential holds at least once. Otherwise the attacker will not be able to distinguish the correct value of the secret key from random values.
2. In the attack the correct value of the key k_r must be suggested at a significantly different rate than other candidate values for k_r.

For the first condition, if we assume the differential holds with probability p and that plaintexts are selected uniformly at random, then if the attack is repeated $\frac{1}{p}$ times one expects the predicted differential to hold once. For the second condition, we need to introduce the concept of the *signal-to-noise ratio*.

Let p_{correct} denote the probability that, in one iteration of a differential attack, the correct value to k_r lies among the candidate values that are proposed. Similarly, let $p_{\text{incorrect}}$ denote the average probability that an incorrect value to k_r lies among the set of proposed values. Then the signal-to-noise ratio, often denoted by S/N, is defined as

$$\text{S/N} = \frac{p_{\text{correct}}}{p_{\text{incorrect}}}.$$

To use this concept we need to define some additional quantities:

- Let λ denote the probability that a randomly chosen pair survives the *filtering* process. Note that λ is always greater than or equal to p, the probability the target differential holds.
- Let γ be the average number of key values suggested by each pair that survives the filtering process.
- Let κ denote the number of bits in the key k_r that we are trying to recover.

To establish an expression for the signal-to-noise ratio, we first note that a pair surviving filtering is a wrong pair with probability $\lambda - p$. We know all right pairs must survive filtering, so after considering N text pairs there must be $N(\lambda - p)$ wrong pairs remaining. We also know that there are $\tau = 2^{\kappa}$ possible values for k_r, including the right one. Each pair surviving the filtering step will suggest γ possibilities. Each wrong pair suggests γ incorrect values for k_r while each right pair suggests $(\gamma - 1)$ incorrect values for k_r as well as the correct value. As we have said previously, we can assume that the incorrect values suggested in the attack are uniformly distributed over all $2^{\kappa} - 1$ incorrect values of the key.

Putting this together, the probability $p_{\text{incorrect}}$ that a specific random (wrong) value of the key is suggested in one iteration of the attack is given by

$$p_{\text{incorrect}} = (\lambda - p) \times \frac{\gamma}{2^\kappa - 1} + p \times \frac{(\gamma - 1)}{2^\kappa - 1} = \frac{\lambda\gamma - p}{2^\kappa - 1},$$

and a formula for S/N is given by

$$\text{S/N} = \frac{p_{\text{correct}}}{p_{\text{incorrect}}} \simeq p \times \left(\frac{2^\kappa - 1}{\lambda\gamma - p}\right) = \frac{(2^\kappa - 1)p}{\lambda\gamma - p}.$$

If S/N is identical to 1 then we cannot hope to distinguish a correct key guess from an incorrect one: correct and incorrect values of the key k_r will be suggested at the same rate. If S/N > 1 then the correct key guess will occur more often than the other values and by examining the key counters at the end of the attack the highest counter value is likely to indicate the correct value to k_r. However, also note that when S/N < 1 the counter for the correct key value will be incremented *less* than the counters corresponding to incorrect key values. Thus by looking for the counter with the smallest value the correct value to k_r can still be recovered.

So a necessary condition for a successful differential attack is that the signal-to-noise ratio be different from 1. The expected success rate of the attack will increase the greater the distance of S/N is from 1. Attacks where we have a signal-to-noise ratio S/N < 1 are, at least in principle, as good as attacks where S/N > 1. However, they do not seem to be easy to find, except in the special case when S/N = 0. These have been named *impossible differentials*; see Sect. 8.1.4.

We typically define the *complexity* of a differential cryptanalytic attack as the number of data pairs required to determine the target key bits plus the number of (off-line) operations required. From experiments on many ciphers over more than a decade, the number of plaintext pairs required in a differential attack appears to remain proportional to p^{-1}. This suffices as an estimator for the complexity of the overall attack, where the constant of proportionality is typically a small integer.

Of course, things might not always be so simple and even with a good S/N a differential attack might fail. First, in the description above it was assumed that the round keys in the iterated cipher were chosen uniformly at random. This is rarely the case. Second, one needs the hypothesis of stochastic equivalence (see Sect. 8.1.2), which might not hold in practice. Despite these hesitations, for the most part things work surprisingly well.

Finally, the asymmetry in effort for the cryptanalyst and the designer is worth observing. To make a successful attack by differential cryptanalysis the existence of one good characteristic is likely to be sufficient. Yet, to guarantee security against a differential attack, the designer must ensure that there are no differentials with high enough probability. This can be hard to do, unless the problem of differential cryptanalysis has been expressly addressed during the design of the cipher.

8.1.3.1 Using Structures to Reduce Data Complexity

It can be advantageous to use several differentials simultaneously. In addition to potentially increasing the success rate for the attacker, distinct differentials can often be used to determine different parts of the secret round keys. However, as we will now show, using more than one differential does not necessarily mean that we need a substantial increase in the number of chosen plaintexts.

Let Δ_0 denote the plaintext difference for some identified differential. During the attack an adversary would repeatedly choose, at random, a plaintext value x and then compute the second value of the pair as x' so that x and x' have the difference Δ_0.

Now consider the case where differentials with two different b-bit additive differences Δ_0^0 and Δ_0^1 are used. For a randomly chosen b-bit value x consider the following four texts:

$$x \quad x \oplus \Delta_0^0 \quad x \oplus \Delta_0^1 \quad x \oplus \Delta_0^0 \oplus \Delta_0^1$$

It is obvious that the first two texts as well as the last two texts have a difference of Δ_0^0, while the first and third, and the second and fourth texts have a difference Δ_0^1. Thus from four texts one can derive four different pairs with the desired differences. More generally, if we wish to generate plaintext pairs for s starting differences, a structure of 2^s plaintexts can be used to form 2^{s-1} pairs of plaintexts for each of the differences Δ_0^0, ..., Δ_0^{s-1}. The use of structures bears some resemblance to higher order differentials that we will see later.

8.1.3.2 Changing the Form of the Data

The standard description of a differential cryptanalytic attack is in terms of chosen messages. While we should always assume that such an attack can be mounted in practice, accumulating vast quantities of chosen messages can be much harder than accumulating vast quantities of known messages.

There is one very simple observation that allows us to convert a differential attack into a known-text attack. Essentially, instead of choosing the target difference we wish the messages to satisfy we wait for the difference to occur at random. To illustrate, suppose that we have identified a differential attack on a b-bit block cipher that requires 2^n chosen pairs. Given a random message pair, the probability that they satisfy the required difference is 2^{-b}. Given a set of t random message pairs we can form $\frac{t(t-1)}{2}$ possible pairs. The number of plaintext pairs satisfying the required difference is then given by

$$\frac{t(t-1)}{2^{b+1}}.$$

Since we require 2^n chosen message pairs for the attack, we can choose t to be sufficiently large so that

$$\frac{t(t-1)}{2^{b+1}} \approx 2^n.$$

We will therefore require around $t \approx 2^{\frac{n+b+1}{2}}$ known message pairs to launch our attack. The limitations and possibilities of this kind of attack were explored in [95] where these observations were extended to consider plaintext sources with patterns of redundancy such as ASCII text.

8.1.4 Enhancements to the Basic Differential Attack

Depending on the cipher under consideration, some slight variations of the basic differential attack can be useful. Their applicability tends to be limited but when they do apply they can often be rather effective.

8.1.4.1 Truncated Differentials

To begin we need to define the *truncation* of a b-bit string. Typically, truncation refers to a shortening of an input, but for the purposes of differential cryptanalysis truncation refers to a relaxation in the specifications of a differential. We will use the symbol $\star$ to denote an unknown value and, for a b-bit string $m_0 m_1 \ldots m_{b-1}$, define

$$n_0 n_1 \ldots n_{b-1} = \text{TRUNC}(m_0 m_1 \ldots m_{b-1})$$

if, and only if, $n_i = m_i$ or $n_i = \star$ for all $0 \le i \le b - 1$. This notion extends naturally to differences. If we have an s-round characteristic

$$\Delta_0 \xrightarrow{\mathscr{R}} \Delta_1 \xrightarrow{\mathscr{R}} \ldots \xrightarrow{\mathscr{R}} \Delta_s$$

then

$$\Delta_0' \xrightarrow{\mathscr{R}} \Delta_1' \xrightarrow{\mathscr{R}} \ldots \xrightarrow{\mathscr{R}} \Delta_s'$$

is a *truncated characteristic* if $\Delta_i' = \text{TRUNC}(\Delta_i)$ for $0 \le i \le s$. In fact, a truncated characteristic is itself a collection of characteristics (and therefore reminiscent of a differential) since a truncated characteristic contains all characteristics of the form

$$\Delta_0'' \xrightarrow{\mathscr{R}} \Delta_1'' \xrightarrow{\mathscr{R}} \ldots \xrightarrow{\mathscr{R}} \Delta_s''$$

for which $\Delta_i'' = \text{TRUNC}(\Delta_i')$. The notion of truncated characteristic extends in a natural way to truncated differentials introduced in [386], and the latter is the term that is often used.

Truncated differentials have proved to be a very useful tool in cryptanalysis and several ciphers which are (or seem to be) secure against basic differential attacks have been attacked successfully using truncated differentials. To illustrate how a truncated differential attack might work we return to CIPHERFOUR, which was introduced in Chap. 6. We recall the description of CIPHERFOUR which was first presented in Fig. 6.5 and is now repeated in Fig. 8.1.

x	0	1	2	3	4	5	6	7	8	9	a	b	c	d	e	f
$S[x]$	6	4	c	5	0	7	2	e	1	f	3	d	8	a	9	b

i	0	1	2	3	4	5	6	7	8	9	10	11	12	13	14	15
$P[i]$	0	4	8	12	1	5	9	13	2	6	10	14	3	7	11	15

1. Set $u_0 = m$.
2. For $i := 1$ to $r - 1$ do:

   a. Combine the round key k_{i-1} with u_{i-1} so that $a_i = u_{i-1} \oplus k_{i-1}$.
   b. Divide a_i into four nibbles $a_i = A_0 \| \ldots \| A_3$.
   c. Compute $S[A_0] \| \ldots \| S[A_3]$ where $S[\cdot]$ is defined above.
   d. Write $S[A_0] \| \ldots \| S[A_3]$ as $y_{15} \ldots y_0 = S[A_0] \| \ldots \| S[A_3]$.
   e. Permute bit y_i to position j according to the permutation $j = P[i]$ defined above.
   f. Set $u_i = y_{15} \| y_{11} \| \ldots \| y_4 \| y_0$.

3. *(Last round)*

   a. Combine the round key k_{r-1} with u_{r-1} so that $a_r = u_{r-1} \oplus k_{r-1}$.
   b. Divide a_r into four nibbles $a_r = A_0 \| \ldots \| A_3$.
   c. Compute $S[A_0] \| \ldots \| S[A_3]$ where $S[\cdot]$ is defined above.
   d. Write $S[A_0] \| \ldots \| S[A_3]$ as $y = y_{15} \ldots y_0 = S[A_0] \| \ldots \| S[A_3]$.
   e. Output $y \oplus k_r$ as ciphertext.

Fig. 8.1 Encryption with CIPHERFOUR (from Chap. 6).

To start we will consider an input difference consisting of four nibbles of the form $(0, 0, 2, 0)$. The difference distribution table for the S-box was given in Table 6.2 and a quick glance confirms that an input difference of 2 to the S-box of CIPHERFOUR can only give four possible output differences, namely 1, 2, 9, and a. Thus, when we take the other components of the round into account, there are only four possible output differences after one round of encryption which are given by $(0, 0, 2, 0)$, $(0, 0, 0, 2)$, $(2, 0, 2, 0)$, and $(2, 0, 0, 2)$.

Since bit-level details will now be required, the differences will be presented as 16-bit quantities. In this notation the four possible one-round characteristics previously identified can be written as

$$(0000\,0000\,0010\,0000) \xrightarrow{\mathcal{R}} \begin{cases} (0000\,0000\,0010\,0000) \text{ or} \\ (0000\,0000\,0000\,0010) \text{ or} \\ (0010\,0000\,0010\,0000) \text{ or} \\ (0010\,0000\,0000\,0010). \end{cases}$$

Now consider adding another round to these cases. By consulting Table 6.2 and tracing the action of the round, we can follow the effect on the input differences $(0, 0, 2, 0)$, $(0, 0, 0, 2)$, $(2, 0, 2, 0)$, and $(2, 0, 0, 2)$. A bit position that might be occupied by either a 1 or a 0 will be indicated with the symbol $\star$. The following interesting patterns emerge for a round of encryption in CIPHERFOUR:

$$(0000000000100000) \xrightarrow{\mathscr{R}} (00\star0000000\star0\,00\star0)$$
$$(0000000000000010) \xrightarrow{\mathscr{R}} (000\star0000000\star\,000\star)$$
$$(0010000000100000) \xrightarrow{\mathscr{R}} (\star0\star00000\star0\star0\,\star0\star0)$$
$$(0010000000000010) \xrightarrow{\mathscr{R}} (\star00\star0000\star00\star\star00\star).$$

To emphasize the notation, as we take all input pairs that satisfy the difference (0000 0000 0010 0000), the 16 bits of output difference after two rounds of encryption will take one of the four forms shown above. In two cases 13 bits will always be 0 and in two cases ten bits will always be 0. By combining these four cases for the second round, we get

$$\left.\begin{array}{l}(0000000000100000) \\ (0000000000000010) \\ (0010000000100000) \\ (0010000000000010)\end{array}\right\} \xrightarrow{\mathscr{R}} (\star0\star\star\;0000\;\star0\star\star\;\star0\star\star).$$

For all the four input differences of interest, seven bits in the output difference are always 0, whereas nine other bits take both values, 1 and 0.

We can continue into another round and we note that the pattern

$$(\star0\star\star\;0000\;\star0\star\star\;\star0\star\star)$$

indicates that the inputs to the second S-box will be the same for both texts in the pair. Therefore the outputs must be the same, with the four output bits being distributed by the action of the permutation P. We therefore have a one-round truncated differential that holds with probability 1 and depends only on the permutation P and not on properties of the S-box:

$$(\star0\star\star\;0000\;\star0\star\star\;\star0\star\star) \xrightarrow{\mathscr{R}} (\star0\star\star\;\star0\star\star\;\star0\star\star\;\star0\star\star).$$

To summarise, we have identified a three-round *truncated differential* of probability 1 for CIPHERFOUR;

$$(0000\,0000\,0010\,0000) \xrightarrow{3\mathscr{R}} (\star0\star\star\;\star0\star\star\;\star0\star\star\;\star0\star\star).$$

Any two inputs with a difference of (0000 0000 0010 0000) will, after three rounds of encryption, provide a pair of partially encrypted texts with equal values in four identified bit positions. We can now appreciate how the term *truncated* draws attention to the fact that only a subset of the entire block of bits is predicted by our analysis.

We will now use this truncated differential to recover key information from the cipher. To illustrate we consider a five-round version of CIPHERFOUR from which we recover key information from the first and last rounds.

Consider the truncated differential that we have just derived and note that it starts with the difference (0000 0000 0010 0000). If this difference occurs at the start of the second round of encryption then the truncated differential would run through the next three rounds with probability 1. To get this difference at the start of the second round, we observe from Fig. 8.2 that the same output difference must be generated by three S-boxes in the first round. The occurrence of 0010 at the second round will therefore depend on the values of the inputs to the third S-box, which in turn depends on the message *and the value of four bits of the first-round key* k_0.

So to get a difference of (0000 0000 0010 0000) after the first round, we will appeal to the *first-round trick* that was first used in [82]. We will use a structure (see page 153) of 16 messages m_i where

$$m_i = (t_0, t_1, i, t_2)$$

for $i = 0, \ldots, 15$ and t_0, t_1, t_2 are randomly chosen, constant, four-bit values.

For any value of the four key bits affecting the third S-box, one can easily find 16 pairs of messages (m_i, m_j) such that the difference in the outputs from the first round S-boxes is (0000 0000 0010 0000). Once we have this the three-round truncated differential with probability 1 will follow. Thus, if the four bits of k_0 are correct, our message pairs will each give a four-round truncated differential of probability 1. The partially-encrypted pairs that result will have a difference of 0 in four identified bit positions. The situation is depicted in Fig. 8.2, where we use c_i and c_j to denote ciphertexts corresponding to m_i and m_j.

For our attack we suppose that for each possible value of k_0 we use s pairs of messages and that these yield the sought-after difference after one round. We will use our s pairs to identify the four target key bits of k_0 as well as all the bits of k_5. Thus 20 bits of key material can potentially be recovered.

For each of the s message pairs (m_i, m_j) corresponding to a guess for k_0 we request the ciphertext pairs (c_i, c_j) generated using the unknown key. We then take each nibble of k_5 in turn and try all possible values. For each guess for k_5 we compute backwards through the corresponding S-box in the last round and check whether a difference with a 0 in the second bit is obtained. As usual we keep counters for each guess and increment the counter by 1 if the condition is satisfied.

Since we use s pairs for each guess to k_0, when we have the correct value of k_0 there will be at least one counter for the first nibble of k_5 with the value s. If we can find at least one input text pair that contradicts the truncated differential for every guess for k_5 then we know that this particular value of k_0 must be wrong and we can move on. Thus, at least in principle, we can recover the correct value to four bits of k_0 and, by working nibble by nibble through k_5, we can recover the entire k_5. Of course, things might not be so straight forward and it is possible that the pairs obtained from one structure of 16 texts are insufficient to uniquely determine the secret keys. If this is the case one can simply generate another structure by using different constant values for t_0, t_1, and/or t_2.

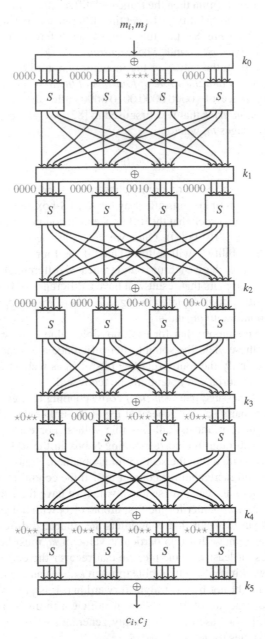

Fig. 8.2 The truncated differential attack on five-round CIPHERFOUR. The (m_i, m_j) are selected from a structure of 16 chosen messages.

The attack has the potential to find a total of 20 key bits. Experimental results show that in 100 tests the four bits of k_0 were correctly identified in 28, 78, and 97 cases when one, two, and three structures were used respectively. This means that pools of 16, 32, and 48 chosen plaintexts were used with 48 chosen plaintexts, giving nearly 100% success in recovering the target nibble from k_0. Recovering bits from k_5 proved to be a little more unreliable, with the number of additional key bits recovered in the three test cases being four, nine, and 12 respectively.

8.1.4.2 Impossible Differentials.

Differentials of probability 0 are often called *impossible differentials*. The technique of combining two differentials of probability 1 so that they conflict when concatenated is called *miss-in-the-middle*. One nice illustration of the application of impossible differentials is given by an attack on a six-round Feistel network [388]. The only requirement for this attack is that the non-linear round function be a bijective mapping for any fixed key. Consider Fig. 8.3, where we will show a five-round differential of probability 0. We will then use this to recover key material from a six-round Feistel cipher.

Suppose we have a pair of input messages where the right-hand halves of both texts have the same value. Such a pair of plaintexts will provide a starting exclusive-or difference of $[\alpha \| 0]$ where the two components of the difference denote the left-hand and the right-hand side of the input difference respectively, and $\alpha \neq 0$. Now suppose, by way of obtaining a contradiction, that the difference after five rounds of encryption and the Feistel swap is given by $[0 \| \alpha]$. If this were the case, then by tracing the action of the Feistel network in Fig. 8.3 we would see that the difference in the inputs to the round function f at both the second and the fourth rounds would have to be α. Yet at the same time the output difference from f in the third round would have to be 0. But since the round function f is a bijective mapping, and an output difference of 0 can only be derived from an input difference of 0, this leads to a contradiction.

Consequently, we have a five-round differential that holds with probability 0. This can be used to mount a differential attack on a six-round Feistel cipher and allows us to find the subkey in the final round in the following way.

Consider a b-bit cipher with subkey k_6 in the sixth round. Choose a structure (see page 153) of $2^{\frac{b}{2}}$ messages m_i of the form

$$m_i = [m_i^L \| m_i^R] = [i \| c]$$

for $i = 0, \ldots, 2^{\frac{b}{2}-1}$, where c is some random and fixed $\frac{b}{2}$-bit value. Let $c_i = [c_i^L \| c_i^R]$ denote the corresponding ciphertexts after six rounds of encryption (with no final swap).

Sort the inputs and find pairs m_i and m_j for $i \neq j$ so that

$$m_i^L \oplus m_j^L = c_i^R \oplus c_j^R = \alpha_{i,j}$$

for nonzero values $\alpha_{i,j}$. Note, for $i \neq j$, that $m_i^L \oplus m_j^L \neq 0$ and so the expected number of pairs for a given $\alpha_{i,j}$ can be estimated by

$$\binom{2^{\frac{b}{2}}}{2} \times \frac{1}{2^{\frac{b}{2}}} \simeq 2^{\frac{b}{2}-1}.$$

For all the pairs that match, we try all possible values of k_6 and decrypt the ciphertexts by one round giving candidate values for the difference after the swap in round 5. If the value of the difference is $[0 \,\|\, \alpha_{i,j}]$ then, since we know this value cannot occur, that particular guess for k_6 must be wrong.

Note that the correct value of k_6 can never yield such a difference after the fifth round since the differential has probability 0. However, for wrong values of k_6, it is reasonable to assume that this difference will occur with probability $2^{-\frac{b}{2}}$ for each analysed pair. Thus with $2^{\frac{b}{2}-1}$ pairs, around half the candidates for k_6 will be discarded. We can now repeat the attack with different structures by changing the value of c. Each time we will halve the remaining candidates for k_6 while always retaining the correct one. Eventually only a few values of k_6 will be left. If we denote the length of k_6 by τ, then the attack requires $\tau \cdot 2^{\frac{b}{2}}$ chosen plaintexts, around $\tau \cdot 2^{\frac{b}{2}}$ words of memory, and an approximate work effort in encryptions given by

$$(2^{\tau} + 2^{\tau-1} + 2^{\tau-2} + \ldots + 2 + 1) \times 2^{\frac{b}{2}} \simeq 2^{\tau+1} \times 2^{\frac{b}{2}} = 2^{\tau+\frac{b}{2}+1}.$$

Once k_6 has been found the attacker can remove the sixth round and attack the remaining five-round Feistel network. A variety of related techniques have been proposed on other ciphers with some of the most impressive results being attacks on Skipjack [57, 59, 58, 407].

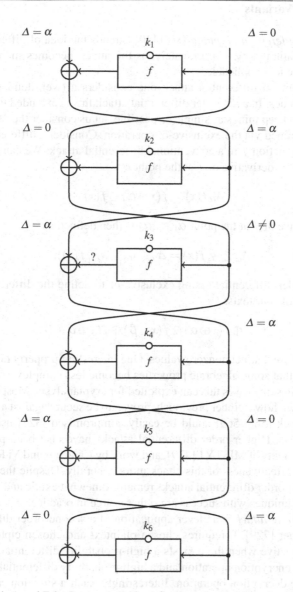

Fig. 8.3 A six-round Feistel network (without final swap) with a bijective round function f for any fixed key. The contradiction in the third round demonstrates that the five-round differential $(\alpha, 0) \to (0, \alpha)$ over the first five rounds can never occur.

8.1.4.3 Other Variants

Higher order differential cryptanalysis [430] extends the idea of differential crypt-analysis in a natural way. Unfortunately, as the attack becomes more complex it tends to become less applicable.

In a conventional differential attack one considers a (well-defined) difference between two values. In a dth-order differential attack this is extended to a collection of 2^d values. To keep things as simple as possible, let us consider the case where the difference is defined via the exclusive-or operation. Consider a difference of $\alpha \neq 0$ through some function f in a conventional differential attack. We can consider this as the (first-order) derivative of f at the point α:

$$\Delta_\alpha f(x) = f(x \oplus \alpha) \oplus f(x).$$

The dth derivative of f at the point $\alpha_1, \ldots, \alpha_d$ is then defined as

$$\Delta_{\alpha_1, \ldots, \alpha_d} f(x) = \Delta_{\alpha_d} (\Delta_{\alpha_1, \ldots, \alpha_{d-1}} f(x)).$$

In a second-order differential (using exclusive-or to define the difference) this extends to tuples of four texts:

$$f(x) \oplus f(x \oplus \alpha) \oplus f(x \oplus \beta) \oplus f(x \oplus \alpha \oplus \beta),$$

where α and β are distinct nonzero values. One interesting property of higher order differentials is that some algebraic properties become less complex as d grows (see page 189) and in some cases this can exploited for cryptanalysis. Most interestingly, it has been shown how a cipher previously proven to be secure against a conventional differential attack [568, 569] could be easily compromised [327] using only 512 chosen plaintexts. Higher order differential attacks have also been proposed [20, 704] against variants of MISTY1 [479], and work by Canteaut and Videau [136] has considered the effectiveness of this attack more generally. Despite these successes, however, higher order differential attacks remain somewhat exotic and opportunities to use these techniques with success are rather scarce in practice.

The *boomerang attack* is a clever application of a second-order differential and is due to Wagner [729]. It requires chosen plaintext and chosen ciphertext and is particularly effective when there exists a high-probability differential covering the first half of the encryption operation and a high-probability differential covering the first half of the decryption operation. Interestingly, such a situation can arise even when there are no useful differentials that cover the whole encryption (or decryption) operation.

To illustrate, let us suppose that some encryption process $\text{ENC}_k(m)$ can be written in the following way:

$$\text{ENC}_k(m) = \text{ENC}^2{}_k(A_k(\text{ENC}^1{}_k(m)))$$

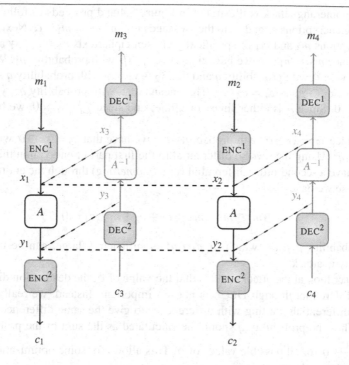

Fig. 8.4 The boomerang attack illustrated with the overall encryption operation being split into two parts. These are joined by a differentially simple operation A. The messages m_1 and m_2 are chosen by the attacker, as are ciphertexts c_3 and c_4.

where A_k is a, potentially key-dependent, linear or affine transformation. For instance, for an r-round block cipher we could consider ENC^1 as representing the first s rounds of encryption while ENC^2 represents encryption with the remaining $r - s$ rounds. A_k would then be the identity function.

Now suppose that there is a differential $\alpha \to \beta$ through $\text{ENC}^1{}_k$ that holds with a high probability p_1. For the decryption direction the reverse differential will hold with probability p_2 (which is not equal to p_1 for all ciphers). We can denote these differentials by

$$\alpha \xrightarrow{\text{ENC}^1} \beta \quad \text{and} \quad \beta \xrightarrow{\text{DEC}^1} \alpha.$$

Further suppose there is a differential through the inverse of the *second* part of encryption that holds with a high probability q and which we denote by

$$\gamma \xrightarrow{\text{DEC}^2} \phi.$$

The boomerang attack is illustrated in Figure 8.4 and proceeds as follows. First obtain the encryptions c_1 and c_2 to the messages m_1 and $m_2 = m_1 \oplus \alpha$. Next, request the decryptions m_3 and m_4 of specifically chosen ciphertexts $c_3 = c_1 \oplus \gamma$ and $c_4 = c_2 \oplus \gamma$. Since $m_2 = m_1 \oplus \alpha$ we have that $x_2 = x_1 \oplus \beta$ with probability p_1. We know that $y_1 \oplus y_3 = \phi$ with probability q and that $y_2 \oplus y_4 = \phi$ with probability q since we chose $c_3 = c_1 \oplus \gamma$ and $c_4 = c_2 \oplus \gamma$. This means that, with probability q^2, $y_1 \oplus y_2 \oplus y_3 \oplus y_4 = 0$. Since A is either linear or affine, and since $\sum_{i=1}^4 y_i = 0$, we have that $\sum_{i=1}^4 x_i = 0$.

But since $x_2 = x_1 \oplus \alpha$ with probability p_1, we have that $x_3 = x_4 \oplus \alpha$ with probability $p_1 q^2$. Using the reverse differential to the first part of encryption this means that we have a second-order differential (or a *boomerang*) through the entire cipher. In this case we have

$$m_1 \oplus m_2 \oplus m_3 \oplus m_4 = c_1 \oplus c_2 \oplus c_3 \oplus c_4 = 0$$

with probability $p_1 p_2 q^2$, which, if $p_1 p_2 q^2$ is sufficiently large, can be used in a cryptanalytic attack.

A closer look at the attack reveals that the value of ϕ, the destination difference for the differential through DEC^2, is not too important. Instead, we really require the two differentials starting with difference γ to give the same difference through DEC^2. Thus the probability q should be calculated as the sum of the probabilities of $\gamma \xrightarrow{\mathrm{DEC}^2} \phi$ for all possible values of ϕ. This allows for some optimisation in the effectiveness of the attack.

The elegance and the delicate construction of the boomerang attack might lead one to suppose that it would remain an academic oddity. However, there are serious block cipher proposals that fall to this attack. For instance, the block cipher Coconut'98 [717] is an iterated 64-bit block cipher which runs over eight Feistel rounds. However, after four rounds a particular non-Feistel mapping is applied. It turns out that the cipher has an ideal form for the boomerang attack where ENC^1 and ENC^2 are two four-round Feistel networks that are joined by a key-dependent affine mapping $A_k(x) = (x \oplus k_1)k_2$ where k_1 and k_2 are derived from the user-supplied key and multiplication is defined over $\mathrm{GF}(2^{64})$. In [729] it was shown that there are differentials through ENC^1 and ENC^2 that hold with a high probability and these allow us to build a boomerang that holds with probability $\frac{1}{1900}$. In turn this leads to a key recovery attack that uses about 2^{16} chosen plaintexts and a total computing time corresponding to 2^{38} encryptions.

One drawback of the boomerang attack is that it requires both chosen messages and chosen ciphertexts. A purely chosen message variant was presented in [364] and the resultant structure termed *amplified boomerangs*. The observation is simple. Instead of choosing ciphertexts c_3 and c_4 with a chosen difference, we once again generate an m_3 and an m_4 related by the chosen difference α so that $m_4 = m_3 \oplus \alpha$. Thus with probability p_1^2 we have that $x_1 = x_2 \oplus \alpha$ and $x_3 = x_4 \oplus \alpha$, which give $\sum_{i=1}^4 x_i = 0$. Then, if we are lucky, we might have $y_1 \oplus y_3 = \gamma$, in which case $y_2 \oplus y_4 = \gamma$ will follow. Typically the "if we are lucky" can be replaced with sufficiently many plaintext pairs so as to ensure a match by the birthday paradox, giving a "match in the middle".

Finally, the *rectangle attack* [62] provides us with an extension to the amplified boomerang attack. The essential observation is that the value of β can be allowed to vary. Thus we only require both message pairs with starting difference α to yield the same output difference after ENC^1. In effect we are asking for four plaintexts that form a second-order differential of value 0 through ENC^1.

Interestingly, it has been shown that there are (non typical) cases where the boomerangs will not come back as described above; indeed the computation of the probabilities above can be inaccurate in certain situations [528]. It is therefore good practice in cryptanalysis to back up any theoretical complexity statements by implementation results whenever possible, as for example in the boomerang attack [118].

Throughout the cryptographic literature there are many adjustments and enhancements to the basic differential attacks. For the most part, the more sophisticated the construction, the less applicable the resultant attack. That said, the idea of viewing the encryption process in a step-by-step fashion as we encrypt chosen inputs is very powerful. So powerful in fact that it is very effective against other cryptographic primitives, including self-synchronising stream ciphers and hash functions. It also remains the most powerful general attack against block ciphers.

8.2 Linear Cryptanalysis Revisited

As we saw in Chap. 7, linear cryptanalysis is a known plaintext attack in which the attacker exploits probabilistic linear relations between bits of the plaintext, the ciphertext and the key. To establish our notation for this section, consider a generic b-bit block cipher and assume that the following relation holds with probability $p \neq \frac{1}{2}$ taken over all possible values of the plaintext m and all values of the key k:

$$(m \cdot \alpha) \oplus (c \cdot \beta) = (k \cdot \gamma),$$

where c represents the ciphertext. We use α, β, and γ to denote b-bit selection vectors, termed *masks*, and '·' to denote the dot or inner product modulo 2. The quantity $\varepsilon = p - \frac{1}{2}$ is called the *bias* and often we are more concerned with the magnitude of the bias, *i.e.*, $|\varepsilon|$, than its exact value. Note that some authors prefer to use the term *imbalance* or *correlation* c when discussing linear cryptanalysis. This is defined as $c = 2p - 1$, and so we have that $c = 2\varepsilon$. While we will continue to use the term bias in what follows, the use of correlation can lead to a very nice formulation of linear cryptanalysis [178].

Let us consider how to find relations for individual rounds in an iterated cipher. It is clear that for linear and affine functions linear relations can be found that hold with probability 1 or 0, offering a bias of $\frac{1}{2}$. The important step in linear cryptanalysis is to establish such relations through the non-linear components of a cipher. For the DES and AES, and in fact most ciphers, it is possible to compute so-called *linear approximation tables* for the different non-linear components. For the AES and the DES these are the S-boxes, and such linear approximation tables contain all possible input and output masks together with the corresponding probability that the parity of input approximation is preserved in the output approximation. Such a table for a function f can be presented as a matrix $A = \{A_{\alpha,\beta}\}$, where

$$A_{\alpha,\beta} = |\{x \in \text{Dom}(f) : (x \cdot \alpha) = (f(x) \cdot \beta)\}|.$$

We have that the bit $x \cdot \alpha$ equals the bit $f(x) \cdot \beta$ with probability

$$p_{\alpha,\beta} = \frac{A_{\alpha,\beta}}{|\text{Dom}(h)|}$$

and the bias for this particular approximation is $\varepsilon = p_{\alpha,\beta} - \frac{1}{2}$ and the correlation c is given by $\varepsilon = 2p_{\alpha,\beta} - 1$.

It is possible to have ciphers where the range or the domain of the non-linear function is too large to allow a complete linear approximation table to be constructed. In such circumstances, a cryptanalyst would likely restrict himself to searching over some useful mask values. These might be ones of low Hamming weight, or perhaps ones with a particular form that would be suitable for approximating other features in the cipher, or particularly suited to iteration.

8.2.1 *Joining Components*

The first stages in a linear cryptanalytic attack are used to establish linear relations for individual components. These are then joined to form approximations to a round which, in turn, are joined to give approximations over more rounds.

Suppose c_i denotes the partially encrypted text after i rounds of encryption for $i = 1$ to r for an r-round block cipher. Further suppose that the per-round relations

$$(c_{i-1} \cdot \alpha_{i-1}) \oplus (c_i \cdot \alpha_i) = (k_i \cdot \gamma_i)$$

hold with probability p_i, where the probability is taken over all possible inputs to the round and for fixed values of the subkeys k_i. By combining these relations with exclusive-or, as in Chap. 7, one obtains an expression over s rounds,

$$(m \cdot \alpha_0) \oplus (c_s \cdot \alpha_s) = \sum_{i=1}^{s} (k_i \cdot \gamma_i),$$

that holds with some probability p when averaged over all plaintexts. The typical approach to estimating p is to use what is often called the *piling-up lemma*, a technique described by Tardy-Corfdir and Gilbert [705] and Matsui [476].

Let Z_i, for $1 \leq i \leq m$, be m independent random variables which take the values $\{0,1\}$, and further suppose that each random variable takes the value 0 with probability p_i. Then we have that

$$\Pr(Z_1 \oplus Z_2 \oplus \ldots \oplus Z_m = 0) = \frac{1}{2} + 2^{m-1} \prod_{i=1}^{m} \left(p_i - \frac{1}{2} \right).$$

The piling-up lemma allows us to compute the bias of a set of combined linear approximations provided that the constituent linear approximations are independent. However, this independence of linear approximations cannot always be guaranteed and there has been interesting work on situations where the piling-up lemma fails to hold [677]. Nevertheless, the piling-up lemma is often the only tool we have and so if one succumbs to the temptation to assume independence, one can estimate the net overall bias of the total approximation.

Note that we can express the piling-up lemma using correlations and we have

$$2 \Pr(Z_1 \oplus Z_2 \oplus \ldots \oplus Z_m = 0) - 1 = \prod_{i=1}^{m} (2p_i - 1).$$

This goes some way to explaining the attractiveness of using correlations instead of biases since correlations can simply be multiplied together.

To introduce the technical notions that follow, we simplify the form of our linear approximation. We consider the linear approximation between message input m and the partially encrypted text after s rounds, c_s, to be

$$(m \cdot \alpha_0) \oplus (c_s \cdot \alpha_s) = 0.$$

Note that if such a linear approximation has bias of magnitude $|\varepsilon|$ then so has the corresponding expression with a 1 on the right-hand side. This then allows us to consider the combination of such approximations without paying regard to any specific choices of key.

To provide a firmer basis for linear cryptanalysis, we can follow much of our treatment of differential cryptanalysis in Section 8.1.

An s-round **linear characteristic** is a series of masks defined as an $(s+1)$-tuple $(\alpha_0, \alpha_1, \ldots, \alpha_{s-1}, \alpha_s)$ with probability p, where α_i is the mask used in the ith round. The characteristic predicts for $i = 1, \ldots, s$ that the sum of certain bits of the input to the ith round will equal the sum of certain bits in the outputs of the ith round.

Most of the linear attacks reported in the literature make use of iterative characteristics. For an iterated block cipher an s-round iterative linear characteristic is an $(s+1)$-tuple $(\alpha_i, \ldots, \alpha_{i+s})$ with $\alpha_i = \alpha_{i+s}$. Their usefulness can easily be seen when we note that an s-round iterative linear characteristic can always be used to build a t-round characteristic for any integer $t \geq s$.

Under the assumption that the s one-round characteristics are independent, we know that we can appeal to the piling-up lemma. The following notion, if applicable to the cipher under analysis, allows us to avoid any ambiguity about the application of the piling-up lemma.

An iterated cipher is called a **Markov cipher** *with respect to linear cryptanalysis if* $|Pr((c_i \cdot \alpha) = (c_{i-1} \cdot \beta) \mid c_{i-1} = \gamma) - 1/2|$ *is independent of γ for all α and β when the round key k is chosen uniformly at random.*

For a Markov cipher (with respect to linear cryptanalysis [577]) we can be sure that if the cipher has independent round keys then the bias of a multiple-round linear approximation, when taken over all values of the round keys, can be calculated from the one-round biases using the piling-up lemma. The probability of a linear characteristic will be calculated as an average over all keys.

While it is often relatively easy to find linear relations through individual rounds, difficulties arise when combining relations over several rounds. While noting the independence of round keys, it is important also to note that linear relations over several rounds can be obtained in different ways. So while analysis might reveal one particular linear approximation through the cipher, other linear relations that involve the same message and ciphertext bits may exist that approximate different key bits with differing biases. Seen from the point of view of an attacker, the masks in the intermediate rounds in an s-round linear characteristic can take different values and this leads to the natural definition of what has become termed a *linear hull*.

An s-round **linear hull** *is a pair of masks (α_0, α_s) with probability p. The linear hull predicts that the sum of certain bits in the plaintexts equals the sum of certain bits in the outputs of the sth round with probability p.*

In some sense an s-round linear hull (α_0, α_s) "covers" the s-round characteristics $(\alpha_0, \alpha_1, \ldots, \alpha_{s-1}, \alpha_s)$ for all values of α_i for $i = 1, \ldots, s-1$. Unsurprisingly, the behavior of a linear hull is both complex and key-dependent. This means that the bias witnessed in practice for a given key may not be that predicted by analysis of a single constituent linear characteristic. In fact, it is rarely easy to fully account for the bias that we might see. However, some block ciphers have a sufficiently regular structure and a sufficiently clean design to allow for some results to be stated. Some of the earliest work in this area is available in [178, 565], but it is an area that is still active and developing [526].

8.2.2 Key Equivalence

In the previous pages we have seen several assumptions that one might use to try and apply linear cryptanalysis. To calculate the probability of linear characteristics for iterated ciphers, one needs to assume that the linear relations of the individual rounds are independent. For Markov ciphers this can be obtained by assuming independent subkeys and by computing the probabilities as averages over all keys.

However, in a typical attack scenario the encryptions all use the same fixed key, and a probability computed over all choices of the plaintext and key may be different from one which is averaged over all choices of the plaintext for an unknown fixed key. With this in mind, and following the example of differential cryptanalysis, we can make the *hypothesis of stochastic equivalence*.

The **hypothesis of stochastic equivalence** *states that, for virtually all high probability s-round linear hulls* (α, β), *we have that*

$$|Pr_{\mathscr{M}}((m \cdot \alpha) \oplus (c_s \cdot \beta) = 0 \mid k = \kappa) - \tfrac{1}{2}| \approx$$
$$|Pr_{\mathscr{M},\mathscr{K}}((m \cdot \alpha) \oplus (c_s \cdot \beta) = 0) - \tfrac{1}{2}|$$

for a substantial fraction of specific key values κ.

In short, the hypothesis of stochastic equivalence states that for the vast majority of keys linear cryptanalysis will behave pretty much like the average case. And with this assumption, we hope to reassure ourselves that the theoretical behavior of characteristics shouldn't be too far from their behavior in reality.

8.2.3 Key Recovery and Data Complexity

An approximation with a sufficiently high bias can be used to determine $k \cdot \gamma$, which corresponds to 1 bit of information about the secret key k. Simultaneously the bias in the message-ciphertext relation allows us to *distinguish* the block cipher, *i.e.*, to observe properties we wouldn't expect to see in a randomly chosen permutation. So considering the basic approximation,

$$(m \cdot \alpha) \oplus (c \cdot \beta) = (k \cdot \gamma),$$

which holds with probability $p = \tfrac{1}{2} + \varepsilon$, assuming, without loss of generality, that $\varepsilon > 0$.

Given N plaintexts m and corresponding ciphertexts c we can recover one key bit as follows. Let T_0 denote the number of times $(m \cdot \alpha) \oplus (c \cdot \beta)$ is equal to 0 while T_1 denotes the number of times $(m \cdot \alpha) \oplus (c \cdot \beta)$ is equal to 1.

The Basic Linear Attack with Characteristic of Bias $\varepsilon > 0$

1. For all N intercepted texts (m, c):

- *Compute $b = (m \cdot \alpha) \oplus (c \cdot \beta)$.*
  - *If $b = 0$ increment counter T_0. Otherwise increment counter T_1.*

2. If $T_0 > \tfrac{N}{2}$ guess that $k \cdot \gamma = 1$. Otherwise guess that $k \cdot \gamma = 0$.

It is easy to see that this basic linear attack allows us to recover one bit of key information. To estimate the success rate of this attack we will appeal to some basic statistical theory. First, recall that our target linear approximation holds with bias $\varepsilon > 0$:

$$(m \cdot \alpha) \oplus (c \cdot \beta) = (k \cdot \gamma).$$

During the attack we keep two counters T_0 and T_1 and increment these according to the value of $(m \cdot \alpha) \oplus (c \cdot \beta)$ for N known plaintexts.

Let X_1, X_2, ... be independent, identically distributed random variables having mean μ and finite nonzero variance σ^2. Set $S_n = X_1 + \cdots + X_n$. Then

$$\lim_{n\to\infty} \Pr\left(\frac{S_n - n\mu}{\sigma\sqrt{n}} \le x \right) = \Phi(x),$$

where Φ denotes the standardised normal distribution function with mean 0 and variance 1.

Fig. 8.5 A statement of the *Central Limit Theorem* which can be found in most text books on statistics, *e.g.*, [239, page 244].

T_1 will follow a binomial distribution with mean $\frac{1}{2} + \varepsilon$ and variance $\frac{1}{4} - \varepsilon^2$. So using the *central limit theorem* (see Fig. 8.5) the behavior of

$$x = \frac{T_1 - N(\frac{1}{2} + \varepsilon)}{\sqrt{N(\frac{1}{4} - \varepsilon^2)}}$$

can be approximated by the standardised normal distribution $\Phi(x)$. There is a similar situation for the case when $k \cdot \gamma = 0$ that is catered for by the counter T_0. By choosing the largest counter value the cryptanalyst is attempting to distinguish between these two cases. To obtain the probability of success when $k \cdot \gamma = 1$, we need to evaluate $1 - \Pr(T_1 \le \frac{n}{2})$. On the assumption that ε is so small we can ignore ε^2,

$$\Pr(\text{success}) = 1 - \Pr(T_1 \le \frac{n}{2}) = 1 - \Phi\left(\frac{2(\frac{N}{2} - \frac{N}{2} - \varepsilon N)}{\sqrt{N}} \right)$$

$$= 1 - \Phi\left(-2\varepsilon\sqrt{N} \right).$$

The same success rate naturally holds for $1 - \Pr(T_0 \le \frac{n}{2})$ when $k \cdot \gamma = 0$, and some example success rates have been tabulated below. The important thing to note is that the number of plaintexts required for a good success rate is proportional to ε^{-2}. So as the bias drops the number of plaintexts needed in an attack goes up as the inverse squared. Some success rates we can expect as we take more texts are tabulated below.

N plaintexts	$\frac{\varepsilon^{-2}}{16}$	$\frac{\varepsilon^{-2}}{8}$	$\frac{\varepsilon^{-2}}{4}$	$\frac{\varepsilon^{-2}}{2}$	ε^{-2}
success rate	69%	76%	84%	92%	98%

A more sophisticated attack allows us to recover more key material. In particular, if we were to consider an r-round cipher then we would expect it to be more advantageous to use a linear approximation over fewer rounds than to use a linear approximation over more rounds. The advantage is two fold; first, we use a shorter approximation that has a higher bias and, second, by using what are termed *1R-* and

2R- linear cryptanalytic attacks against a cipher we have the chance to recover more key information.

In what follows, we restrict ourselves to removing a single round from a cipher, but in an attack on DES, for instance, one can find the values of the round keys in the first and last rounds at the same time [475].

To avoid the complexity of dealing with the key information being approximated, let us use (α_0, α_{r-1}) to denote a linear hull over the first $r-1$ rounds of some block cipher. We will assume that this linear hull can be reasonably expected to exhibit a bias of $\varepsilon > 0$. Since an attacker knows the ciphertexts, he is at liberty to guess some (or all) of the key bits that might have been used in the last round. With each key guess, the attacker can at least partially compute back one round and recover the partially encrypted text c_{r-1} that may have occurred after $r-1$ rounds. The attacker can then compute the value of (α_0, α_{r-1}), which analysis has already shown should exhibit a bias.

To formalize the description of this more advanced attack, denote the computation of c_{r-1} by $g^{-1}(c,t)$, where t is a potential value for the key used in the last round. Assume that there are τ possible values for t, namely 0 to $\tau - 1$, and set up 2τ counters numbered $T_{0,0} \ldots T_{0,\tau-1}$ and $T_{1,0} \ldots T_{1,\tau-1}$ where the subscript t is interpreted as the value of k_r and $T_{0,t}/T_{1,t}$ as its associated counter-pair. Given N messages and ciphertexts, and for each value of t, let $T_{0,t}$ denote the number of times $(m \cdot \alpha_0) \oplus (g^{-1}(c,t) \cdot \alpha_{r-1}) = 0$ and let T_1 denote the number of times $(m \cdot \alpha_0) \oplus (g^{-1}(c,t) \cdot \alpha_{r-1}) = 1$.

The Advanced Linear 1R-Attack with Bias $\varepsilon > 0$

1. *For all N intercepted text pairs (m, c):*

   - *For all τ values $t = 0, \ldots, \tau - 1$:*
     - *Compute $b = (m \cdot \alpha_0) \oplus (g^{-1}(c,t) \cdot \alpha_{r-1})$.*
     - *If $b = 0$ increment $T_{0,t}$; otherwise increment $T_{1,t}$.*

2. *Identify the counter $T_{i,s}$ for $0 \le i \le 1$ and $0 \le s \le \tau - 1$ with the largest value.*
3. *Guess that $k_r = s$.*

There are two observations to make.

First, we have made no effort here to recover the internal bit of key information. Instead, we restrict ourselves to recovering $\log_2(\tau)$ bits of key information. Attacks on DES [475], and most other descriptions of this attack, will probably include recovery of the internal key bit. However, by ignoring the internal key bit we have broadened our notion to the linear hull from a single linear approximation. Further, exactly the same algorithm can be used without change even if the bias $\varepsilon < 0$ since we would still recover the correct value for $t = j$ (only using the other counter in a given pair).

Second, in attacks where we try to determine part of the key from the outer rounds, we need to understand the behavior of an incorrect guess for the key. The simplest approach is to assume that wrong guesses to the key (or parts of the key) will result in essentially random values to c_{r-1}. This assumption is termed the *hy-*

pothesis of wrong key randomisation [281]. Without this assumption, and more importantly without something that closely approximates this in practice, it would become very difficult to mount the attack. In reality, tests and experiments on most good ciphers reveal this to be a reasonable assumption in practice.

The general outline of a linear cryptanalytic attack is nicely described in [92] and can be summarised as a three-phase process:

1. A *distillation* phase, where the *effective* data is collected, sorted, and efficiently stored.
2. An *analysis* phase, where all the possibilities for the *effective* key bits are matched to the effective data, with the correct value to the effective key bits being derived.
3. A *search* phase, where any remaining key bits not discovered in the analysis phase are recovered by exhaustive search.

The evaluation of the success rate of the advanced linear cryptanalytic attack is not so straightforward. One of the best treatments can be found in [678], which also takes into account a variety of optimisations to the basic attacks. In general terms, however, the practical value of linear cryptanalysis on a given cipher will depend on the number of known messages that are required and on the number of effective key bits that are being recovered from the outer rounds. As we increase the number of effective key bits, we will need to increase the amount of data to identify the right key value, we will need to increase the work effort to test all the effective key bits, and we will need to increase the memory requirements for processing the data.

8.2.4 Enhancements to the Basic Linear Attack

In some ways linear cryptanalysis is less versatile than differential cryptanalysis. Certainly, we don't see the same range of effective variants such as truncated, higher order, and other differential-based attacks. There have nevertheless been a few proposals for enhancements. In the practical execution of a 1R- or 2R-linear attack one might recover a single guess for part of the key from the final counter values. However, one can increase the success probability by taking (say) the top ten guesses and moving on with the rest of the analysis with these ten. This is referred to as *key ranking* [475] and by doing this we reduce the chance that we miss the correct key value at the first stage of cryptanalysis. The potential of some other enhancements may be more far-reaching.

8.2.4.1 Multiple Linear Approximations

When mounting a linear cryptanalytic attack we typically concentrate on using a single linear characteristic. We collect message-ciphertext pairs and by computing the sum of certain bits from the text pairs we obtain an approximation to a bit of key information.

In Sect. 8.2.1 we introduced the concept of linear hulls and we saw that other, unknown, approximations may impact the effective bias of a linear approximation. These other approximations have the same effective message and ciphertext bits, while approximating different effective key bits. But this idea can be turned around, as was first done in [350]. There it was shown how n linear approximations, each approximating the same internal key bit of the cipher with the same probability, could be used to provide an n-fold reduction in the amount of text for an equally successful attack. It is easy to see why this would be helpful; each evaluation of one of the n linear approximations gives us another guess for the internal key bit. Instead of calling for more text to improve the reliability of our guess, we reuse the data we have and use additional approximations to generate more guesses.

In a good block cipher it is unlikely that there will be many approximations to the same internal bit of key information. So we can relax these exacting conditions in several ways. We can deal with the case when the linear approximations to the same internal key bit have different biases. We can introduce a system of *weighting* so that we pay more attention to the guess arising from more reliable approximations. Under certain conditions [350], it can be shown that if we use approximations A_0, ..., A_{n-1} with biases ε_0, ..., ε_{n-1} instead of A_0 alone, then the amount of data to recover the internal key bit with the same success rate can be reduced by a factor of

$$\frac{\sum_{i=0}^{n-1} \varepsilon_i^2}{\varepsilon_0^2}.$$

Having dealt with different biases we can turn to using approximations to different key bits, and here there are some obvious [350, 351] and some more advanced techniques [92]. These allow us to simultaneously use linear approximations that approximate different internal key bits. One particularly interesting case is that of using a set of linear approximations for which the masks are linearly dependent. That one can still gain an advantage has at times been viewed as somewhat counter intuitive. But this phenomenon was first observed in 1994 [351] and has been examined in a body of work that continues today [92, 527].

8.2.4.2 Other Variants

The use of parity relations in forming our approximations might appear to be somewhat limited. It is therefore natural to ask if we can use other functions in a similar way so as to offer the cryptanalyst some advantage. There has, however, only been sporadic progress on alternative formulations to linear cryptanalysis and the conjectured improvements have tended to be limited in scope.

Harpes *et al.* [281] consider what are termed *I/O sums* and demonstrate a cipher for which this type of analysis would be more successful than linear cryptanalysis. They go on to broaden the framework for analysis to what is termed *partitioning cryptanalysis* [282], an approach which echoes to some extent earlier unpublished work by Murphy *et al.* [530].

Elsewhere [406], the role of non-linear approximations was considered. For most ciphers the use of such approximations within the internal rounds of a cipher is limited, but there are opportunities to improve the practical impact of an attack by improving analysis in the outer rounds of the cipher. This issue was revisited in [169] when *bilinear cryptanalysis* was introduced. Such analysis depends strongly on the form of the internal rounds to the cipher and provides a complement to the work of [406]. However, the net result is that if one is to achieve good results with this kind of cryptanalysis, then one needs to invent a block cipher that is particularly susceptible. Such block ciphers do not arise very often in practice.

Finally we mention *mod n cryptanalysis* [367], which applies to a limited group of cipher variants, but which nevertheless makes an interesting addition to the literature. The notion of a biased two-valued function is extended to a t-valued function for $t > 2$, for instance, the result of reduction modulo t. This allows for some progress to be made on the open status of some pre-existing proposals in the literature [367].

8.3 Differential-Linear Cryptanalysis

One intriguing extension of differential and linear cryptanalysis has been their combination in *differential-linear cryptanalysis*. Due to Langford and Hellman [437], this attack uses a differential of high probability to set up a known relation between the bits internal to two parallel instantiations of a block cipher. These bits are then used at the start of a parallel linear approximation in both block ciphers and a range of linear cryptanalytic techniques can be used to recover key information.

This approach provides an effective and elegant attack on eight rounds of DES, which we will now illustrate. The attack is a 2R-attack and so we refer the reader to [437] for details on the key recovery. Here we restrict our attention to the six-round differential-linear structure that lies at the heart of the attack. The necessary components are illustrated in Figure 8.6.

Like differential cryptanalysis, the attack is a chosen plaintext attack that requires pairs of messages with a specific difference between them. For the attack we illustrate we will denote the starting difference as $(\Delta, 0)$. Thus, over three rounds of a Feistel cipher, the difference evolves in the following way with probability 1:

$$(\Delta, 0) \xrightarrow{\mathscr{R}} (0, \Delta) \xrightarrow{\mathscr{R}} (\Delta, \Delta_1) \xrightarrow{\mathscr{R}} (\Delta_1, \Delta_3).$$

If we choose the difference Δ so that only 1 (or two) carefully chosen bits are involved, then we can identify the bits of Δ_1 that might change value. In particular, if the active bits in Δ are those bits that are input to DES S-box 1, then only the output bits from S-box 1 can change and so we know the active bits in Δ_1. In our attack we avoid these bits. What we want are bits of Δ_1 that never change. Then if we take a linear approximation that starts with these fixed bits, which therefore applies to both texts in the chosen-plaintext pair, then we know that the starting value for the mask α must be the same for both texts in the pair.

Assuming that we can choose a good set of starting bits for the linear approximation, which is the case for DES, we can use a good linear approximation to cover the next three rounds of the cipher. If this approximation holds with a bias ε then we have the following probability that the output from the linear approximation is the same for both messages:

$$\left(\frac{1}{2}+\varepsilon\right)^2 + \left(\frac{1}{2}-\varepsilon\right)^2 = \frac{1}{2}+2\varepsilon^2.$$

So if we were to take both outputs and to exclusive-or the bits indicated by the linear approximation, the value would be 0 with probability $\frac{1}{2}+2\varepsilon^2$ and 1 with probability $\frac{1}{2}-2\varepsilon^2$.

Typical techniques that we have described elsewhere can be used to provide 1R- and 2R-attacks. Note, however, that the amount of text required for a differential-linear attack is proportional to ε^{-4}, where ε is the bias of the underlying linear approximation. Thus the effectiveness of this attack quickly falls as we attempt to cover more rounds. That said, the attack given in [437] remains the most efficient on eight-round DES and it has been experimentally confirmed that ten bits of key can be recovered with 80% success rate from only 512 chosen message pairs.

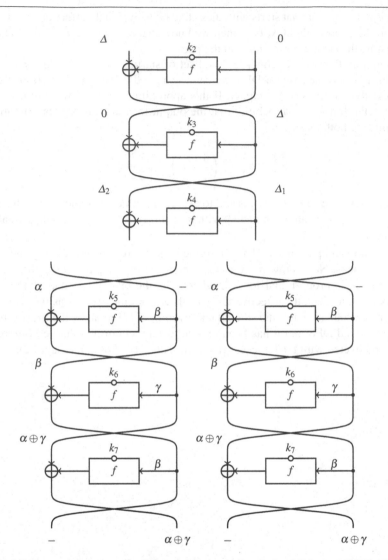

Fig. 8.6 Two components to the differential-linear cryptanalysis of eight-round DES. The second component consists of the three-round linear cryptanalysis of two simultaneous encryptions. The first component shows a three-round differential. If the bits α are guaranteed to be 0 for Δ_1, then the parity bits denoted by α in the text encryptions are guaranteed to be the same.

8.4 The Interpolation Attack

The interpolation attack was proposed in [327, 328]. In the simple version, the ciphertexts are expressed as polynomials $p(m)$ of the messages m. If such a polynomial has a sufficiently low degree then it can be reconstructed from a collection of plaintexts and their corresponding ciphertexts. In fact, the important property is that the polynomial should have a low number of coefficients. If an attacker can construct this polynomial then he has, in effect, a representation of the encryption function without necessarily having the exact value of the secret key used. There are variants of the attack which can also be used for key recovery, and while the simple attack uses a univariate polynomial, a more general version of the interpolation attack allows the ciphertext to be expressed as multivariate polynomials of several (sub) words of the message.

As an example, consider an r-round iterated b-bit cipher with subkeys $k_1, \ldots, k_r$ where the round function in round i is

$$g(x_i, k_i) = (x_i \oplus k_i)^3$$

and x_1 is the message m. The function g has inputs and output over $GF(2^b)$, a finite field with 2^b elements. To this description let us add some post-whitening, and we will define the ciphertext c for this unusual, but perfectly valid, b-bit block cipher as

$$c = g(x_r, k_r) \oplus k_{r+1}.$$

For odd values of b the cubing function is bijective so the cipher is well defined. Further, because of the good properties of the cubing function it can be shown that the cipher is highly resistant to (basic) differential and linear attacks.

To begin, let us assume that $r = 2$ and let c denote the ciphertext for a message m. Then we can write

$$c = m^9 + m^8 k_1 + m^6 k_2 + m^4 k_1^2 k_2 + m^3 k_2^2 + m^2 (k_1 k_2^2 + k_1^4 k_2)$$
$$+ m(k_1^2 k_2^2 + k_1^8) + k_1^3 k_2^2 + k_1^6 k_2 + k_1^9 + k_2^3 + k_3.$$

Given m and c this equation seems hard to solve for the unknown key (k_1, k_2, k_3). However, it is fairly easy to break this cipher in a known message attack. To see why, note that the terms involving k_1, k_2, and k_3 in the equation above are constants since the key is fixed. This means we can rewrite the equation as

$$c = m^9 + m^8 c_1 + m^6 c_2 + m^4 c_3 + m^3 c_4 + m^2 c_5 + m c_6 + c_7$$

for a set of constants $c_1, \ldots, c_7$. So given a message m and its ciphertext c we have an equation in seven unknowns. For any other value of m and c we get a different equation but in the same seven unknowns. Consequently, given the encryptions of seven different messages we can determine the constants $c_1, \ldots, c_7$. With these, one has enough information to encrypt any plaintext message even though the exact value of the key has not been determined.

In general, if one can express the ciphertext c as a polynomial of degree d in the plaintext message m, then one needs no more than $d + 1$ known plaintext messages and corresponding ciphertexts to find a representation equivalent to encryption with the secret key. The attack extends in a natural way to give a key recovery attack on iterated ciphers. The interpolation step is repeated for some guesses of some key bits until the interpolation succeeds as expected.

There is a variant of the interpolation attack that is often even more efficient. Consider an r-round iterated cipher and let z be the intermediate text after some $s < r$ rounds of encryption. Express z as a polynomial $p(m)$ of the message m. Then express z as a polynomial $q(c)$ of the ciphertext c. We then have that $z = p(m)$ and $z = q(c)$, and consequently

$$p(m) = q(c).$$

If both p and q are polynomials with a low number of coefficients then one can solve this equation for the unknown coefficients. This variant of the interpolation attack is reminiscent of the so-called algebraic attacks proposed on AES; see Sect. 3.2.3.

The attacks described above all assume that the polynomials in question exist with probability 1. In [326] a probabilistic variant of the interpolation attack was given. It was further shown that there are ciphers for which this attack is superior to the deterministic attack.

8.5 The Key Schedule

In iterated ciphers the ciphertext is computed over a number of rounds and it is a typical design policy that each round use a fresh round key. These are computed from the user-provided key using what is called the *key schedule*.

While many block ciphers share features in the design of the round, few share similar features in the key schedule. The design of a key schedule design is somewhat *ad hoc*, and while there are certainly sensible things to do, there are few guidelines or design criteria in this area. One interesting aspect is the performance of the key schedule. For the encryption of short messages or packets the key schedule is a fixed, and potentially significant, overhead. However, for many applications, when long messages are being encrypted, the cost of the key schedule is amortised over the entire encryption process. Another issue that often features is that of *on-the-fly* key generation, whereby successive subkeys are derived from one another, and one needn't generate and store all the subkeys at the same time. This can be helpful in very constrained devices.

One might classify the key schedules of popular ciphers in the following way:

- *Affine key schedules.* Subkeys are derived as an affine transformation of (parts of) the user-supplied key. Examples include the key schedules of DES [545], Triple-DES [549], and KASUMI [228].

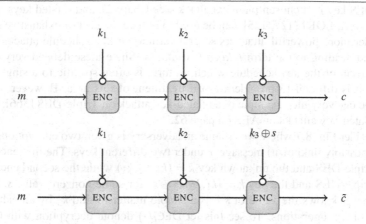

Fig. 8.7 The set-up for the related-key attack on Triple-DES [366]. The same message is encrypted under the two key-triples (k_1, k_2, k_3) and $(k_1, k_2, k_3 \oplus s)$.

- *Non-linear key schedules.* Subkeys are generated as (simple) non-linear transformations of (parts of) the user-supplied key. Examples include the key schedules of AES [551] and Serpent [11].
- *Complex key schedules.* The subkeys are generated as (complex) non-linear transformations of (parts of) the user-supplied key. Examples include the key schedules of DEAL [388] and RC5 [638], among many others.

Since there are no systematic design approaches to the key schedule, there are few generally applicable attacks. However, one preoccupation is often that of avoiding the kind of *weak-key* or *related-key* phenomenon we saw with DES (see Chap. 2).

The term *weak key* is used widely but rarely consistently. Often one talks of *weak keys* for some block cipher if there exists a subset of keys for which an attack can be mounted successfully (and which is less successful on other keys). The practical importance of weak keys can sometimes be disputed and a small number of weak keys need pose no problem if the keyspace is sufficiently large and encryption keys are chosen at random. However, when block ciphers are used within another construction, for example, in hashing (see Chap. 4), then weak keys can play an important role [190, 613].

An important concept in this area is that of *related keys*, when two (or more) keys k and k^* that are somehow related give rise to some structured phenomenon during encryption [46, 380]. The most well-known example of related keys can be found in DES where a key k and its bitwise complemented value $\bar{k}$ give rise to the complementation property. If c is the encryption of m using k then $\bar{c}$ will be the encrypted value of $\bar{m}$ using the key $\bar{k}$. This can be used to speed up exhaustive search for a DES key in a chosen plaintext attack (see Chap. 2), and related keys in other ciphers, *e.g.*, LOKI [127, 125], can be used to reduce the cost of exhaustive search.

Under more powerful attackers several variants of key schedule attacks can be mounted, leading to the term *related-key attacks*. Since these depend very closely on the form of the key schedule which, in turn, is often specific to a single block cipher, it is difficult to provide a broad treatment of this area. However, we will describe one very nice example of a related-key attack on Triple-DES [366], but see also related-key attacks on AES on page 62.

Consider Fig. 8.7 where we assume an adversary is given two encryptions of the same (possibly unknown) message m under two different keys. The first encryption uses Triple-DES and the unknown key $k = (k_1, k_2, k_3)$ while the second encryption uses Triple-DES and the key $k = (k_1, k_2, k_3 \oplus s)$ for some nonzero value s. Now if the attacker knows the value of s it is a simple matter to find k_3 by an exhaustive search of 2^{56} operations. To see this let $\text{DEC}_\ell(\cdot)$ denote decryption with (single) DES using the key ℓ. Then as we move through the different possible values for ℓ we have that $\text{DEC}_\ell(c) = \text{DEC}_{\ell \oplus s}(\tilde{c})$ when $\ell = k_3$. However, for any other value of ℓ this equality is unlikely to hold and so we can recover the value of the third key in only 2^{56} operations. Of course this requires that s be known. If this is not the case then the attacker faces a cryptanalytic problem that requires computational resources equivalent to those required for compromising double-DES.

Finally, we observe that the *slide attack* [98, 99] is, in effect, a variant of the related-key attack and it can be applied to ciphers where the sequence of round keys has repeated patterns. The perfect situation for the attacker would be that all round functions were identical, in which case there could be very efficient attacks. More sophisticated attacks exploit situations where the sequence of round keys in the decryption direction has similarities to the round keys in the encryption direction. In such situations the so-called *slide with a twist* can be useful, and in extreme (but thankfully rare) cases, the slide attack (and its twist) can be applied, independently of the number of rounds.

8.6 The Impact of Analysis on Design

Despite the large variety of block ciphers, there is a body of work that aims to put design on a consistent and sound footing. Indeed, it is possible to prove the resistance of a block cipher to specific attacks. However, block ciphers with particularly broad proofs of security are often of limited practical use. Instead, it is more common for designers to make the argument—perhaps incorporating elements of provable security—that a given cipher resists some well-known cryptanalytic approaches.

There are numerous ciphers for which absolute resistance to certain attacks has been claimed, yet the same ciphers have been vulnerable to unforeseen analysis. It is therefore worth being particularly wary of sweeping claims for cipher security. To illustrate, the group properties of cryptosystems based on permutation groups have been studied [464] and some state that the ability of a system to generate the symmetric group [162, see, *e.g.*, page 48] the message space is "one of the strongest security conditions that can be offered". Yet other researchers [529] have designed a cipher that generates the symmetric group that is particularly vulnerable to a known message attack.

But despite such unfortunate examples, and setting aside group theoretic properties, there remains considerable hope for block cipher designers. The area has progressed tremendously over the last several years and, at the very least, there are block cipher constructions that are believed by many cryptographers and cryptanalysts to be strong and hard to attack. Although Shannon's design principles are informal, they are natural enough to still provide the basis for all practical secret-key cipher designs.[1] More formal approaches, rather than some of the ad hoc methods from the past, now help us obtain very good diffusion and confusion for the components of a block cipher. However, there still remains no known provably secure construction for a complete block cipher.

8.6.1 Block Cipher Topology

In attempting to justify the design of a block cipher, one might consider the overall construction independently of any specific components. It is conceivable that if we were to assume some very general conditions on a round function (say), then we would be able to say something about the overall cipher. Of course one has to be careful in practice. It is not clear that a theoretical framework requiring ideal behavior in some component will be realistic when we instantiate the cipher. But it can still provide useful validation of a sound approach.

The pioneers in this field were Luby and Rackoff [453]. In their work they showed how the Feistel network [235] could be used to construct $2n$-bit pseudo-random permutations from n-bit random functions. As we recall from Chap. 2, the Feistel network uses the left and right halves of the cipher state during encryption. As before, we will denote the left and right halves of the input to round i for $1 \le i \le r$ by a_{i-1} and b_{i-1}, and round i of the Feistel network is defined as

$$b_i = a_{i-1} \oplus f(b_{i-1}, k_i), \text{ and } a_i = b_{i-1}.$$

In reality $f(\cdot, \cdot)$ is a carefully designed round function. But in the work of Luby and Rackoff $f(\cdot, \cdot)$ is chosen at random from the set of all functions on $\frac{b}{2}$ bits.

Luby and Rackoff were able to show several things. First they demonstrated that to be able to distinguish a three-round Feistel construction from a randomly chosen b-bit function, and to do so with a probability of success that is very close to 1, an attacker needs about $2^{\frac{b}{4}}$ chosen messages and their corresponding ciphertexts. A permutation achieving this bound is called a *pseudorandom* permutation [453]. They also showed that in a combined chosen plaintext and chosen ciphertext attack,

---

[1] Shannon also suggested constructing ciphers whose security would depend on a known, difficult problem. This is exactly how many public-key ciphers are constructed today.

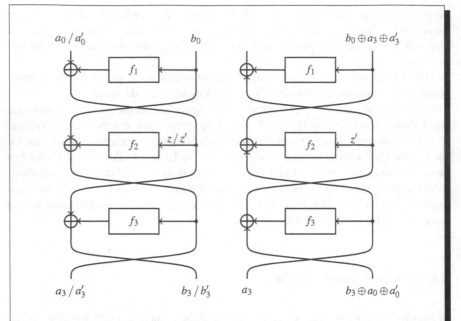

Fig. 8.8 Two three-round encryptions acting as a distinguisher for a three-round Feistel network. Encryption with two chosen messages is illustrated on the left while decryption with the chosen ciphertext on the right completes the distinguisher.

an attacker would need roughly $2^{\frac{b}{4}}$ chosen texts to distinguish a four-round Feistel construction from a random function with probability close to 1. Such a permutation is called *super pseudorandom*.

It is interesting to see what happens when we give the attacker more freedom. In particular, if we allow a ciphertext to be chosen adaptively in response to some of the data that has already been demanded, then it becomes very easy to distinguish a three-round Feistel network from a random function [453]. In fact we can distinguish such a three-round Feistel construction from a randomly chosen b-bit function with only two chosen plaintexts and one chosen ciphertext.

Consider Fig. 8.8 where we choose two messages of the form $(a_0 \| b_0)$ and $(a_0' \| b_0)$ with $a_0 \neq a_0'$. Encrypt these and denote the corresponding ciphertexts $(a_3 \| b_3)$ and $(a_3' \| b_3')$. Next, form a chosen ciphertext with the value $(a_3 \| b_3 \oplus a_0 \oplus a_0')$ and ask for the corresponding message.

For both messages the same input was used to the same round function in the first round. So we must have that $z \oplus z' = a_0 \oplus a_0'$. Now consider the decryption of our chosen ciphertext. The input to the third iteration of the round function is a_3, for which we know, from the two chosen messages, the output must be $b_3 \oplus z$. This means that the input to the round function in the second round must be

$$(b_3 \oplus a_0 \oplus a_0') \oplus (b_3 \oplus z) = a_0 \oplus a_0' \oplus z.$$

From our chosen messages we know that $z \oplus z' = a_0 \oplus a_0'$ and so the input to the second round of the ciphertext decryption is equal to z'. With this input the output from the second round function must be $b_0 \oplus a_3'$. Combining this with a_3 from the ciphertext, we have that the right side of the message that results from decrypting our chosen ciphertext must be $b_0 \oplus a_3 \oplus a_3'$. This holds with certainty. Yet for a function chosen at random this should only happen with probability $2^{-\frac{b}{2}}$. We can therefore distinguish between the two cases.

Moving back to the original setting, the results of Luby and Rackoff tell us that to distinguish three- and four-round Feistel networks from random b-bit functions, we need at least $2^{\frac{b}{4}}$ chosen plaintexts or chosen plaintexts and ciphertexts. The next two results tell us that if we have that much data available then we can efficiently distinguish between the Feistel network and the random permutation. In other words the bounds of Luby and Rackoff [453] are tight.

First we show that with q chosen plaintexts we can distinguish a three-round Feistel network from a random function with a probability

$$p = 1 - e^{-q(q-1)/2^{\frac{b}{2}+1}}$$

that is very close to 1 when $q \simeq 2^{\frac{b}{4}}$ [4, 590]. Imagine choosing a set of q messages $(a_0^i \| b_0^i)$ for $1 \leq i \leq q$, where all the a_0^i are (pair-wise) distinct and all b_0^i are fixed to the same, arbitrary, value. As before we denote the corresponding ciphertexts by $(a_3^i \| b_3^i)$. Then if q is close to $2^{\frac{b}{4}}$, with the probability p given above one expects to find at least one pair (with $i \neq j$) for which $a_0^i \oplus a_0^j = b_3^i \oplus b_3^j$ and $a_3^i = a_3^j$. For a random b-bit function and q close to $2^{\frac{b}{4}}$ this would happen with only a very small probability.

Now consider distinguishing a four-round Feistel construction from a random function. Here we choose a set of $q = n2^{\frac{b}{4}}$ messages (a_0^i, b_0^i) for $1 \leq i \leq q$ and some integer n, with all the a_0^i (pairwise) distinct and b_0^i fixed to some arbitrary value. Then it can be shown that one expects to find n pairs of messages for which $a_0^i \oplus a_0^j = b_3^i \oplus b_3^j$ for the four-round construction, whereas for a random b-bit function one expects to find only $\frac{n}{2}$ such pairs [590].

Since the work of Luby and Rackoff there have been many results published in the area and a good survey is provided by [538]. A few results are mentioned here. For instance, it has been shown [591, 652] that four-round super pseudorandom permutations can be constructed from just one or two (pseudo) random $\frac{b}{2}$-bit functions. Furthermore, in the four-round construction, the first and fourth functions can be replaced by simpler "combinatorial" functions while the same level of security as the original construction is maintained [538]. Coppersmith [168] has shown that round functions in a four-round construction can be identified (up to symmetry) using $\frac{b}{2}2^{\frac{b}{2}}$ chosen messages, and that with $8 \times 2^{\frac{b}{2}}$ texts 99.9% of the functions can be identified.

Note that for any number of rounds r there is a trivial upper bound for distinguishing an r-round Feistel construction from a randomly chosen b-bit function. This is because the Luby-Rackoff construction is a permutation. This means that with $n2^{\frac{b}{2}}$ chosen distinct messages, for some integer $n \geq 1$, the generated ciphertexts will all be distinct. However, for a truly random function one would expect collisions (*i.e.*, pairs of distinct inputs with equal outputs) [4, 590]. Thus it is can be more interesting to see how Luby-Rackoff constructions can be distinguished from randomly chosen b-bit *permutations* [725]. More recent results [400, 594, 595] indicate that with a larger number of rounds, the lower bound for the security of the Luby-Rackoff constructions approaches $2^{\frac{b}{2}}$. Other results extend the work of Luby and Rackoff to other block cipher topologies such as that found in IDEA [719] or those used in the AES finalists [518].

8.6.2 Resistance to Differential and Linear Cryptanalysis

While Shannon's theory of perfect secrecy and results such as those by Luby and Rackoff are major contributions to the area of block cipher theory, neither provides a practical method to construct block ciphers. Instead the main driving force behind block cipher design has been in countering attacks such as differential and linear cryptanalysis. Over the years this approach has become more systematic and designers tend to deploy two general strategies:

1. Improve the overall structure of the block cipher so as to better resist differential and linear cryptanalysis.
2. Improve the resistance provided by individual cipher components.

Many results on differential and linear cryptanalysis are quoted using the unified notions of differential probability and correlation that we introduced at the close of Chaps. 6 and 7. In what follows we will let p denote the highest probability of a one-round differential and we will let q denote the highest squared correlation of a one-round linear hull.

The Structure Approach

Given results on the differential or linear cryptanalysis of one round of a cipher, typically expressed in terms of the quantities p and q, it is possible to bound the performance of any differential or linear hull over r rounds of an iterated cipher.

Consider an r-round iterated cipher which has independent round keys. Then any s-round differential with $s \geq 1$ has a probability of at most p, and any s-round linear hull has a squared correlation of at most q [382].

Unfortunately, in the special case of Feistel ciphers, this result is not particularly useful. For a Feistel cipher there are trivial one-round differentials and characteristics and so the bound above is not particularly helpful. However, we can deliberately exclude these trivial cases and build our results on the best nontrivial one-round differential (holding with probability p_{max}) or the best nontrivial one-round linear hull (holding with a squared correlation of q_{max}). Then we have the following result for Feistel ciphers.

Consider an r-round Feistel cipher with independent round keys. Then any s-round differential for $s \geq 4$ has a probability of at most $2p_{max}^2$ and any s-round linear hull for $s \geq 4$ has a squared correlation of at most $2q_{max}^2$ [478, 569].

A substantial body of work has been built on this result and this confirms that round functions in an iterated cipher can be chosen so that the probabilities of all (nontrivial) differentials and the squared correlations of all (nontrivial) linear hulls can be made arbitrarily small [568, 569, 564]. These are not just abstract theoretical results. Instead, they have been used directly in the design of successful block ciphers such as MISTY [479] and KASUMI [228].

That said, this approach is not without its problems. One important issue is that the bounds are derived as an *average* over all keys for the block cipher. Yet in a given attack the secret key is (usually) fixed and there is no guarantee that the bounds hold in this case. In fact, it is easy to construct an example cipher demonstrating this. Consider a rather strange b-bit block cipher with a nonzero b-bit key k. We will define encryption as the operation

$$c = \text{ENC}_k(x) = k \times x$$

where multiplication is defined over the finite field $GF(2^b)$. Then, since we have

$$\text{ENC}_k(x) + \text{ENC}_k(x + \alpha) = k \times x + k \times (x + \alpha) = k \times \alpha,$$

the probability of any differential for this block cipher, when taken as an average over all keys, is given by $\frac{1}{2^b - 1}$. However for a *fixed* value of the key, and for two inputs x and $x + \alpha$, the difference in the ciphertexts will be constant and independent of the input x. Thus differentials for this cipher will hold with probability that is either 0 or 1 for any fixed key.

It therefore comes as no surprise that a number of pitfalls await if we try to apply theoretical results without some care. On a more positive note, many theoretical results on block ciphers do not give tight bounds and the practical resistance of a cipher to some attack might in fact be better than the bounds suggest. To help with this, more accurate bounds are typically gained by looking closer at the structure of the round, and this in turn leads to the idea of bounding the effectiveness of the best differential or linear characteristic in terms of the number of active S-boxes that must be encountered.

One interesting approach in the design of block ciphers is termed *decorrelation theory* [717]. Here an adversary attempts to distinguish a b-bit block cipher from an ideal b-bit permutation. Using well-defined measures of distance, this approach can be used to compare the distribution of messages and ciphertexts for a block cipher with those from a uniform distribution. The approach is well developed and one can speak about decorrelations of certain orders depending on the type of attack under consideration. It has been shown [717, 725] that this technique can be used to prove resistance against elementary versions of differential and linear cryptanalysis. In addition, such an approach allows for relatively easy proofs of the results by Luby and Rackoff and can provide bounds on security for the most well-known cipher structures.

However, the approach cannot be used to give security bounds for practical iterated ciphers "unless we approximate them to an ideal model" [725]. Once again we find ourselves stuck between a theoretically attractive analysis and its practical instantiation. It is only natural then that designers have turned to considering specific components and most obviously to best exploiting the fundamental role of the S-boxes.

The Component Approach

As we saw in Chaps. 6 and 7, when building a differential or linear attack one considers the differences and correlations of many components of the cipher. For those ciphers that use S-boxes, this is a component that can easily be adapted to help resist attack.

We can illustrate with the case of differential cryptanalysis. One speaks of *active* S-boxes in a differential characteristic when the characteristic requires a nonzero input difference to the S-box. If the maximum differential probability through a single S-box is p, and if the number of active S-boxes in a differential characteristic is t, then if we are willing to assume S-box independence we can bound the probability of the differential characteristic by p^t.

This is a powerful result. Provided we design the cipher in such a way as to guarantee that a certain number of active S-boxes must be involved in any differential characteristic, and provided we design the S-box with sufficient resistance to differential cryptanalysis locally, we can derive bounds for the security of the cipher. Thus the designer is aiming to maximise t and to minimise p so that p^t is sufficiently low. We can obviously adopt a similar approach for linear cryptanalysis where the designer will try to force the cryptanalyst to use linear characteristics with many active S-boxes, each with a low squared correlation q. There is therefore some justified hope that the cipher might resist differential cryptanalysis, though we note that such an approach typically bounds the effectiveness of single characteristics and approximations. It does not necessarily take into account the role of differentials, for instance.

But how low can we make p and q? Consider a function that maps n bits to n bits and let p_{max} and q_{max} denote the maximum probability of a differential characteristic and the maximum squared correlation of a linear relation across the function. It can be shown for all n-bit mappings that

$$p_{max} \geq 2^{1-n} \text{ and } q_{max} \geq 2^{1-n}$$

Table 8.1 The maximum differential probabilities and squared correlations for some power mappings over $GF(2^n)$. The most important case is that given by x^{-1} for even n, which is used in the AES S-box.

$f(x)$	p_{max}	q_{max}	Conditions
x^{2^k+1}	2^{s-n}	–	$s = \gcd(k,n)$
x^{2^k+1}	–	2^{s-n}	$s = \gcd(k,n)$ and $\frac{n}{s}$ odd
x^{-1}	2^{1-n}	–	n odd
x^{-1}	2^{2-n}	2^{2-n}	n even

where the probability and correlation are computed over all inputs. Mappings that attain these minimum values are said to be *almost perfect non-linear functions* [568] in the differential case and *almost bent functions* [140] for the linear case. Such functions are known to exist for odd n and not to exist for even n. Examples attaining the minimum values for odd n and low values for even n are given in Table 8.1. These are a special class of functions that are defined by raising a given input to different powers over the field $GF(2^n)$, so-called *power mappings*.

One of the most elegant examples of this approach is the design of the AES (see Chap. 3). It can be shown [189] that for any four-round differential or linear characteristic there must be at least 25 active S-boxes. Immediately therefore, there must be at least 50 active S-boxes over eight of the ten rounds of encryption. Since the AES S-box has been chosen to provide low differential probabilities and linear correlations, the resistance of the AES to these basic techniques is assured.

8.6.3 S-Box Properties

Soon after the publication of DES it was realised that the DES S-boxes had been carefully chosen. Later, with the advent of differential cryptanalysis, we found out just how inspired their design had been. Throughout the 1980s, and often in response to the discovery of a new cryptanalytic technique, additional criteria were proposed in the design of good S-boxes. It has been a productive area of research, but the practical results have been somewhat mixed. In this section we provide a quick overview of some of the properties that have been proposed, and we find it easy to classify them into two groups:

1. There are properties that reflect the propagation of difference that is induced by an S-box. These include the *strict avalanche criterion (SAC)* which stipulates that if we complement any single input bit to an S-box, then any of the output bits should flip with probability $\frac{1}{2}$.

2. There are properties that reflect the algebraic structure of an S-box. These include issues such as the *algebraic degree*, the *non-linear degree*, and what has been termed the *IO degree*.

Most, if not all of these properties, can be explained and manipulated using combinatorial techniques and the study of Boolean functions. After all, an S-box is essentially a look-up table or a compact representation of an ensemble of particularly useful Boolean functions.

Let $f(x)$ be a Boolean function $f : \{0,1\}^n \to \{0,1\}$ that maps n-bit inputs to a single bit of output. We will denote the binary representation of an input x by $x_{n-1}x_{n-2}\ldots x_1x_0$. It is well known that we can write the output of the function as a (unique) multivariate expression in terms of the n input bits:

$$f(x) = \sum_{b \in \{0,1\}^n} A_f(b) \quad x_{n-1}^{b_{n-1}} x_{n-2}^{b_{n-2}} \cdots x_1^{b_1} x_0^{b_0}.$$

Here $b_{n-1}b_{n-2}\ldots b_1b_0$ denotes the binary representation of b and determines which input bits are involved in a particular summand. The sum itself is computed modulo 2 and A_f is another Boolean function. If we consider this representation we see that the constant term of this summation is given by $A_f(0,\ldots,0)$ and the coefficient to the term $x_{n-1}^{b_{n-1}} x_{n-2}^{b_{n-2}} \ldots x_1^{b_1} x_0^{b_0}$ is given by $A_f(b_{n-1}b_{n-2}\ldots b_1b_0)$.

This representation of the Boolean function $f(x)$ is known as the *algebraic normal form*. An important quantity is the maximum Hamming weight of $b \in \{0,1\}^n$ for which $A_f(b) = 1$. This is known as the (algebraic) degree of f and will be denoted by $\mathrm{AD}(f)$. The algebraic degree gives a measure of how many input bits might simultaneously have an impact on the value of the output. Intuitively we might expect a high algebraic degree to be a desirable attribute for an S-box, but things are rarely so clear-cut and trade-offs may well come into play. Here is an example of the algebraic normal form.

Let $f : \{0,1\}^3 \to \{0,1\}$ be a Boolean function defined by the look-up table:

x	0	1	2	3	4	5	6	7
$f(x)$	1	0	1	0	1	0	1	1

The algebraic normal form can be derived as $f(x) = x_2x_1x_0 + x_0 + 1$ and so the algebraic degree $\mathrm{AD}(f)$ is 3.

We can extend these ideas to a general n-bit to m-bit function as instantiated by an S-box S. One can view such a function as the concatenation of m Boolean functions, so that the evaluation of $S[x]$ with input x is given by the evaluation of m component n-bit Boolean functions f_{m-1} to f_0 with

$$S[x] = f_{m-1}(x)\|\ldots\|f_1(x)\|f_0(x).$$

These component Boolean functions f_i are called the *coordinate functions* of the S-box.

Table 8.2 The values of the algebraic, non-linear, and IO degree for selected power mappings $f(x)$ over $GF(2^n)$. In the first line k and n are chosen so that $3 \leq 2^k + 1 < n - 2$.

$f(x)$	AD(f)	NL(f)	IO(f)
x^{2^k+1}	2	2	2
x^{-1}	$n-1$	$n-1$	2

We now have a very natural definition for the *algebraic degree* of an S-box (or any n-bit to m-bit function f), and we define the *algebraic degree* of the S-box to be the maximum algebraic degree of the coordinate functions of the S-box. Again, it seems that a high algebraic degree would be a very desirable property since it implies that at least one of the output bits depends on many input bits.

Related to the algebraic degree is another quantity called the *non-linear degree*. It is also called the *non-linear order* by some authors. Given an n-bit to m-bit function f with $f : \{0,1\}^n \rightarrow \{0,1\}^m$, the non-linear degree of f, denoted by NL(f), is the *minimum* value of the algebraic degree of all linear combinations of the coordinate functions of f. This is a more demanding criterion than the algebraic degree itself and a high value to this quantity would again, at least intuitively, appear to be a desirable attribute. Loosely, it implies that not only are individual output bits from the S-box complicated, but so are some simple algebraic combinations.

In the special case of n-bit to n-bit functions, for $n > 1$, the highest possible non-linear degree is $n - 1$ while the highest possible algebraic degree is n. In Table 8.2 some power mappings $f : \{0,1\}^n \rightarrow \{0,1\}^n$ are shown for which the algebraic and the non-linear degrees are both $n - 1$.

There is an interesting connection between the non-linear degree and higher order differentials as presented in Sect. 8.1.4.3. Consider a function f with non-linear degree s. Then the non-linear degree of a first-order derivative of f is at most $s - 1$. Moreover, the non-linear degree of a d-order derivative of f is at most $s - d$, where $s \geq d$. Here is an example using the function f from before.

Let $f : \{0,1\}^3 \rightarrow \{0,1\}$ be a Boolean function with the algebraic normal form $f(x) = x_2 x_1 x_0 + x_0 + 1$. The algebraic degree AD(f) is 3. The first-order derivative at the point 1 (001 in binary notation) is

$$\Delta_1 f(x) = x_2 x_1 + 1.$$

The algebraic degree of $\Delta_1 f(x)$ is therefore 2. Consider the second-order derivative of f at the point $(1, 2)$:

$$\Delta_{1,2} f(x) = x_2.$$

The algebraic degree of $\Delta_{2,1} f(x)$ is therefore 1. Consider the third-order derivative of f at the point $(1, 2, 4)$:

$$\Delta_{1,2,4}f(x) = 1.$$

The algebraic degree of $\Delta_{1,2,4}f(x)$ is therefore 0.

The *input-output degree* or IO degree is of interest for resisting certain kinds of algebraic attacks [172]. In many ways, it provides a natural extension of the ideas embodied in the algebraic degree. Suppose we have an n-bit to m-bit function f. Then we can represent all the input bits as $x_{n-1}\dots x_0$ and all the output bits as $y_{m-1}\dots y_0$. We can then consider deriving a multivariate equation that holds over all inputs and has the following form:

$$\sum_{b\in\{0,1\}^{n+m}} B_f(b)\, y_{m-1}^{b_{n+m-1}} y_{n+m-2}^{b_{2n-2}} \cdots y_1^{b_{n+1}} y_0^{b_n} x_{n-1}^{b_{n-1}} x_{n-2}^{b_{n-2}} \cdots x_1^{b_1} x_0^{b_0} = 0.$$

The sum is computed modulo 2 and B_f is a Boolean function on $(m+n)$-bit inputs that sets each coefficient for the 2^{n+m} terms in the summation. If we define d to be equal to the maximum Hamming weight of $b\in\{0,1\}^{m+n}$ for which $B_f(b)=1$, then d is termed the *degree* of the multivariate expression. (This is an exact analogy of the algebraic degree, but involves all the input and output bits for a non-Boolean function.) However, in contrast to the case of the algebraic degree the sum considered here is not unique, and so we imagine searching over all multivariate expressions of the form above and keeping the lowest value of all *degrees*. This is called the *IO degree* of the function f.

It will not be a surprise that the quantities we have introduced are related. For an n-bit to m-bit function f the IO degree is less than or equal to the non-linear degree, which in turn is less than or equal to the algebraic degree. More concisely,

$$AD(f) \ge NL(f) \ge IO(f).$$

Unlike in the algebraic and non-linear degrees, bounds on the IO degree depend closely on the size of the input and output. Consider an n-bit to m-bit function f with input x and corresponding output $y = f(x)$. In determining the value $d = IO(f)$ we are finding multivariate expressions of degree d in the bits of x and y. And this expression must hold with certainty. A little reflection shows that d cannot get so large, even if we significantly increase the size of our S-box. The number of possible terms of degree t in $p(x,y)$ is

$$\binom{n+m}{t}$$

and so once we have d that satisfies

$$\sum_{t=0}^{d}\binom{n+m}{t} > 2^n$$

there must be a multivariate expression $p(x,y) = 0$ of degree at most d that holds with certainty. The lowest possible value of d gives an upper-bound to the IO degree and Table 8.3 gives the maximum values of d for different values of n and m that appear in common S-box designs.

Table 8.3 The maximum IO degree for n-bit to m-bit mappings f. As expected, the maximum I/O-degree increases for larger n and m, but only moderately.

n	m	Maximum IO degree
4	4	2
6	4	3
8	8	3
16	16	5
32	32	8
64	64	16
128	128	30

8.6.3.1 Choosing S-boxes

Given the research devoted to S-box properties one might expect consensus on how best to choose an S-box. However, there is still healthy debate on the topic.

One school of thought suggests that we construct an S-box from a (simple) mathematical function with known properties. In this way we know exactly what we are getting. The downside is that the box will inevitably contain some structure, though if said structure cannot be exploited then it is hard to see what risks it might pose. The most frequently used mathematical functions in S-boxes have been the power functions over a Galois field. For the algebraic and non-linear degree of these power mappings, it can be shown [137] that for $f(x) = x^d$ in $GF(2^n)$, the algebraic degree as well as the non-linear degree of f equals the Hamming weight of d (see Table 8.2). The S-boxes in the cipher MISTY [479] are constructed from power mappings in $GF(2^7)$ and $GF(2^9)$ while the S-box in the AES is constructed from an affine transformation (over $GF(2)$) of the power mapping $f(x) = x^{-1} = x^{254}$ in $GF(2^8)$. In this way it is guaranteed that the AES S-box has the highest possible algebraic and non-linear degree over $GF(2^8)$ for a bijection, namely 7.

The opposing design approach is to choose an S-box at random, or rather to generate S-boxes repeatedly in a pseudorandom and verifiable way until we find an S-box that satisfies all the criteria that we seek [11].

In addition to the security that is offered by an S-box, we are likely to be interested in the efficiency of its implementation, particularly in hardware. Examples where this was a concern include DES [545] and MISTY [479], where the S-boxes were the most efficient that could be identified from within a relatively small set of possible S-boxes. The block ciphers Twofish [667], and Anubis [28] use S-boxes that are constructed from smaller four-bit S-boxes, which can lead to compact implementations. In fact, hardware efficiency has been the fundamental constraint in the design of several recent block ciphers and the cipher PRESENT [116] was optimised with the most extremely limited hardware environments in mind.

8.7 Getting to the Source

Given the range of topics for this chapter, there are far too many interesting publications for us to list them all here. However, the following will make a good start for those interested in reading about some of the more advanced block cipher topics in more detail.

Differential-based attacks	[59, 81, 328, 391, 433, 531, 729]
Linear-based attacks	[437, 476]
Block cipher structure	[184, 453, 478]
S-box design	[147, 375, 563, 564, 571]
Other cryptanalysis	[13, 17, 18, 19, 27, 31, 47, 64, 67, 70, 72, 83, 86]
	[89, 90, 91, 113, 130, 143, 156, 170, 177, 182]
	[211, 212, 214, 220, 225, 244, 251, 258, 260, 271]
	[276, 292, 313, 319, 337, 339, 342, 365, 379, 390]
	[401, 402, 404, 409, 410, 421, 424, 434, 440, 447]
	[452, 456, 458, 482, 502, 516, 520, 523, 560, 586]
	[593, 596, 597, 608, 625, 633, 668, 702, 718, 730]
	[731, 736, 748, 750, 755, 756, 757, 759]
Other design aspects	[2, 3, 14, 22, 32, 34, 35, 44, 52, 84, 88, 108, 109]
	[126, 131, 134, 146, 147, 149, 154, 188, 207, 210]
	[213, 229, 230, 245, 248, 249, 254, 262, 265, 273]
	[283, 308, 311, 313, 317, 321, 322, 323, 338, 343]
	[346, 376, 425, 443, 444, 448, 449, 454, 459, 461]
	[472, 473, 474, 485, 486, 487, 488, 492, 500, 508]
	[512, 561, 562, 563, 566, 567, 570, 571, 572, 573]
	[580, 592, 598, 599, 603, 607, 609, 610, 611, 614]
	[616, 623, 628, 635, 645, 646, 653, 658, 665, 666]
	[671, 672, 673, 674, 684, 685, 686, 688, 690, 695]
	[699, 720, 721, 722, 723, 737, 764, 766, 767, 768]
	[769]
Implementation aspects	[150, 192, 304, 442, 480, 481, 601, 618, 619, 752]

Chapter 9
A Short Survey and Six Prominent Ciphers

Four events provided much of the inspiration for the design and the cryptanalysis of block ciphers.

The first and most significant development was the invention and publication of DES. This was such an intriguing design, particularly given its provenance, that it lead to the development and foundation of the increasingly sophisticated science of block cipher design and cryptanalysis in the open community.

Differential cryptanalysis was a second major spur, and this was described openly for the first time in the early 1990s. It was the first method to propose the recovery of a DES key by analytic methods and understanding this technique gave birth to many new design criteria and block cipher proposals. This was further advanced with the advent of linear cryptanalysis. When applied to DES it is superior to differential cryptanalysis; not only is less data required but it need "only" be known rather than chosen by the attacker. Unsurprisingly the advent of linear cryptanalysis spawned new design criteria for block ciphers, although many (but not all) design criteria for resisting differential cryptanalysis sometimes help resist linear cryptanalysis.

The fourth advance in block cipher design came with the NIST-organised competition to establish the Advanced Encryption Standard (AES). During this process, and also in the years leading up to it, many new block ciphers were designed. By understanding the level of security that they provided, the community was able to substantially extend its knowledge of block ciphers.

In what follows we will give a short survey of the surprisingly large number of block ciphers that have been proposed over the past 30 years. We will then close the chapter, and the book, by highlighting six of these in detail.

9.1 From DES to the AES

The DES [545] is one of the earliest and most important block ciphers; as such, Chap. 2 is devoted solely to this remarkable cipher. Close relatives to DES include Lucifer, a name which appears to refer to several ciphers developed by IBM but which evolved into the DES [234, 235]. Over the years there have been many attempts to improve DES by changing or rearranging components, for example, in GDES [659, 661] and RDES [416] which sports a modified key schedule[128]. These variants have been shown to be weaker than DES [81, 38, 99]. By contrast, Matsui [477] and Biham and Biryukov [53] considered DES variants where the order of the S-boxes is different from that specified in the original. Some variants offer increased security from various attacks, but none have gained popularity as a serious replacement of DES. It is interesting to include in the ciphers of the DES era the cipher GOST [760], which is a 64-bit block cipher with a 256-bit key. The cipher, occasionally referred to as "the Russian DES", was developed in the 1970s but only made public in 1990 as a Soviet standard, GOST 28147-89. This was translated into English by Pieprzyk and Tombak and made public in 1994.

One early alternative to DES was Madygra [463, 95], which was proposed in 1984 and, even then, was designed with an eye to improved software performance. The cipher MULTI2, a Feistel cipher with a 64-bit block and a 64-bit key, was developed by Hitachi in 1988 and is still of interest today [18]. In 1989 Miyaguchi presented the block cipher FEAL [504]. After a series of strong cryptanalytic attacks, FEAL was modified and FEAL-N [505] was presented the following year. Despite its unfortunate history, FEAL remains an important contribution to the field of block ciphers and is considered in more detail later in the chapter. It is also interesting to note that one of the most widely used "non-DES" block ciphers during the early 1990's dates from this time. The cipher RC2 was a confidential and proprietary block cipher designed in 1989 by Rivest for RSA Data Security, Inc. Later, in 1997, RC2 was published as an Internet draft [641] and described at FSE 1998 along with an accompanying analysis [405].

LOKI [127] is a 64-bit block cipher with a 64-bit key and was proposed as an alternative to DES. However some problems with the key schedule were soon reported [380, 46] and a modified version, called LOKI91 [125], was presented the following year and in turn seems to have triggered the invention of so-called related-key attacks [380, 46]. Also in 1990, Lai and Massey proposed PES, the Proposed Encryption Standard, which received much attention. Its redesign, IDEA [433], appeared the following year, and its continued importance is reflected in this chapter.

The 1990s were important years for block cipher designers and analysts. At the 1990 Crypto conference, one of the three flagship conferences in academic cryptography, three block ciphers were presented. (This would be unheard of today, so much has the field of cryptography changed in the past 20 years.) These were REDOC II [176] by Cusick and Wood and the ciphers Khufu and Khafre by Merkle [496]. Of these, it is fair to say that Khufu and Khafre have had a more lasting impact. Khufu is a 64-bit Feistel cipher that uses a secret key of 512 bits. Perhaps the most interesting part of the design is the use of secret S-boxes that are generated from the user-supplied key. These S-boxes take an eight-bit input and return a 32-bit output. The ideas of both larger and key-dependent S-boxes later reappeared in designs that were submitted to the AES competition. Khafre is similar to Khufu, but the S-boxes are fixed and were generated by Merkle. The two ciphers have now been fairly well studied and a number of cryptanalytic results have been established [81, 257, 58, 729].

To provide an outlet for this new work in symmetric key cryptography, the workshop series *Fast Software Encryption* was started in December 1993 by Anderson. In the years that followed, a large number of block ciphers were presented, some of which in turn evolved into AES submissions. In 1993 Schneier presented the 64-bit block cipher Blowfish [662, 716], which is heavily inspired by Merkle's Khufu design. Kaliski and Robshaw [349] presented an experimental block cipher with very large blocks based around the hash function MD5 [651]. Massey introduced SAFER K-64 [470, 471], a 64-bit block cipher with a 64-bit key. In a departure from Feistel-like designs, SAFER K-64 is a traditional SP-network with all operations being byte-oriented. Two years later the cipher was redesigned with the inclusion of a new key schedule proposed in [384]. The resultant cipher was named SAFER SK-64 [468]. Early cryptanalytic results on SAFER can be found in [391, 525], and Massey later designed several extended variants of the cipher, though all built around the same core principles.

1993 and 1994 saw the publication of several block ciphers that reflected the growing philosophy of using simple internal operations during encryption. Daemen proposed 3-way [180, 178], a lightweight block cipher with a block size of 96 bits, an unusual block cipher length. Wheeler proposed Wake [740], a variable-width block cipher based on an auto-keyed table lookup, and also the lightweight cipher TEA [742]. Due to the discovery of weak keys in TEA [366] the cipher XTEA (eXtended TEA) was developed in unpublished work [741]. It was further tweaked in the ensuing years. However the most strikingly simple design of that period was the block cipher RC5, designed by Rivest [638]. This prominent cipher will be considered in more detail later in the chapter.

Two important predecessors to Rijndael are SHARK [632] and SQUARE [183] which were proposed by Rijmen *et al.* in 1996 and 1997 respectively. Anderson and Biham are behind the designs BEAR, LION, and LIONESS [10], which are large block ciphers built from a stream cipher and a hash function. Matsui proposed MISTY1 [479] for which bounds for the best differentials and linear hulls could be established and which built on earlier work by Nyberg and Knudsen [568, 569]. Kwan proposed the block cipher ICE [428, 649], which used key-dependent permutations, while M'Raihi *et al.* proposed XMX [522, 119], a cipher that used only modular multiplications and the exclusive-or operation. CAST-128 is a block cipher constructed according to the CAST procedure by Adams and Tavares [1] and which successfully reflected some of the same design choices in Merkle's Khufu. The block cipher Akelarre [466] combined elements of IDEA and RC5, though unfortunately with less success than either of the original ciphers [403].

Some designers tried to address the occasionally cited need for block ciphers that operate on different block lengths. UES [280] was designed by Handschuh and Vaudenay as a method for doubling the block length of a block cipher while Daemen [178] proposed a number of block ciphers in his Ph.D. dissertation, including BaseKing which is a large-block variant of 3-way (see above). SPEED [765, 274] by Zheng is a block cipher with a variable block length. However, while all this

Table 9.1 The 15 candidates for the AES. All use 128-bit blocks and 128-, 192-, and 256-bit keys.

Block cipher	Designers
CAST-256	Adams, Tavares, Heys, Wiener
CRYPTON	Lim
DEAL	Knudsen (submitted by Outerbridge)
DFC	Gilbert, Girault, Hoogvorst, Noilhan, Pornin, Poupard, Stern, Vaudenay
E2	Kanda, Moriai, Aoki, Ueda
FROG	Georgoudis, Leroux, Chaves
Hasty Pudding Cipher	Schroeppel
LOKI97	Brown, Pieprzyk, Seberry
MARS	Burwick, Coppersmith, D'Avignon, Gennaro, Halevi, Jutla, Matyas Jr., O'Connor, Peyravian, Safford, Zunic
Magenta	Jacobson Jr., Huber
RC6	Rivest, Robshaw, Sidney, Yin
Rijndael	Daemen, Rijmen
SAFER+	Massey, Khachatrian, Kuregian
Serpent	Anderson, Biham, Knudsen
Twofish	Schneier, Kelsey, Whiting, Wagner, Hall, Ferguson

work is interesting and raised intriguing design issues, the availability of standardised modes of operation means that the need to support unusual block lengths is not always so clear.

9.2 The AES Process and Finalists

The AES competition was, in some sense, the culmination of two decades of work on block ciphers. While more background to the competition and, of course, full details on the eventual winner were given in Chap. 3, Table 9.1 gives the 15 first-round ciphers which were accepted from an initial field of 21 submissions. Rijndael was chosen as the AES and deservedly gets Chap. 3 to itself. The other four finalists, MARS, RC6, Twofish, and Serpent, will always live in its considerable shadow.

MARS [132] was a rather unusual cipher with different types of rounds. It used eight *forward mixing* rounds, 16 rounds of *core encryption*, and finally eight rounds of *backward mixing*. The mixing rounds were related so that the inverse of the backwards mixing rounds is almost identical to the forward mixing rounds. MARS used a substantial mix of operations such as exclusive-ors, additions modulo 2^{32}, multiplications modulo 2^{32}, fixed bitwise shifts, data-dependent bitwise shifts, and S-box table lookups. Its performance suffered accordingly, though it was widely viewed as a particularly secure option.

RC6 [642] was an evolutionary development of RC5. Like RC5, RC6 made essential use of data-dependent rotations. RC6 was an iterated cipher running over 20 rounds, with each round having a Feistel-like structure. The operations used in RC6 were exclusive-ors, additions modulo 2^{32}, multiplications modulo 2^{32}, fixed bitwise shifts, and data-dependent bitwise shifts. While offering very good software performance, the use of multiplication was less easily accommodated on smart cards and in hardware.

Serpent [11] was a classical SP-network. Each of the 32 rounds used 32 applications of a four-bit S-box and the S-box layer was designed so that so-called *bit-slice* techniques would permit an efficient implementation. The operations used in the encryption routine of Serpent were exclusive-ors, fixed bitwise shifts, and table lookups. Serpent's similarities with DES were appealing, particularly similarities that extended to an exceptional performance in hardware. But the large security margin built into the cipher hampered the throughput in software.

Finally, Twofish [667] was a 16-round cipher with a Feistel-like structure. The cipher built on a variety of techniques that had been used with success elsewhere, including components that were inspired by the ciphers Khufu [496], Square [183], and SAFER [471]. The operations used within Twofish included exclusive-ors, additions modulo 2^{32}, fixed bitwise shifts, and table lookups. Some very impressive implementation efforts by the designers gave Twofish an appealing performance profile and, like the other unsuccessful candidates, was widely viewed as secure.

Over the years of the AES process a substantial body of cryptanalytic and performance analysis was completed by many different commentators. Much of this remains interesting and relevant today and a concise summary was provided in the NIST final report on the AES process [550]. In a far-sighted initiative, all materials that were submitted to NIST during the AES process have been archived and are intended to remain available via www.csrc.nist.gov.

9.3 After the AES

For many commentators the end of the AES process was supposed to bring the design of new block ciphers to an end. However, this has not been the case and while there was a short lull in activity, the field of block ciphers has once again become very active.

9.3.1 Other Competitions and Standardisation Efforts

After the AES there were other large-scale efforts to compare and analyse block ciphers. However, these have not had as great an impact as the AES process.

New European Schemes for Signatures, Integrity and Encryption, NESSIE, was a research project sponsored by the European Commission for a three-year period that

Table 9.2 The NESSIE [558] and CRYPTREC [175] block ciphers.

NESSIE Block Ciphers		
Name	Block size	Key size
CS-Cipher	64	128
Hierocrypt-L1	64	128
IDEA	64	128
Khazad	64	128
MISTY1	64	128
Nimbus	64	128
Anubis	128	128–320
Camellia	128	128, 192, 256
Grand Cru	128	128
Hierocrypt-3	128	128, 192, 256
Noekeon	128	128
Q	128	128, 192, 256
RC6	128	128, 192, 256
SC2000	128	128, 192, 256
SHACAL	160	512
NUSH	64,128,256	128, 192, 256
SAFER++	64,128	128, 256
CRYPTREC Block Ciphers		
Name	Block size	Key size
CIPHERUNICORN-E	64	128
FEAL-NX	64	128
Hierocrypt-L1	64	128
MISTY1	64	128
Camellia	128	128, 192, 256
CIPHERUNICORN-A	128	128, 192, 256
Hierocrypt-3	128	128, 192, 256
MARS	128	128, 192, 256
RC6	128	128, 192, 256
SC2000	128	128, 192, 256

ended in 2003. As part of this project there was an open call for candidate algorithms within many areas of cryptology. A total of 17 block ciphers were proposed (see Table 9.2) of which many were variants of AES candidates or heavily inspired by these designs. More details can be found in the final NESSIE report [558].

The *Cryptography Research and Evaluation Committee*, CRYPTREC, was a Japanese initiative to evaluate a range of cryptographic primitives. In many aspects it was the Japanese equivalent of NESSIE and several block ciphers that were not submitted to either the AES competition or to NESSIE were considered within CRYPTREC; see [175] for more details.

9.3.2 Niche Proposals

With the standardisation of the AES and its very versatile performance profile, it is rare to find situations where the AES is not a good choice for implementation. Certainly, when the AES can be used there are no compelling reasons to look elsewhere or to turn to a less well-studied algorithm.

Nevertheless, new ciphers have been proposed for specific applications. The cipher Mercy [174] has a large block size and is intended for disk encryption. FOX [340] is a cipher from a family of constructions that use techniques seen in the cipher IDEA. With regards to provable security, DFC v2 [269] is the successor to the AES candidate DFC while KFC [21] offers a proof of security and is claimed, by the authors, to be the first such practical cipher. SMS4 is a 128-bit block cipher that appears as part of WAPI, the Chinese WLAN standard [450], and its simple design has received some independent analysis [227, 451, 762]

From time to time proprietary ciphers from the 1980s appear zombie-like as a result of reverse-engineering. Not all proprietary ciphers are weak but many are. The *Cellular Message Encryption Algorithm (CMEA)* was proposed for mobile communications in the US while Keeloq is used in remote keyless entry systems. Both use small block sizes, both were designed in the mid-1980s, and both offer poor security [731, 313].

9.3.3 Lightweight Block Ciphers

One important area of research has become known as *lightweight* block ciphers. Technological advances mean that the widespread deployment of very simple devices such as passive RFID tags is becoming a reality. Yet these devices are very constrained with regards to the space available for security features and the power that might be available for a computation. This means that ciphers such as the AES, while they can be implemented in a surprisingly compact manner [236, 237], will likely remain unsuitable for passive RFID deployment in the short term.

More details on hardware implementation can be found in [587] though a comparison between different implementations of different ciphers is difficult. The typical base point for a comparison of size is given by a unit called a *gate-equivalent* (GE). This is computed as the number of NAND gates that would fit into the same physical area occupied by the implementation. Unfortunately, given its name, it is not quite a standard unit and the same implementation of the same cipher in two different fabrication technologies can have different GE counts. Nevertheless, the notion of the GE allows at least a crude comparison of the areas required for different implementations. It is important to note, however, that the area occupied by an implementation is not the only important factor and issues such as the power profile, energy consumption, and processing time can all impact the suitability of an algorithm for a given application.

Given the prominence of the AES most other cipher implementations are measured against implementations of this cipher [237, 275] and a typical goal for lightweight ciphers is to have an area less than 3 000 GE. Several promising designs can be implemented under 2 000 GE and some proposals even break the 1 000 GE barrier. However this can sometimes be at the price of a long processing time and/or restricted functionality.

The area is moving quickly and any survey here would soon be out of date. The reader is therefore referred to the current literature for more details, though some background and indications of early trends can be found in [587, 612]. As we saw in Chap. 2, DES was designed with hardware efficiency in mind and DES offers very competitive hardware implementations from around 3 000 GE [726] down to 2 300 GE [438]. Since the key length of DES limits its usefulness, other proposals such as DESXL (2 168 GE) are of some interest [438]. Estimates in [587] for the *Tiny Encryption Algorithm, TEA* [742] suggest that it will occupy around 2 100 GE while XTEA needs more. These estimates have been confirmed by work elsewhere [357, 758]. Some dedicated block cipher proposals for low-cost implementation include Clefia [689], a generalised Feistel cipher with elements borrowed from the AES and an implementation under 3 000 GE, mCrypton [446] (2 681 GE), HIGHT [302] (3 048 GE), and the *Scaleable Encryption Algorithm (SEA)* [694].

One of the most prominent lightweight block cipher is PRESENT [116] with two implementation reference points being given by 1 075 and 1 570 GE respectively [116, 648]. The design of PRESENT is based on techniques seen in both DES and the AES candidate Serpent [11] and this association is further confirmed by the name PRESENT being an anagram of Serpent. Sect. 9.4.6 gives more details.

9.4 Six Prominent Block Ciphers

We now focus our attention on six block ciphers that have, for one reason or another, become particularly prominent in the field of block ciphers. Setting aside FEAL, they are all considered secure today. We have nevertheless included FEAL as a prominent cipher since its design anticipated many future trends, even though its success was somewhat limited.

- FEAL is notable for using S-boxes built from simple arithmetic operations and for being one of the first ciphers to attempt to address the software performance of DES. It is unfortunately also notable as a test bed for cryptanalysis.
- IDEA has a very different design philosophy from many other block ciphers and is widely considered to be very strong.
- KASUMI is, in many ways, the logical conclusion of much of the work on DES cryptanalysis. It retains many classical block cipher features, yet it offers some elements of assured security. KASUMI features in third generation mobile phone standards.
- RC5 is a strikingly simple and elegant cipher. This positively encourages cryptanalysis which, when unsuccessful, increases trust. RC5 appeared within some early mobile phone application standards.
- Skipjack is a conventional cipher, supposedly designed by the NSA (National Security Agency) of the US. If so, it is the only NSA-designed block cipher design to be released in the public domain.
- PRESENT was specifically designed to be suited for lightweight hardware implementation. The cipher builds on a variety of techniques that first appeared during the AES process.

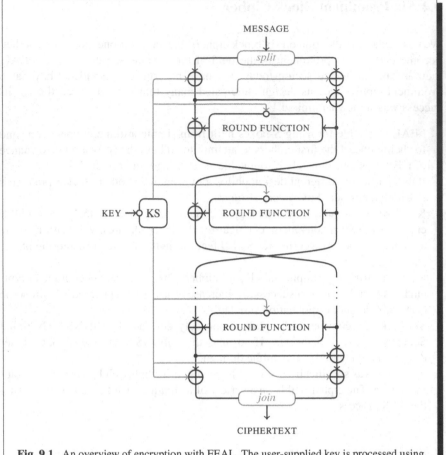

MESSAGE

split

ROUND FUNCTION

KEY → KS   ROUND FUNCTION

ROUND FUNCTION

join

CIPHERTEXT

Fig. 9.1 An overview of encryption with FEAL. The user-supplied key is processed using a *key schedule* (*KS*) to derive a set of *round keys* and what have since become termed *pre-* and *post-whitening* keys.

9.4.1 FEAL

While it is unlikely to be used, the *Fast Data Encipherment Algorithm* (FEAL) [504] has an important place in block cipher development. Somewhat unfortunately, it is the cryptanalysis of the cipher that is remembered rather than some of the innovative design features that have appeared in some other, more recent, ciphers.

FEAL is a 64-bit Feistel cipher. At the start of the encryption process, the plaintext is exclusive-ored with key material, a process called *pre-whitening*, and the result is processed using a classical Feistel network. The output from the Feistel network is *post-whitened* to give the ciphertext. Initial versions of FEAL had four

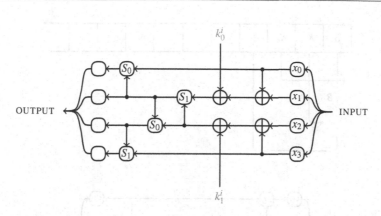

Fig. 9.2 An overview of the round function for FEAL where the input is a 32-bit word partitioned into four bytes $x_0 \| x_1 \| x_2 \| x_3$. The user-supplied key is processed using a *key schedule* (KS) to deliver the 16-bit round keys $k_0^i \| k_1^i$.

rounds [681], but as cryptanalysis progressed this increased to eight [504] and later to an arbitrary number of rounds [505]. The first two versions used 64-bit keys while later versions used 128-bit keys. Fig. 9.3 gives the key schedule for 64-bit keys, though attacks on FEAL are more or less independent of this.

The round function is relatively simple and the cipher was designed to be fast on eight-bit processors; all operations are byte-oriented. See Figs. 9.1 and 9.2. The functions S_0 and S_1 are defined as follows:

$$S_0(x,y) = (x+y \bmod 256) \lll 2, \text{ and}$$
$$S_1(x,y) = (x+y+1 \bmod 256) \lll 2,$$

where $z \lll 2$ denotes the bitwise rotation of z by two positions to the left.

Unfortunately, the security of FEAL was not as high as expected by the designers and the appearance of FEAL coincided with the remarkable development in cryptanalysis that took place during the 1990s. The first version of the original proposal, with only four rounds, was broken in 1988 using an attack that required between 100 and 10,000 chosen texts [206]. This was reduced by Murphy to around 20 in 1990 [524] and then later to five known texts by Matsui [483]. For FEAL with eight rounds an attack in 1990 used 20,000 chosen texts [705], which decreased to 256 chosen texts in 1993 and to 12 chosen texts three years later [16]. It has since been shown that there are attacks on up to 31 rounds of FEAL which would allow the key to be recovered faster than by exhaustive search.

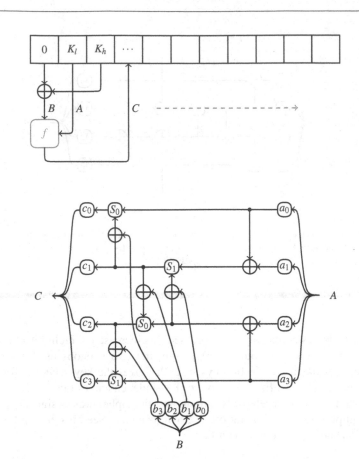

Fig. 9.3 An overview of the FEAL key schedule for an eight-round version of FEAL. The input is a 64-bit user-supplied key, partitioned into two 32-bit words K_l and K_h. A simple recurrence is used to generate eight 32-bit round keys, where the 32-bit inputs to the function f (also illustrated) are split into bytes $a_0\|a_1\|a_2\|a_3$ and $b_0\|b_1\|b_2\|b_3$. Each four bytes of output $c_0\|c_1\|c_2\|c_3$ provide two different 16-bit subkeys $k_0\|k_1$ that are used during successive rounds of encryption, as well as the pre- and post-whitening keys.

9.4.2 IDEA

The block cipher IDEA, which stands for *International Data Encryption Algorithm*, was proposed by Lai and Massey and published in a 1991 paper co-authored with Murphy [433]. IDEA is a strengthened version of the *Proposed Encryption Standard*, PES [431], that was cryptanalysed by Murphy [433], and IDEA was originally, and briefly, named *Improved Proposed Encryption Standard (IPES)* .

IDEA is a 64-bit iterated block cipher that takes 128-bit keys. It is based on the design concept of "mixing operations from different algebraic groups". The cipher was developed specifically to resist differential cryptanalysis and in [429] it is argued that there are no useful differentials for three rounds of IDEA and that IDEA would resist a differential attack after only four rounds.

IDEA has eight rounds followed by an output transformation. Researchers typically count the output transformation as an extra half round. The complete first round and the output transformation are shown in Fig. 9.4. The two multiplications and the two additions in the middle of the figure are often identified as the *MA-structure*. The key schedule takes the 128-bit user-defined key and returns 52 16-bit subkeys. The key schedule is very simple and illustrated in Fig. 9.5.

Two rounds of IDEA were cryptanalysed by Meier in a differential-like attack using a partial distributive law [491] and required a time of about 2^{42} operations. It was estimated that an extension of this attack to two and a half rounds would require around 2^{106} operations. In separate research Daemen found large classes of weak keys for IDEA [181] and described an attack on two and a half rounds that applied to all keys [179]. Differential-like attacks for IDEA up to three and a half rounds using *truncated differentials* was described in [121] while [57] details attacks up to four and a half rounds using *impossible differentials*. An extended series of papers [203, 205] exploited what has become known as the *Biryukov-Demirci distinguisher*, and the best current attack comprises six rounds of the cipher [71].

IDEA is a remarkable cipher that has withstood all cryptanalysis since its invention in 1991. Unfortunately, however, IDEA is not very fast, particularly when compared to more recent proposals.

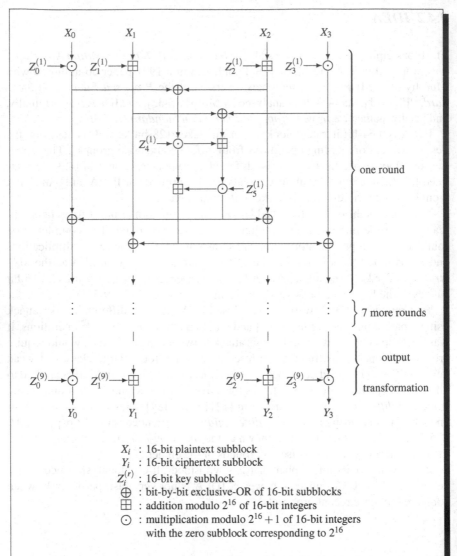

Fig. 9.4 Computational graph for the encryption process of the IDEA cipher.

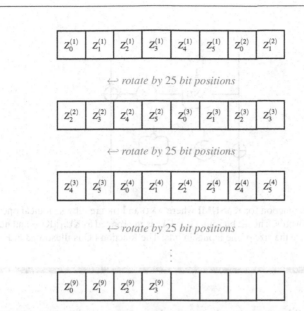

Fig. 9.5 The key schedule for IDEA generates 52 subkeys of 16 bits from a 128-bit user-supplied key. At each round the current contents of the 128-bit key register are considered as eight subkeys of 16 bits. At the end of the round the entirety of the 128-bit key register is rotated by 25 bit positions to the left.

9.4.3 KASUMI

KASUMI is a 64-bit block cipher that takes a key of size 128 bits. Essentially it is a Feistel cipher with eight rounds, though it has an elegant recursive structure and subcomponents of the cipher are, themselves, of a Feistel-like form.

Designed by the ETSI *Security Algorithms Group of Experts (SAGE)* for use within third generation telecommunications networks, KASUMI is a variant[1] of the block cipher MISTY [479] and was slightly modified for better hardware implementation [228]. The overall form of encryption with KASUMI is illustrated in Figs. 9.6, 9.7, and 9.8.

The recursive structure of KASUMI serves several purposes. Since all subcomponents are relatively small it allows for efficient and compact implementations, one of the design goals of the cipher. The two S-boxes, S7 and S9, were chosen to provide optimum security against differential and linear cryptanalysis. The S-box specifications are not given here but they can be found in [479] along with a justification for their choice and a description of their properties.

---

[1] In a play on words, KASUMI is the Japanese word for "misty".

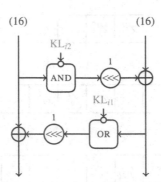

Fig. 9.6 The FL function for KASUMI where AND and OR are bitwise logical operations applied to 16-bit words. The 32-bit round key KL_i is interpreted as $KL_{i1} \| KL_{i2}$ and numbers in brackets indicate the size of the inputs in bits. The function FO is illustrated in Fig. 9.8.

Although the S-boxes themselves are relatively small, the recursive structure of KASUMI allows one to use the properties of the S-boxes to derive good immunity against differential and linear attacks. This builds on other work [568, 569] and provides bounds on the performance of these attacks when *averaged* over all choices of the key. The key schedule of KASUMI is described in Fig. 9.9.

Several attacks on reduced-round KASUMI have been proposed [417, 418] and a related-key attack on the full eight rounds of KASUMI is given in [69]. At Crypto 2010 Dunkelman *et al* presented a related-key attack on KASUMI which finds a 128-bit key [222] with a time complexity that permits experimental verification. However the attack requires encryptions under related-keys and, when used in its intended environment, KASUMI is immune to this cryptanalysis.

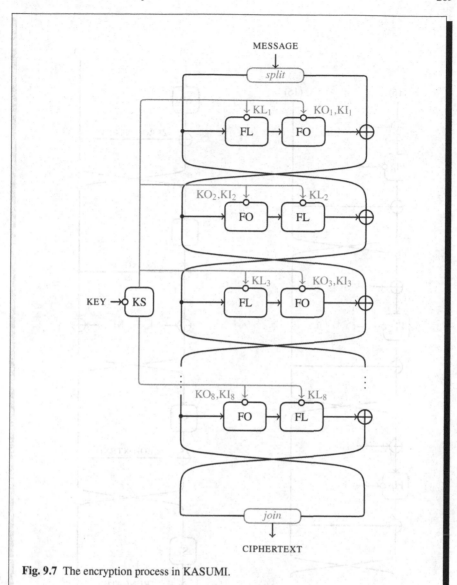

Fig. 9.7 The encryption process in KASUMI.

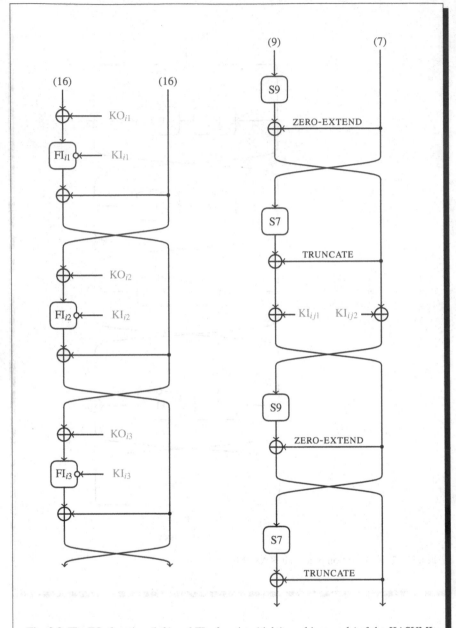

Fig. 9.8 The FO_i function (left) and FI_{ij} function (right) used in round i of the KASUMI cipher. Numbers in brackets indicate the size of the inputs in bits. The 48-bit round key KO_i is interpreted as $KO_{i1} \| KO_{i2} \| KO_{i3}$ while the 48-bit round key KI_i is interpreted as $KI_{i1} \| KI_{i2} \| KI_{i3}$. Within FI_{ij} the 16-bit KI_{ij} is interpreted as $KI_{ij1} \| KI_{ij1}$.

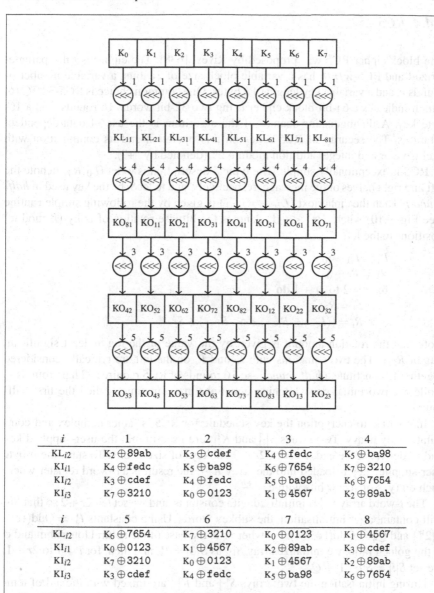

i	1	2	3	4
KL_{i2}	$K_2 \oplus \texttt{89ab}$	$K_3 \oplus \texttt{cdef}$	$K_4 \oplus \texttt{fedc}$	$K_5 \oplus \texttt{ba98}$
KI_{i1}	$K_4 \oplus \texttt{fedc}$	$K_5 \oplus \texttt{ba98}$	$K_6 \oplus \texttt{7654}$	$K_7 \oplus \texttt{3210}$
KI_{i2}	$K_3 \oplus \texttt{cdef}$	$K_4 \oplus \texttt{fedc}$	$K_5 \oplus \texttt{ba98}$	$K_6 \oplus \texttt{7654}$
KI_{i3}	$K_7 \oplus \texttt{3210}$	$K_0 \oplus \texttt{0123}$	$K_1 \oplus \texttt{4567}$	$K_2 \oplus \texttt{89ab}$

i	5	6	7	8
KL_{i2}	$K_6 \oplus \texttt{7654}$	$K_7 \oplus \texttt{3210}$	$K_0 \oplus \texttt{0123}$	$K_1 \oplus \texttt{4567}$
KI_{i1}	$K_0 \oplus \texttt{0123}$	$K_1 \oplus \texttt{4567}$	$K_2 \oplus \texttt{89ab}$	$K_3 \oplus \texttt{cdef}$
KI_{i2}	$K_7 \oplus \texttt{3210}$	$K_0 \oplus \texttt{0123}$	$K_1 \oplus \texttt{4567}$	$K_2 \oplus \texttt{89ab}$
KI_{i3}	$K_3 \oplus \texttt{cdef}$	$K_4 \oplus \texttt{fedc}$	$K_5 \oplus \texttt{ba98}$	$K_6 \oplus \texttt{7654}$

Fig. 9.9 The key schedule for KASUMI. The 128-bit user-provided key K is partitioned into 16-bit words K_0, K_1, ..., K_7. The subkeys $KL_i = KL_{i1} \| KL_{i2}$, $KO_i = KO_{i1} \| KO_{i2} \| KO_{i3}$, and $KI_i = KI_{i1} \| KI_{i2} \| KI_{i3}$ are derived as above.

9.4.4 RC5

The block cipher RC5 was proposed by Rivest [638]. The cipher is fully parameterized and RC5-$w/r/b$ has a variable block size of $2w$ bits, a variable number of rounds r, and a variable key length of b bytes. One common choice is RC5-32/16/16, which indicates a 64-bit block cipher using two 32-bit words, 16 rounds, and a 16-byte key. A distinguishing feature of the algorithm is the use of data-dependent rotations. The security of RC5 relies on this operation and its combination with exclusive-or and integer addition modulo 2^w (denoted by "+").

RC5 is exceptionally simple. The typical description is to let (L_0, R_0) denote the left and right halves of the message, respectively, and to let S_i be the key used in *half-round* i. Then the ciphertext (L_{2r+1}, R_{2r+1}) is given by the following simple routine (see Fig. 9.10) where $(\alpha \lll \beta)$ denotes the bitwise rotation of α by $(\beta \bmod w)$ positions to the left:

$$L_1 = L_0 + S_0$$
$$R_1 = R_0 + S_1$$
for $i = 2$ **to** $2r + 1$ **do**
$$L_i = R_{i-1}$$
$$R_i = ((L_{i-1} \oplus R_{i-1}) \lll R_{i-1}) + S_i$$

Note that the rotation amount is given by the value of the $\log_2 w$ least significant bits of R_{i-1}. The two equations with L_i and R_i on the left are typically considered together to constitute a *half-round*—so 16 rounds of RC5 requires 32 half-rounds—while the two initial pre-whitening equations are sometimes called the first half-round.

In contrast to encryption the key schedule for RC5 is rather complex and computationally heavy. Two arrays $S[\cdot]$ and $K[\cdot]$ are used to mix the user-supplied key and to derive, at the end, a set of $2r + 2$ subkeys of size w bits. To start, the b-byte user-supplied key is loaded into the array $K[\cdot]$; the result is a c-word register where each entry is of size w bits with $c = \frac{8b}{w}$.

The t-word array $S[\cdot]$ is initialised with constants, and we set $t = 2r + 2$ so that $S[\cdot]$ will contain, after initialisation, the subkey words. Using constants $P_w = \text{Odd}((e - 2)2^w)$ and $Q_w = \text{Odd}((\phi - 1)2^w)$, where e is the base of the natural logarithm and ϕ is the golden ratio, we initialize array $S[\cdot]$ as $S[0] = P_w$ and then, for $i = 1$ to $2r + 1$, we set $S[i] = S[i-1] + Q_w$.

During initialisation the two arrays $S[\cdot]$ and $K[\cdot]$ are mixed with the aid of temporary variables A and B and positions $S[0]$ and $K[0]$ of the respective arrays are considered the leftmost.

$$A = B = 0$$
for $i = 1$ **to** $3 \times \max(t, c)$ **do**
$$A = S[0] = (A + B + S[0]) \lll 3$$
$$B = K[0] = (A + B + K[0]) \lll (A + B)$$
left-rotate (wordwise) register $S[0] \cdots S[t-1]$
left-rotate (wordwise) register $K[0] \cdots K[c-1]$

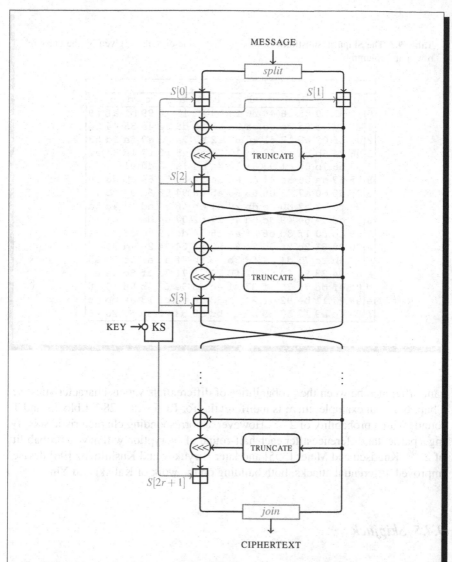

Fig. 9.10 An overview of encryption with RC5 where each Feistel-like round is referred to as a *half-round*. The user-supplied key is processed using a *key schedule* (KS) to derive a set of *round keys*.

The round subkeys S_0 to S_{2r+1} are taken from the final state of $S[\cdot]$.

Kaliski and Yin evaluated RC5 with respect to differential and linear cryptanalysis [352]. Their analysis showed that the cipher resists the two attacks relatively well, but perhaps most notably, the results illustrated for the first time a signifi-

Table 9.3 The Skipjack substitution box $S[\cdot]$. The value of $S[\text{ab}]$ is given by the entry in row a and column b.

$S[\cdot]$	0	1	2	3	4	5	6	7	8	9	a	b	c	d	e	f
0	a3	d7	09	83	f8	48	f6	f4	b3	21	15	78	99	b1	af	f9
1	e7	2d	4d	8a	ce	4c	ca	2e	52	95	d9	1e	4e	38	44	28
2	0a	df	02	a0	17	f1	60	68	12	b7	7a	c3	e9	fa	3d	53
3	96	84	6b	ba	f2	63	9a	19	7c	ae	e5	f5	f7	16	6a	a2
4	39	b6	7b	0f	c1	93	81	1b	ee	b4	1a	ea	d0	91	2f	b8
5	55	b9	da	85	3f	41	bf	e0	5a	58	80	5f	66	0b	d8	90
6	35	d5	c0	a7	33	06	65	69	45	00	94	56	6d	98	9b	76
7	97	fc	b2	c2	b0	fe	db	20	e1	eb	d6	e4	dd	47	4a	1d
8	42	ed	9e	6e	49	3c	cd	43	27	d2	07	d4	de	c7	67	18
9	89	cb	30	1f	8d	c6	8f	aa	c8	74	dc	c9	5d	5c	31	a4
a	70	88	61	2c	9f	0d	2b	87	50	82	54	64	26	7d	03	40
b	34	4b	1c	73	d1	c4	fd	3b	cc	fb	7f	ab	e6	3e	5b	a5
c	ad	04	23	9c	14	51	22	f0	29	79	71	7e	ff	8c	0e	e2
d	0c	ef	bc	72	75	6f	37	a1	ec	d3	8e	62	8b	86	10	e8
e	08	77	11	be	92	4f	24	c5	32	36	9d	cf	f3	a6	bb	ac
f	5e	6c	a9	13	57	25	b5	e3	bd	a8	3a	01	05	59	2a	46

cant difference between the probabilities of differentials versus characteristics; see Chap. 6. As an example, there is a differential for RC5 with 128-bit blocks and 12 rounds with a probability of 2^{-53}. However, a corresponding characteristic specifying a particular difference after each half-round of encryption will have a probability of 2^{-96}. Knudsen and Meier [395] and later Biryukov and Kushilevitz [96] devised improved differential attacks, both building on the work of Kaliski and Yin.

9.4.5 Skipjack

Skipjack is a 64-bit block cipher that uses an 80-bit key. The cipher was proposed for use in the ill-fated *Clipper Chip* [664, 216] and made public in 1998 [557]. It is a simple cipher and one interesting feature is the use of two different types of rounds; see Fig. 9.11. These are typically referred to as A-rounds and B-rounds, and encryption with Skipjack consists of first applying eight A-rounds, then eight B-rounds, then once again eight A-rounds and finally eight B-rounds. Each round takes an input of 64 bits which is split into four 16-bit words a, b, c, and d. The rounds are defined as follows, where s_i is a round counter that is incremented from 1, and G_k is a key-dependent permutation:

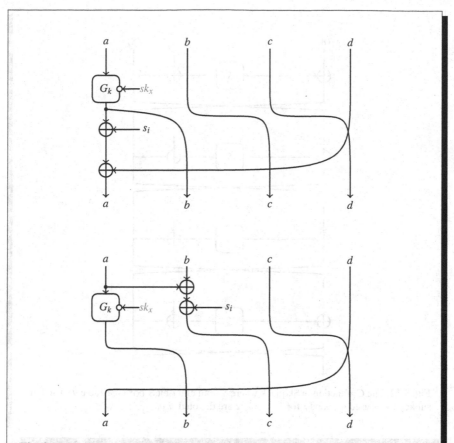

Fig. 9.11 The A-round (top) and B-round (bottom) used in Skipjack. The variable s_i is a round-dependent constant and G_k is a key-dependent permutation.

$$A[a, b, c, d] \rightarrow [d \oplus G_k(a) \oplus s_i, G_k(a), b, c] \text{ and}$$
$$B[a, b, c, d] \rightarrow [d, G_k(a), a \oplus b \oplus s_i, c]$$

In each round of Skipjack, one of the 16-bit words passes through the keyed permutation G_k and we observe that at most two 16-bit words are modified during a single round of Skipjack. G_k itself is a four-round Feistel network on 16-bit words which uses four key bytes; see Fig. 9.12. At each round one half of the 16-bit data is exclusive-ored with a byte of key material and the resulting byte is transformed using an S-box. The specifications of the Skipjack S-box are given in Table 9.3.

Interestingly, Skipjack has one of the simplest key schedules. At each round the four key bytes used in G_k are derived from the user-supplied key. Denoting the four key bytes used in round r, for $1 \leq r \leq 32$, by $sk_{4(r-1)}$, $sk_{4(r-1)+1}$, $sk_{4(r-1)+2}$, and

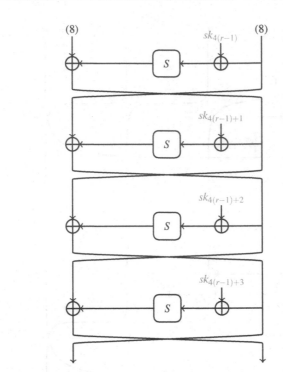

Fig. 9.12 The G_k function in Skipjack where S is an eight-bit S-box (see Table 9.3) and the subkey bytes used in round r for $1 \leq r \leq 32$ are denoted $sk_{4(r-1)}, \ldots, sk_{4(r-1)+3}$.

$sk_{4(r-1)+3}$, these are derived from the ten-byte user-supplied key $K = k_0 \| \cdots \| k_9$ as

$$sk_{4(r-1)+i} = k_{(4(r-1)+i) \bmod 10}$$

for $1 \leq r \leq 32$ and for $0 \leq i \leq 3$. That is, the bytes of the key are taken in rotation with byte alignment being preserved throughout.

The best known cryptanalytic attack on Skipjack is a truncated, impossible differential attack that is faster than exhaustive key search on a variant of Skipjack reduced to 31 instead of 32 rounds [57, 59]. There have been a large number of other interesting observations on the cipher [56, 407, 703, 268, 408, 309, 630].

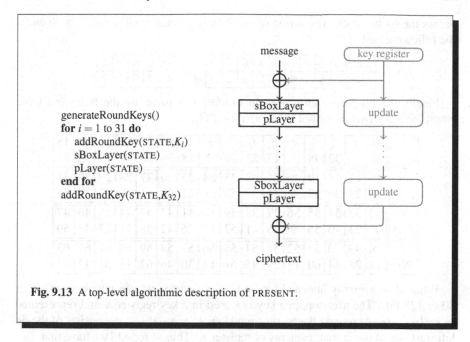

```
generateRoundKeys()
for i = 1 to 31 do
    addRoundKey(STATE,Kᵢ)
    sBoxLayer(STATE)
    pLayer(STATE)
end for
addRoundKey(STATE,K₃₂)
```

Fig. 9.13 A top-level algorithmic description of PRESENT.

9.4.6 PRESENT

The block cipher PRESENT was designed to be particularly compact when implemented in hardware [116]. Designed by a combined team from DTU (Denmark), Orange Labs (France), and Bochum University (Germany), the cipher has become prominent as one of the earliest examples of what is sometimes referred to as *lightweight cryptography*. The phrase doesn't imply that security has been reduced; rather that the design of the cipher from the first steps is intended to reduce the hardware footprint for a given level of security. The designers of PRESENT deliberately borrowed features seen in both DES and Serpent [11] since these are well known to be particularly suited to hardware implementation.

PRESENT is an example of an SP-network [493] and consists of 31 rounds. The block length is 64 bits and the primary objective was to support a key, and hence a security level, of 80 bits. Each of the 31 rounds consists of an XOR operation to introduce a round key K_i for $1 \leq i \leq 32$, where K_{32} is used for post-whitening, a linear bitwise permutation and a non-linear substitution layer. The non-linear layer uses a single four-bit S-box S which is applied 16 times in parallel in each round. The cipher is described in pseudo-code in Fig. 9.13, while each stage is described as follows.

The step addRoundKey takes a round key and the current cipher state and combines them together using bitwise exclusive-or. The S-box used in PRESENT is a four- to four-bit S-box with an S-box operation being performed 16 times in parallel across the 64-bit block. The action of this box in hexadecimal notation is given by the following table.

x	0	1	2	3	4	5	6	7	8	9	a	b	c	d	e	f
$S[x]$	c	5	6	b	9	0	a	d	3	e	f	8	4	7	1	2

Finally the bit permutation used in PRESENT is given by the following table where bit i of STATE is moved to bit position $P(i)$.

i	0	1	2	3	4	5	6	7	8	9	10	11	12	13	14	15
$P(i)$	0	16	32	48	1	17	33	49	2	18	34	50	3	19	35	51
i	16	17	18	19	20	21	22	23	24	25	26	27	28	29	30	31
$P(i)$	4	20	36	52	5	21	37	53	6	22	38	54	7	23	39	55
i	32	33	34	35	36	37	38	39	40	41	42	43	44	45	46	47
$P(i)$	8	24	40	56	9	25	41	57	10	26	42	58	11	27	43	59
i	48	49	50	51	52	53	54	55	56	57	58	59	60	61	62	63
$P(i)$	12	28	44	60	13	29	45	61	14	30	46	62	15	31	47	63

While it is primarily intended for 80-bit keys, PRESENT can take keys of either 80 or 128 bits. The user-supplied key is stored in a key register K and represented as $k_{79}k_{78}\ldots k_0$. At round i the 64-bit round key $K_i = \kappa_{63}\kappa_{62}\ldots\kappa_0$ consists of the 64 leftmost bits of the current contents of register K. Thus at round i we have that

$$K_i = \kappa_{63}\kappa_{62}\ldots\kappa_0 = k_{79}k_{78}\ldots k_{16}.$$

After extracting the round key K_i, the key register $K = k_{79}k_{78}\ldots k_0$ is updated as follows:

1. $[k_{79}k_{78}\ldots k_1 k_0] = [k_{18}k_{17}\ldots k_{20}k_{19}]$
2. $[k_{79}k_{78}k_{77}k_{76}] = S[k_{79}k_{78}k_{77}k_{76}]$
3. $[k_{19}k_{18}k_{17}k_{16}k_{15}] = [k_{19}k_{18}k_{17}k_{16}k_{15}] \oplus \texttt{round\_counter}$

Thus, the key register is rotated by 61 bit positions to the left, the leftmost four bits are passed through the PRESENT S-box, and the `round_counter` value i is exclusive-ored with bits $k_{19}k_{18}k_{17}k_{16}k_{15}$ of K, with the least significant bit of `round_counter` on the right.

Since its publication PRESENT has received considerable attention and a variety of papers have been published [759, 733, 586, 6, 579, 163, 536, 151]. Results seem to confirm the widely held view that up to around 25 out of 31 encryption rounds might be theoretically compromised if the entire codebook is available to the cryptanalyst. In real terms this seems to confirm that the cipher has been designed with a narrow but sufficient security margin that matches its intended applications.

9.5 Getting to the Source

An exhaustive summary of block cipher proposals and their cryptanalysis is not easy. As a result we have shied away from providing a comprehensive survey and instead we have tried to identify a few notable proposals along the way. This chapter, therefore, is nicely complemented by other summaries such as those to be found in [493, Chap. 7], [663, Chaps. 13 and 14] and [189].

In closing, we note that the research literature on block cipher continues to develop. Not only is there continued work on the main proposals, but hardware implementation demands have opened up a completely new area of design and analysis. In addition, much contemporary work on hash functions is a natural extension of block cipher design and, as a result of the NIST SHA-3 competition, block ciphers, especially their key schedules, have come under renewed scrutiny.

The field is as lively as it has ever been.

9.3 Getting to the Source

Index

References

1. C. M. Adams. Constructing symmetric ciphers using the CAST design procedure. *Designs, Codes, and Cryptography*, 12(3):283–316, 1997.
2. C. M. Adams and S. E. Tavares. Good S-boxes are easy to find. In G. Brassard, editor, *Advances in Cryptology - CRYPTO '89*, volume 435 of *Lecture Notes in Computer Science*, Springer, pages 612–615, 1990.
3. C. M. Adams and S. E. Tavares. The structured design of cryptographically good S-boxes. *J. Cryptology*, 3(1):27–41, 1990.
4. W. Aiello and R. Venkatesan. Foiling birthday attacks in length-doubling transformations - Benes: A non-reversible alternative to Feistel. In U. Maurer, editor, *Advances in Cryptology - EUROCRYPT '96*, volume 1070 of *Lecture Notes in Computer Science*, Springer, pages 307–320, 1996.
5. M.-L. Akkar and C. Giraud. An implementation of DES and AES, secure against some attacks. In Ç.K. Koç, D. Naccache and C. Paar, editors, *Cryptographic Hardware and Embedded Systems - CHES 2001*, volume 2162 of *Lecture Notes in Computer Science*, Springer, 2001, pages 309–318, 2001.
6. M. Albrecht and C. Cid. Algebraic techniques in differential cryptanalysis. In O. Dunkelman, editor, *Fast Software Encryption, FSE 2009*, volume 5665 of *Lecture Notes in Computer Science*, Springer, pages 193–208, 2009.
7. American Bankers Association. American national standard - financial institution key management (wholesale), ANSI X9.17. ASC X9 Secretariat, 1985.
8. American Bankers Association. American national standard - financial institution retail message authentication, ANSI X9.19. ASC X9 Secretariat, 1985.
9. H. Amirazizi and M. E. Hellman. Time-memory-processor trade-offs. *IEEE Trans. Information Theory*, 34(3):505–512, May 1988.
10. R. J. Anderson and E. Biham. Two practical and provably secure block ciphers BEAR and LION. In D. Gollmann, editor, *Fast Software Encryption, FSE 1996*, volume 1039 of *Lecture Notes in Computer Science*, Springer, pages 113–120, 1996.
11. R. J. Anderson, E. Biham, and L. R. Knudsen. SERPENT - a 128-bit block cipher. A candidate for the Advanced Encryption Standard. Available via www.nist.gov/aes.
12. K. Aoki. On maximum non-averaged differential probability. In S.E. Tavares and H. Meijer, editors, *Selected Areas in Cryptography, SAC 1998*, volume 1556 of *Lecture Notes in Computer Science*, Springer, pages 118–130, 1999.
13. K. Aoki. Efficient evaluation of security against generalized interpolation attack. In H.M. Heys and C.M. Adams, editors, *Selected Areas in Cryptography, SAC 1999*, volume 1758 of *Lecture Notes in Computer Science*, Springer, pages 135–146, 2000.
14. K. Aoki, T. Ichikawa, M. Kanda, M. Matsui, S. Moriai, J. Nakajima, and T. Tokita. Camellia: A 128-bit block cipher suitable for multiple platforms - design and analysis. In D.R. Stin-

son and S.E. Tavares, editors, *Selected Areas in Cryptography, SAC 2000,* volume 2012 of *Lecture Notes in Computer Science,* Springer, pages 39–56, 2001.

15. K. Aoki, K. Kobayashi, and S. Moriai. Best differential characteristic search of FEAL. In E. Biham, editor, *Fast Software Encryption, FSE 1997,* volume 1267 of *Lecture Notes in Computer Science,* Springer, pages 41–53, 1997.

16. K. Aoki and K. Ohta. Differential-linear cryptanalysis of FEAL-8. In *IEICE Transactions: Fundamentals of Electronics, Communications, and Computer Sciences (Japan),* volume E79-A, 1, pages 20–27, 1996.

17. G. Ars, J.-C. Faugére, M. K. Hideki Imai, and M. Sugita. Comparison between XL and Gröbner basis algorithms. In P.J. Lee, editor, *Advances in Cryptology - ASIACRYPT 2004,* volume 3329 of *Lecture Notes in Computer Science,* Springer, pages 338–353, 2004.

18. J.-P. Aumasson, J. Nakahara, and P. Sepehrdad. Cryptanalysis of the ISDB scrambling algorithm (MULTI2). In O. Dunkelman, editor, *Fast Software Encryption, FSE 2009,* volume 5665 of *Lecture Notes in Computer Science,* Springer, pages 296–307, 2009.

19. E. S. Ayaz and A. A. Selçuk. Improved DST cryptanalysis of IDEA. In E. Biham and A.M. Youssef, editors, *Selected Areas in Cryptography, SAC 2006,* volume 4356 of *Lecture Notes in Computer Science,* Springer, pages 1–14, 2007.

20. S. Babbage and L. Frisch. On MISTY1 higher order differential cryptanalysis. In D. Won, editor, 3rd *International Conference on Information Security and Cryptology (ICISC 2000),* volume 2015 of *Lecture Notes in Computer Science,* pages 22–36. Springer, 2001.

21. T. Baignéres and M. Finiasz. KFC - the krazy Feistel cipher. In X. Lai and K. Chen, editors, *Advances in Cryptology - ASIACRYPT 2006,* volume 4284 of *Lecture Notes in Computer Science,* Springer, pages 380–395, 2006.

22. T. Baignères and M. Finiasz. Dial C for cipher. In E. Biham and A.M. Youssef, editors, *Selected Areas in Cryptography, SAC 2006,* volume 4356 of *Lecture Notes in Computer Science,* Springer, pages 76–95, 2007.

23. T. Baignères, P. Junod, and S. Vaudenay. How far can we go beyond linear cryptanalysis. In P.J. Lee, editor, *Advances in Cryptology - ASIACRYPT 2004,* volume 3329 of *Lecture Notes in Computer Science,* Springer, pages 432–450, 2004.

24. T. Baignères, J. Stern, and S. Vaudenay. Linear cryptanalysis of non binary ciphers. In C.M. Adams, A. Miri, and M.J. Wiener, editors, *Selected Areas in Cryptography, SAC 2007,* volume 4876 of *Lecture Notes in Computer Science,* Springer, pages 184–211, 2007.

25. T. Baignères and S. Vaudenay. Proving the security of AES substitution-permutation network. In B. Preneel and S.E. Tavares, editors, *Selected Areas in Cryptography, SAC 2005,* volume 3897 of *Lecture Notes in Computer Science,* Springer, pages 65–81, 2006.

26. E. Barkan and E. Biham. In how many ways can you write Rijndael? In Y. Zheng, editor, *Advances in Cryptology - ASIACRYPT 2002,* volume 2501 of *Lecture Notes in Computer Science,* Springer, pages 160–175, 2002.

27. E. Barkan, E. Biham, and A. Shamir. Rigorous bounds on cryptanalytic time/memory trade-offs. In C. Dwork, editor, *Advances in Cryptology - CRYPTO 2006,* volume 4117 of *Lecture Notes in Computer Science,* Springer, pages 1–21, 2006.

28. P. S. L. M. Barreto and V. Rijmen. The Anubis block cipher. Available via www. cryptonessie.org, May 2003.

29. P. S. L. M. Barreto and V. Rijmen. The Khazad block cipher. Available via www. cryptonessie.org, May 2003.

30. P. S. L. M. Barreto and V. Rijmen. The Whirlpool hashing function. Available via www. cryptonessie.org, May 2003.

31. P. S. L. M. Barreto, V. Rijmen, J. Nakahara Jr., B. Preneel, J. Vandewalle, and H. Y. Kim. Improved Square attacks against reduced-round HIEROCRYPT. In M. Matsui, editor, *Fast Software Encryption, FSE 2001,* volume 2355 of *Lecture Notes in Computer Science,* Springer, pages 165–173, 2001.

32. M. Bellare, A. Boldyreva, L. R. Knudsen, and C. Namprempre. Online ciphers and the Hash-CBC construction. In J. Kilian, editor, *Advances in Cryptology - CRYPTO 2001,* volume 2139 of *Lecture Notes in Computer Science,* Springer, pages 292–309, 2001.

33. M. Bellare, J. Kilian, and P. Rogaway. The security of cipher block chaining. In Y.G. Desmedt, editor, *Advances in Cryptology - CRYPTO '94*, volume 839 of *Lecture Notes in Computer Science*, Springer, pages 341–358, 1994.

34. M. Bellare, T. Krovetz, and P. Rogaway. Luby-Rackoff backwards: Increasing security by making block ciphers non-invertible. In K. Nyberg, editor, *Advances in Cryptology - EUROCRYPT '98*, volume 1403 of *Lecture Notes in Computer Science*, Springer, pages 266–280, 1998.

35. M. Bellare and P. Rogaway. Encode-then-encipher encryption: How to exploit nonces or redundancy in plaintexts for efficient cryptography. In T. Okamoto, editor, *Advances in Cryptology - ASIACRYPT 2000*, volume 1976 of *Lecture Notes in Computer Science*, Springer, pages 317–330, 2000.

36. M. Bellare and P. Rogaway. The security of triple encryption and a framework for code-based game-playing proofs. In S. Vaudenay, editor, *Advances in Cryptology - EUROCRYPT 2006*, volume 4004 of *Lecture Notes in Computer Science*, Springer, pages 409–426, 2006.

37. M. Bellare, P. Rogaway, and D. Wagner. The EAX mode of operation. In W. Meier and B. Roy, editors, *Fast Software Encryption, FSE 2004*, volume 3017 of *Lecture Notes in Computer Science*, Springer, pages 389–407, 2004.

38. I. Ben-Aroya and E. Biham. Differtial cryptanalysis of Lucifer. In D. Stinson, editor, *Advances in Cryptology - CRYPTO '93*, volume 773 of *Lecture Notes in Computer Science*, Springer, pages 187–199, 1994.

39. I. Ben-Aroya and E. Biham. Differential cryptanalysis of Lucifer. *J. Cryptology*, 9(1):21–34, 1996.

40. R. Benadjila, O. Billet, H. Gilbert, G. Macario-Rat, T. Peyrin, M. Robshaw, and Y. Seurin. SHA-3 proposal: ECHO. Available from [542].

41. D. J. Bernstein. Cache timing attacks on AES. Available via cr.yp.to/antiforgery/.

42. T. A. Berson. Long key variants of DES. In D. Chaum, R.L. Rivest, and A.T. Sherman, editors, *Advances in Cryptology: Proceedings of CRYPTO '82*, Plenum, pages 311–313, 1983.

43. G. Bertoni, L. Breveglieri, P. Fragneto, M. Macchetti, and S. Marchesin. Efficient software implementation of AES on 32-bit platforms. In B.S. Kaliski Jr., Ç.K. Koç and C. Paar, editors, *Cryptographic Hardware and Embedded Systems - CHES 2002*, volume 2523 of *Lecture Notes in Computer Science*, Springer, 2003, pages 159–171, 2003.

44. T. Beth and C. Ding. On almost perfect nonlinear permutations. In T. Helleseth, editor, *Advances in Cryptology - EUROCRYPT '93*, volume 765 of *Lecture Notes in Computer Science*, Springer, pages 65–76, 1994.

45. E. Biham. New types of cryptanalytic attacks using related keys. *J. Cryptology*, 7(4):229–246, 1994.

46. E. Biham. New types of cryptanalytic attacks using related keys (extended abstract). In T. Helleseth, editor, *Advances in Cryptology - EUROCRYPT '93*, volume 765 of *Lecture Notes in Computer Science*, Springer, pages 398–409, 1994.

47. E. Biham. Cryptanalysis of multiple modes of operation. In J. Pieprzyk and R. Safavi-Naini, editors, *Advances in Cryptology - ASIACRYPT '94*, volume 917 of *Lecture Notes in Computer Science*, Springer, pages 278–292, 1995.

48. E. Biham. On Matsui's linear cryptanalysis. In A. De Santis, editor, *Advances in Cryptology - EUROCRYPT '94*, volume 950 of *Lecture Notes in Computer Science*, Springer, pages 341–355, 1995.

49. E. Biham. Cryptanalysis of Ladder-DES. In E. Biham, editor, *Fast Software Encryption, FSE 1997*, volume 1267 of *Lecture Notes in Computer Science*, Springer, pages 134 138, 1997.

50. E. Biham. Cryptanalysis of multiple modes of operation. *J. Cryptology*, 11(1):45–58, 1998.

51. E. Biham. Cryptanalysis of triple modes of operation. *J. Cryptology*, 12(3):161–184, 1999.

52. E. Biham, R. J. Anderson, and L. R. Knudsen. Serpent: A new block cipher proposal. In S. Vaudenay, editor, *Fast Software Encryption, FSE 1998*, volume 1372 of *Lecture Notes in Computer Science*, Springer, pages 222–238, 1998.

53. E. Biham and A. Biryukov. How to strengthen DES using existing hardware. In J. Pieprzyk and R. Safavi-Naini, editors, *Advances in Cryptology - ASIACRYPT '94*, volume 917 of *Lecture Notes in Computer Science*, Springer, pages 398–412, 1995.

54. E. Biham and A. Biryukov. An improvement of Davies' attack on DES. In A. De Santis, editor, *Advances in Cryptology - EUROCRYPT '94*, volume 950 of *Lecture Notes in Computer Science*, Springer, pages 461–467, 1995.

55. E. Biham and A. Biryukov. An improvement of Davies' attack on DES. *J. Cryptology*, 10(3):195–206, 1997.

56. E. Biham, A. Biryukov, O. Dunkelman, E. Richardson, and A. Shamir. Initial observations on Skipjack: Cryptanalysis of Skipjack-3xor. In S.E. Tavares and H. Meijer, editors, *Selected Areas in Cryptography, SAC 1998*, volume 1556 of *Lecture Notes in Computer Science*, Springer, pages 362–376, 1999.

57. E. Biham, A. Biryukov, and A. Shamir. Cryptanalysis of Skipjack reduced to 31 rounds using impossible differentials. In J. Stern, editor, *Advances in Cryptology - EUROCRYPT '99*, volume 1592 of *Lecture Notes in Computer Science*, Springer, pages 12–23, 1999.

58. E. Biham, A. Biryukov, and A. Shamir. Miss in the middle attacks on IDEA and Khufu. In L.R. Knudsen, editor, *Fast Software Encryption, FSE 1999*, volume 1636 of *Lecture Notes in Computer Science*, Springer, pages 124–138, 1999.

59. E. Biham, A. Biryukov, and A. Shamir. Cryptanalysis of Skipjack reduced to 31 rounds using impossible differentials. *J. Cryptology*, 18(4):291–311, 2005.

60. E. Biham and O. Dunkelman. The SHAvite-3 hash function. Available from [542].

61. E. Biham, O. Dunkelman, and N. Keller. Linear cryptanalysis of reduced round Serpent. In M. Matsui, editor, *Fast Software Encryption, FSE 2001*, volume 2355 of *Lecture Notes in Computer Science*, Springer, pages 16–27, 2001.

62. E. Biham, O. Dunkelman, and N. Keller. The rectangle attack - rectangling the Serpent. In B. Pfitzmann, editor, *Advances in Cryptology - EUROCRYPT 2001*, volume 2045 of *Lecture Notes in Computer Science*, Springer, pages 340–357, 2001.

63. E. Biham, O. Dunkelman, and N. Keller. Enhancing differential-linear cryptanalysis. In Y. Zheng, editor, *Advances in Cryptology - ASIACRYPT 2002*, volume 2501 of *Lecture Notes in Computer Science*, Springer, pages 254–266, 2002.

64. E. Biham, O. Dunkelman, and N. Keller. New results on boomerang and rectangle attacks. In J. Daemen and V. Rijmen, editors, *Fast Software Encryption, FSE 2002, February 2002*, volume 2365 of *Lecture Notes in Computer Science*, Springer, pages 1–16, 2002.

65. E. Biham, O. Dunkelman, and N. Keller. Differential-linear cryptanalysis of Serpent. In T. Johansson, editor, *Fast Software Encryption, FSE 2003*, volume 2887 of *Lecture Notes in Computer Science*, Springer, pages 9–21, 2003.

66. E. Biham, O. Dunkelman, and N. Keller. Rectangle attacks on 49-round SHACAL-1. In T. Johansson, editor, *Fast Software Encryption, FSE 2003*, volume 2887 of *Lecture Notes in Computer Science*, Springer, pages 22–35, 2003.

67. E. Biham, O. Dunkelman, and N. Keller. New combined attacks on block ciphers. In H. Gilbert and H. Handschuh, editors, *Fast Software Encryption, FSE 2005*, volume 3557 of *Lecture Notes in Computer Science*, Springer, pages 126–144, 2005.

68. E. Biham, O. Dunkelman, and N. Keller. Related-key boomerang and rectangle attacks. In R. Cramer, editor, *Advances in Cryptology - EUROCRYPT 2005*, volume 3494 of *Lecture Notes in Computer Science*, Springer, pages 507–525, 2005.

69. E. Biham, O. Dunkelman, and N. Keller. A related-key rectangle attack on the full KASUMI. In B.K. Roy, editor, *Advances in Cryptology - ASIACRYPT 2005*, volume 3788 of *Lecture Notes in Computer Science*, Springer, pages 443–461, 2005.

70. E. Biham, O. Dunkelman, and N. Keller. New cryptanalytic results on IDEA. In X. Lai and K. Chen, editors, *Advances in Cryptology - ASIACRYPT 2006*, volume 4284 of *Lecture Notes in Computer Science*, Springer, pages 412–427, 2006.

71. E. Biham, O. Dunkelman, and N. Keller. A new attack on 6-round IDEA. In A. Biryukov, editor, *Fast Software Encryption, FSE 2007*, volume 4593 of *Lecture Notes in Computer Science*, Springer, pages 211–224, 2007.

72. E. Biham, O. Dunkelman, and N. Keller. A unified approach to related-key attacks. In K. Nyberg, editor, *Fast Software Encryption, FSE 2008*, volume 5086 of *Lecture Notes in Computer Science*, Springer, pages 73–96, 2008.
73. E. Biham, V. Furman, M. Misztal, and V. Rijmen. Differential cryptanalysis of Q. In M. Matsui, editor, *Fast Software Encryption, FSE 2001*, volume 2355 of *Lecture Notes in Computer Science*, Springer, pages 174–186, 2001.
74. E. Biham and N. Keller. Cryptanalysis of reduced variants of Rijndael. Available via www. nist.gov/aes.
75. E. Biham and L. R. Knudsen. Cryptanalysis of the ANSI X9.52 CBCM mode. In K. Nyberg, editor, *Advances in Cryptology - EUROCRYPT '98*, volume 1403 of *Lecture Notes in Computer Science*, Springer, pages 100–111, 1998.
76. E. Biham and L. R. Knudsen. Cryptanalysis of the ANSI X9.52 CBCM mode. *J. Cryptology*, 15(1):47–59, 2002.
77. E. Biham and A. Shamir. Differential cryptanalysis of DES-like cryptosystems. In A.J. Menezes and S.A. Vanstone, editors, *Advances in Cryptology - CRYPTO '90*, volume 537 of *Lecture Notes in Computer Science*, Springer, pages 2–21, 1991.
78. E. Biham and A. Shamir. Differential cryptanalysis of DES-like cryptosystems. *J. Cryptology*, 4(1):3–72, 1991.
79. E. Biham and A. Shamir. Differential cryptanalysis of Feal and N-Hash. In D.W. Davies, editor, *Advances in Cryptology - EUROCRYPT '91*, volume 547 of *Lecture Notes in Computer Science*, Springer, pages 1–16, 1992.
80. E. Biham and A. Shamir. Differential cryptanalysis of Snefru, Khafre, REDOC-II, LOKI and Lucifer. In J. Feigenbaum, editor, *Advances in Cryptology - CRYPTO '91*, volume 576 of *Lecture Notes in Computer Science*, Springer, pages 156–171, 1992.
81. E. Biham and A. Shamir. *Differential Cryptanalysis of the Data Encryption Standard*. Springer, 1993.
82. E. Biham and A. Shamir. Differential cryptanalysis of the full 16-round DES. In E.F. Brickell, editor, *Advances in Cryptology - CRYPTO '92*, volume 740 of *Lecture Notes in Computer Science*, Springer, pages 487–496, 1993.
83. E. Biham and A. Shamir. Differential fault analysis of secret key cryptosystems. In B.S. Kaliski Jr., editor, *Advances in Cryptology - CRYPTO '97*, volume 1294 of *Lecture Notes in Computer Science*, Springer, pages 513–525, 1997.
84. O. Billet and H. Gilbert. A traceable block cipher. In C-S. Laih, editor, *Advances in Cryptology - ASIACRYPT 2003*, volume 2894 of *Lecture Notes in Computer Science*, Springer, pages 331–346, 2003.
85. O. Billet, H. Gilbert, and C. Ech-Chatbi. Cryptanalysis of a white box AES implementation. In H. Handschuh and M.A. Hasan, editors, *Selected Areas in Cryptography, SAC 2004*, volume 3357 of *Lecture Notes in Computer Science*, Springer, pages 227–240, 2004.
86. A. Biryukov. Analysis of involutional ciphers: Khazad and Anubis. In T. Johansson, editor, *Fast Software Encryption, FSE 2003*, volume 2887 of *Lecture Notes in Computer Science*, Springer, pages 45–53, 2003.
87. A. Biryukov. The boomerang attack on 5 and 6-round reduced AES. In V. Rijmen, H. Dobbertin, and A. Sowa, editors, *Advanced Encryption Standard - AES, Fourth International Conference*, volume 3373 of *Lecture Notes in Computer Science*, Springer, pages 11–15, 2005.
88. A. Biryukov. The design of a stream cipher LEX. In E. Biham and A.M. Youssef, editors, *Selected Areas in Cryptography, SAC 2006*, volume 4356 of *Lecture Notes in Computer Science*, Springer, pages 67–75, 2007.
89. A. Biryukov and C. De Cannière. Block ciphers and systems of quadratic equations. In T. Johansson, editor, *Fast Software Encryption, FSE 2003*, volume 2887 of *Lecture Notes in Computer Science*, Springer, pages 274–289, 2003.
90. A. Biryukov, C. De Cannière, A. Braeken, and B. Preneel. A toolbox for cryptanalysis: Linear and affine equivalence algorithms. In E. Biham, editor, *Advances in Cryptology - EUROCRYPT 2003*, volume 2656 of *Lecture Notes in Computer Science*, Springer, pages 33–50, 2003.

91. A. Biryukov, C. De Canniére, and G. Dellkrantz. Cryptanalysis of SAFER++. In D. Boneh, editor, *Advances in Cryptology - CRYPTO 2003*, volume 2729 of *Lecture Notes in Computer Science*, Springer, pages 195–211, 2003.

92. A. Biryukov, C. De Canniére, and M. Quisquater. On multiple linear approximations. In M. Franklin, editor, *Advances in Cryptology - CRYPTO 2004*, volume 3152 of *Lecture Notes in Computer Science*, Springer, pages 1–22, 2004.

93. A. Biryukov, O. Dunkelman, N. Keller, D. Khovratovich, and A. Shamir. Key recovery attacks of practical complexity on AES variants with up to 10 rounds. Available via eprint.iacr.org/2009/374.

94. A. Biryukov and D. Khovratovich. Related-key cryptanalysis of the full AES-192 and AES-256. In M. Matsui, editor, *Advances in Cryptology - ASIACRYPT 2009*, volume 5912 of *Lecture Notes in Computer Science*, Springer, pages 1–18, 2009.

95. A. Biryukov and E. Kushilevitz. From differential cryptanalysis to ciphertext-only attacks. In H. Krawczyk, editor, *Advances in Cryptology - CRYPTO '98*, volume 1462 of *Lecture Notes in Computer Science*, Springer, pages 72–88, 1998.

96. A. Biryukov and E. Kushilevitz. Improved cryptanalysis of RC5. In K. Nyberg, editor, *Advances in Cryptology - EUROCRYPT '98*, volume 1403 of *Lecture Notes in Computer Science*, Springer, pages 85–99, 1998.

97. A. Biryukov and A. Shamir. Structural cryptanalysis of SASAS. In B. Pfitzmann, editor, *Advances in Cryptology - EUROCRYPT 2001*, volume 2045 of *Lecture Notes in Computer Science*, Springer, pages 394–405, 2001.

98. A. Biryukov and D. Wagner. Slide attacks. In L.R. Knudsen, editor, *Fast Software Encryption, FSE 1999*, volume 1636 of *Lecture Notes in Computer Science*, Springer, pages 245–259, 1999.

99. A. Biryukov and D. Wagner. Advanced slide attacks. In B. Preneel, editor, *Advances in Cryptology - EUROCRYPT 2000*, volume 1807 of *Lecture Notes in Computer Science*, Springer, pages 589–606, 2000.

100. J. Black. The ideal-cipher model, revisited: An uninstantiable blockcipher-based hash function. In M.J.B. Robshaw, editor, *Fast Software Encryption, FSE 2006*, volume 4047 of *Lecture Notes in Computer Science*, Springer, pages 328–340, 2006.

101. J. Black, M. Cochran, and T. Shrimpton. On the impossibility of highly-efficient blockcipher-based hash functions. *J. Cryptology*, 22(3):311–329, 2009.

102. J. Black and P. Rogaway. CBC MACs for arbitrary-length messages: The three-key constructions. In M. Bellare, editor, *Advances in Cryptology - CRYPTO 2000*, volume 1880 of *Lecture Notes in Computer Science*, Springer, pages 197–215, 2000.

103. J. Black and P. Rogaway. A block-cipher mode of operation for parallelizable message authentication. In L.R. Knudsen, editor, *Advances in Cryptology - EUROCRYPT 2002*, volume 2332 of *Lecture Notes in Computer Science*, Springer, pages 384–397, 2002.

104. J. Black, P. Rogaway, and T. Shrimpton. Black-box analysis of the block-cipher-based hash-function constructions from PGV. In M. Yung, editor, *Advances in Cryptology - CRYPTO 2002*, volume 2442 of *Lecture Notes in Computer Science*, Springer, pages 320–335, 2002.

105. J. Black and H. Urtubia. Side-channel attacks on symmetric encryption schemes: The case for authenticated encryption. In *Proceedings of the 11th USENIX Security Symposium*, pages 327–338, August 2002.

106. G. R. Blakley. Information theory without the finiteness assumption, II: Unfolding the DES. In H.C. Williams, editor, *Advances in Cryptology - CRYPTO '85*, volume 218 of *Lecture Notes in Computer Science*, Springer, pages 282–337, 1986.

107. M. Blaze, W. Diffie, R. L. Rivest, B. Schneier, T. Shimomura, E. Thompson, and M. Wiener. Minimal key lengths for symmetric ciphers to provide adequate commercial security. A Report by an Ad Hoc Group of Cryptographers and Computer Scientists. Available via people.csail.mit.edu/rivest/bsa-final-report.txt, January 1996.

108. M. Blaze, J. Feigenbaum, and M. Naor. A formal treatment of remotely keyed encryption. In K. Nyberg, editor, *Advances in Cryptology - EUROCRYPT '98*, volume 1403 of *Lecture Notes in Computer Science*, Springer, pages 251–265, 1998.

109. M. Blaze and B. Schneier. The MacGuffin block cipher algorithm. In B. Preneel, editor, *Fast Software Encryption, FSE 1994*, volume 1008 of *Lecture Notes in Computer Science*, Springer, pages 97–110, 1995.

110. U. Blöcher and M. Dichtl. Problems with the linear cryptanalysis of DES using more than one active S-box per round. In B. Preneel, editor, *Fast Software Encryption, FSE 1994*, volume 1008 of *Lecture Notes in Computer Science*, Springer, pages 265–274, 1995.

111. J. Blömer, J. Guajardo, and V. Krummel. Provably secure masking of AES. In H. Handschuh and M.A. Hasan, editors, *Selected Areas in Cryptography, SAC 2004*, volume 3357 of *Lecture Notes in Computer Science*, Springer, pages 69–83, 2004.

112. J. Blömer and V. Krummel. Analysis of countermeasures against access driven cache attacks on AES. In C.M. Adams, A. Miri, and M.J. Wiener, editors, *Selected Areas in Cryptography, SAC 2007*, volume 4876 of *Lecture Notes in Computer Science*, Springer, pages 96–109, 2007.

113. M. Blunden and A. Escott. Related key attacks on reduced round KASUMI. In M. Matsui, editor, *Fast Software Encryption, FSE 2001*, volume 2355 of *Lecture Notes in Computer Science*, Springer, pages 277–285, 2001.

114. A. Bogdanov. Improved side-channel collision attacks on AES. In C.M. Adams, A. Miri, and M.J. Wiener, editors, *Selected Areas in Cryptography, SAC 2007*, volume 4876 of *Lecture Notes in Computer Science*, Springer, pages 84–95, 2007.

115. A. Bogdanov. Multiple-differential side-channel collision attacks on AES. In E. Oswald and P. Rohatgi, editors, *Cryptographic Hardware and Embedded Systems - CHES 2008*, volume 5154 of *Lecture Notes in Computer Science*, Springer, 2008, pages 30–44, 2008.

116. A. Bogdanov, L. R. Knudsen, G. Leander, C. Paar, A. Poschmann, M. J. B. Robshaw, Y. Seurin, and C. Vikkelsoe. PRESENT: An ultra-lightweight block cipher. In P. Paillier and I. Verbauwhede, editors, *Cryptographic Hardware and Embedded Systems - CHES 2007*, volume 4727 of *Lecture Notes in Computer Science*, Springer, 2007, pages 450–466, 2007.

117. J. Bonneau and I. Mironov. Cache-collision timing attacks against AES. In L. Goubin and M. Matsui, editors, *Cryptographic Hardware and Embedded Systems - CHES 2006*, volume 4249 of *Lecture Notes in Computer Science*, Springer, 2006, pages 201–215, 2006.

118. J. Borghoff, L. R. Knudsen, G. Leander, and K. Matusiewicz. Cryptanalysis of C2. In S. Halevi, editor, *Advances in Cryptology - CRYPTO 2009*, volume 5677 of *Lecture Notes in Computer Science*, Springer, pages 250–266, 2009.

119. N. Borisov, M. Chew, R. Johnson, and D. Wagner. Multiplicative differentials. In J. Daemen and V. Rijmen, editors, *Fast Software Encryption, FSE 2002, February 2002*, volume 2365 of *Lecture Notes in Computer Science*, Springer, pages 17–33, 2002.

120. J. Borst. The block cipher: Grand Cru. Available via www.cryptonessie.org.

121. J. Borst, L. R. Knudsen, and V. Rijmen. Two attacks on reduced IDEA. In W. Fumy, editor, *Advances in Cryptology - EUROCRYPT '97*, volume 1233 of *Lecture Notes in Computer Science*, Springer, pages 1–13, 1997.

122. J. Borst, B. Preneel, and J. Vandewalle. On time-memory trade-off between exhaustive key search and table pre-computation. In P. H. N. de With and M. van der Schaar-Mitrea, editors, *Proceedings of the 19th Symposium on Information Theory in the Benelux*, pages 111–118. Werkgemeenschap Informatie- en Communicatietheorie, Enschede, The Netherlands, 1998.

123. J. Borst, B. Preneel, and J. Vandewalle. Linear cryptanalysis of RC5 and RC6. In L.R. Knudsen, editor, *Fast Software Encryption, FSE 1999*, volume 1636 of *Lecture Notes in Computer Science*, Springer, pages 16–30, 1999.

124. E. F. Brickell, J. H. Moore, and M. R. Purtill. Structure in the S-boxes of the DES. In A.M. Odlyzko, editor, *Advances in Cryptology - CRYPTO '86*, volume 263 of *Lecture Notes in Computer Science*, Springer, pages 3–8, 1988.

125. L. Brown, M. Kwan, J. Pieprzyk, and J. Seberry. Improving resistance to differential cryptanalysis and the redesign of LOKI. In H. Imai, R.L. Rivest and T. Matsumoto, editors, *Advances in Cryptology - ASIACRYPT '91*, volume 739 of *Lecture Notes in Computer Science*, Springer, pages 36–50, 1993.

126. L. Brown and J. Pieprzyk. Introducing the new LOKI97 block cipher. Submitted as candidate for AES. Available via www.nist.gov/aes, 1997.

127. L. Brown, J. Pieprzyk, and J. Seberry. LOKI - a cryptographic primitive for authentication and secrecy applications. In J. Seberry and J. Pieprzyk, editors, *Advances in Cryptology - AUSCRYPT '90*, volume 453 of *Lecture Notes in Computer Science*, Springer, pages 229–236, 1990.

128. L. Brown and J. Seberry. Key scheduling in DES type cryptosystems. In J. Seberry and J. Pieprzyk, editors, *Advances in Cryptology - AUSCRYPT '90*, volume 453 of *Lecture Notes in Computer Science*, Springer, pages 221–228, 1990.

129. L. Brown and J. Seberry. On the design of permutation P in DES type cryptosystems. In J.-J. Quisquater and J. Vandewalle, editors, *Advances in Cryptology - EUROCRYPT '89*, volume 434 of *Lecture Notes in Computer Science*, Springer, pages 696–705, 1990.

130. J. Buchmann, A. Pyshkin, and R.-P. Weinmann. A zero-dimensional Gröbner basis for AES-128. In M.J.B. Robshaw, editor, *Fast Software Encryption, FSE 2006*, volume 4047 of *Lecture Notes in Computer Science*, Springer, pages 78–88, 2006.

131. L. Burnett, G. Carter, E. Dawson, and W. Millan. Efficient methods for generating MARS-like S-boxes. In B. Schneier, editor, *Fast Software Encryption, FSE 2000*, volume 1978 of *Lecture Notes in Computer Science*, Springer, pages 300–314, 2000.

132. C. Burwick, D. Coppersmith, E. D. Avignon, R. Gennaro, S. Halevi, C. Jutla, S. M. Matyas Jr., L. O'Connor, M. Peyravian, D. Safford, and N. Zunic. MARS - a candidate cipher for AES. Submitted as candidate for AES. Available via www.nist.gov/aes, 1997.

133. K. W. Campbell and M. J. Wiener. DES is not a group. In E.F. Brickell, editor, *Advances in Cryptology - CRYPTO '92*, volume 740 of *Lecture Notes in Computer Science*, Springer, pages 512–520, 1993.

134. V. Canda, T. van Trung, S. S. Magliveras, and T. Horváth. Symmetric block ciphers based on group bases. In D.R. Stinson and S.E. Tavares, editors, *Selected Areas in Cryptography, SAC 2000*, volume 2012 of *Lecture Notes in Computer Science*, Springer, pages 89–105, 2001.

135. D. Canright. A very compact S-box for AES. In J.R. Rao and B. Sunar, editors, *Cryptographic Hardware and Embedded Systems - CHES 2005*, volume 3659 of *Lecture Notes in Computer Science*, Springer, 2005, pages 441–455, 2005.

136. A. Canteaut and M. Videau. Degree of composition of highly nonlinear functions and applications to higher order differential cryptanalysis. In L.R. Knudsen, editor, *Advances in Cryptology - EUROCRYPT 2002*, volume 2332 of *Lecture Notes in Computer Science*, Springer, pages 518–533, 2002.

137. C. Carlet. *Codes de Reed-Muller, codes de Kerdock et de Preparata*. PhD thesis, Publication of LITP, Institut Blaise Pascal, Université 6, 90.59, 1990.

138. G. Carter, A. Clark, and L. Nielsen. DESV-1: A variation of the Data Encryption Standard (DES). In J. Pieprzyk and R. Safavi-Naini, editors, *Advances in Cryptology - ASIACRYPT '94*, volume 917 of *Lecture Notes in Computer Science*, Springer, pages 427–430, 1995.

139. G. Carter, E. Dawson, and L. Nielsen. DESV: A Latin square variation of DES. In *Workshop Record of SAC'95 (Selected Areas in Cryptology), May 18-19, 1995, Carleton Univ, Canada*, 1995.

140. F. Chabaud and S. Vaudenay. Links between differential and linear cryptanalysis. In A. De Santis, editor, *Advances in Cryptology - EUROCRYPT '94*, volume 950 of *Lecture Notes in Computer Science*, Springer, pages 356–365, 1995.

141. D. Chang, S. Hong, C. Kang, J. Kang, J. Kim, C. Lee, J. Lee, J. Lee, S. Lee, Y. Lee, J. Lim, and J. Sung. Arirang. Available from [542].

142. D. Chang, J. Sung, S. H. Sung, S. Lee, and J. Lim. Full-round differential attack on the original version of the hash function proposed at PKC'98. In K. Nyberg and H.M. Heys, editors, *Selected Areas in Cryptography, SAC 2002*, volume 2595 of *Lecture Notes in Computer Science*, Springer, pages 160–174, 2003.

143. C. Charnes, L. O'Connor, J. Pieprzyk, R. Safavi-Naini, and Y. Zheng. Comments on Soviet encryption algorithm. In A. De Santis, editor, *Advances in Cryptology - EUROCRYPT '94*, volume 950 of *Lecture Notes in Computer Science*, Springer, pages 433–438, 1995.

144. C. Charnes and J. Pieprzyk. Linear nonequivalence versus nonlinearity. In J. Seberry and Y. Zheng, editors, *Advances in Cryptology - ASIACRYPT '92*, volume 718 of *Lecture Notes in Computer Science*, Springer, pages 156–164, 1993.

145. D. Chaum and J.-H. Evertse. Crytanalysis of DES with a reduced number of rounds: Sequences of linear factors in block ciphers. In H.C. Williams, editor, *Advances in Cryptology - CRYPTO '85*, volume 218 of *Lecture Notes in Computer Science*, Springer, pages 192–211, 1986.

146. Z.-G. Chen and S. E. Tavares. Toward provable security of substitution-permutation encryption networks. In S.E. Tavares and H. Meijer, editors, *Selected Areas in Cryptography, SAC 1998*, volume 1556 of *Lecture Notes in Computer Science*, Springer, pages 43–56, 1999.

147. J. H. Cheon, S. Chee, and C. Park. S-boxes with controllable nonlinearity. In J. Stern, editor, *Advances in Cryptology - EUROCRYPT '99*, volume 1592 of *Lecture Notes in Computer Science*, Springer, pages 286–294, 1999.

148. J. H. Cheon, M. Kim, K. Kim, J.-Y. Lee, and S. Kang. Improved impossible differential cryptanalysis of Rijndael and Crypton. In K. Kim, editor, *4th International Conference on Information Security and Cryptology (ICISC 2001)*, volume 2288 of *Lecture Notes in Computer Science*, pages 39–49. Springer, 2002.

149. J. H. Cheon and D. H. Lee. Resistance of S-boxes against algebraic attacks. In W. Meier and B. Roy, editors, *Fast Software Encryption, FSE 2004*, volume 3017 of *Lecture Notes in Computer Science*, Springer, pages 83–94, 2004.

150. O. Y. H. Cheung, K. H. Tsoi, P. H. W. Leong, and M. P. Leong. Tradeoffs in parallel and serial implementations of the international data encryption algorithm IDEA. In Ç.K. Koç, D. Naccache and C. Paar, editors, *Cryptographic Hardware and Embedded Systems - CHES 2001*, volume 2162 of *Lecture Notes in Computer Science*, Springer, 2001, pages 333–347, 2001.

151. J. Y. Cho. Linear cryptanalysis of reduced-round PRESENT. In J. Pieprzyk, editor, *Topics in Cryptology - CT-RSA 2010*, volume 5985 of *Lecture Notes in Computer Science*, pages 302–317. Springer, 2010.

152. P. Chodowiec and K. Gaj. Very compact FPGA implementation of the AES algorithm. In C.D. Walter, Ç.K. Koç and C. Paar, editors, *Cryptographic Hardware and Embedded Systems - CHES 2003*, volume 2779 of *Lecture Notes in Computer Science*, Springer, 2003, pages 319–333, 2003.

153. S. Chow, P. A. Eisen, H. Johnson, and P. C. V. Oorschot. White-box cryptography and an AES implementation. In K. Nyberg and H.M. Heys, editors, *Selected Areas in Cryptography, SAC 2002*, volume 2595 of *Lecture Notes in Computer Science*, Springer, pages 250–270, 2003.

154. J. Choy, G. Chew, K. Khoo, and H. Yap. Cryptographic properties and application of a generalized unbalanced Feistel network structure. In C. Boyd and J.G. Nieto, editors, *Proceedings of ACISP 2009*, volume 5594 of *Lecture Notes in Computer Science*, Springer, pages 73–89, 2009.

155. K. Chun, S. Kim, S. Lee, S. Sung, and S. Yoon. Differential and linear cryptanalysis for 2-round SPNs. *Information Processing Letters*, 87:277–282, 2003.

156. C. Cid and G. Leurent. An analysis of the XSL algorithm. In B.K. Roy, editor, *Advances in Cryptology - ASIACRYPT 2005*, volume 3788 of *Lecture Notes in Computer Science*, Springer, pages 333–352, 2005.

157. C. Cid, S. Murphy, and M. J. B. Robshaw. Computational and algebraic aspects of the Advanced Encryption Standard. In V. G. Ganzha, E. W. Mayr, and E. V. Vorozhtsov, editors, *Seventh International Workshop on Computer Algebra in Scientific Computing - CASC 2004*, pages 93–103, 2004.

158. C. Cid, S. Murphy, and M. J. B. Robshaw. Small scale variants of the AES. In H. Gilbert and H. Handschuh, editors, *Fast Software Encryption, FSE 2005*, volume 3557 of *Lecture Notes in Computer Science*, Springer, pages 145–162, 2005.

159. C. Cid, S. Murphy, and M. J. B. Robshaw. *Algebraic Aspects of the Advanced Encryption Standard*. Springer, 2006.

160. R. Clayton and M. Bond. Experience using a low-cost FPGA design to crack DES keys. In B.S. Kaliski Jr., Ç.K. Koç and C. Paar, editors, *Cryptographic Hardware and Embedded Systems - CHES 2002*, volume 2523 of *Lecture Notes in Computer Science*, Springer, 2003, pages 579–592, 2003.

161. H. Cloetens, Y. Desmedt, L. Bierens, and R. G. Joos Vandewalle. Additional properties in the S-boxes of the DES. In I. Ingemarsson, editor, *Abstracts of Papers: EUROCRYPT 1986*, *Linköping, Sweden*, page 20, 1986.

162. P. M. Cohn. *Algebra, Volume 1*. John Wiley & Sons, 1982.

163. B. Collard and F.-X. Standaert. A statistical saturation attack against the block cipher PRESENT. In M. Fischlin, editor, *Topics in Cryptology - CT-RSA 2009*, volume 5473 of *Lecture Notes in Computer Science*, pages 195–210. Springer, 2009.

164. B. Collard, F.-X. Standaert, and J.-J. Quisquater. Experiments on the multiple linear cryptanalysis of reduced round Serpent. In K. Nyberg, editor, *Fast Software Encryption, FSE 2008*, volume 5086 of *Lecture Notes in Computer Science*, Springer, pages 382–397, 2008.

165. S. Contini, R. L. Rivest, M. J. B. Robshaw, and Y. L. Yin. Improved analysis of some simplified variants of RC6. In L.R. Knudsen, editor, *Fast Software Encryption, FSE 1999*, volume 1636 of *Lecture Notes in Computer Science*, Springer, pages 1–15, 1999.

166. D. Coppersmith. The real reason for Rivest's phenomenon. In H.C. Williams, editor, *Advances in Cryptology - CRYPTO '85*, volume 218 of *Lecture Notes in Computer Science*, Springer, pages 535–536, 1986.

167. D. Coppersmith. The Data Encryption Standard and its strength against attacks. IBM Technical Report, RC18613 (81421), December 1992.

168. D. Coppersmith. Luby-Rackoff: Four rounds is not enough. Technical Report RC 20674, IBM, December 1996.

169. N. Courtois. Feistel schemes and bi-linear cryptanalysis. In M. Franklin, editor, *Advances in Cryptology - CRYPTO 2004*, volume 3152 of *Lecture Notes in Computer Science*, Springer, pages 23–40, 2004.

170. N. Courtois, G. V. Bard, and D. Wagner. Algebraic and slide attacks on KeeLoq. In K. Nyberg, editor, *Fast Software Encryption, FSE 2008*, volume 5086 of *Lecture Notes in Computer Science*, Springer, pages 97–115, 2008.

171. N. Courtois, A. Klimov, J. Patarin, and A. Shamir. Efficient algorithms for solving overdefined systems of multivariate polynomial equations. In B. Preneel, editor, *Advances in Cryptology - EUROCRYPT 2000*, volume 1807 of *Lecture Notes in Computer Science*, Springer, pages 392–407, 2000.

172. N. Courtois and J. Pieprzyk. Cryptanalysis of block ciphers with overdefined systems of equations. In Y. Zheng, editor, *Advances in Cryptology - ASIACRYPT 2002*, volume 2501 of *Lecture Notes in Computer Science*, Springer, pages 267–287, 2002.

173. D. Cox, J. Little, and D. O'Shea. *Ideals, Varieties, and Algorithms*. Undergraduate Texts in Mathematics. Springer, 2nd edition, 1997.

174. P. Crowley. Mercy: A fast large block cipher for disk sector encryption. In B. Schneier, editor, *Fast Software Encryption, FSE 2000*, volume 1978 of *Lecture Notes in Computer Science*, Springer, pages 49–63, 2000.

175. CRYPTREC. Cryptography research and evaluation committee. Available via www.ipa. go.jp/security/.

176. T. W. Cusick and M. C. Wood. The REDOC II cryptosystem. In A.J. Menezes and S.A. Vanstone, editors, *Advances in Cryptology - CRYPTO '90*, volume 537 of *Lecture Notes in Computer Science*, Springer, pages 545–563, 1991.

177. J. Daemen. Limitations of the Even-Mansour construction. In H. Imai, R.L. Rivest and T. Matsumoto, editors, *Advances in Cryptology - ASIACRYPT '91*, volume 739 of *Lecture Notes in Computer Science*, Springer, pages 495–498, 1993.

178. J. Daemen. *Cipher and Hash Function Design*. PhD thesis, Katholieke Universiteit Leuven, March 1995.

179. J. Daemen, R. Govaerts, and J. Vandewalle. Cryptanalysis of 2,5 rounds of IDEA. Technical Report ESAT-COSIC Report 94-1, Department of Electrical Engineering, Katholieke Universiteit Leuven, March 1994.

180. J. Daemen, R. Govaerts, and J. Vandewalle. A new approach to block cipher design. In R. Anderson, editor, *Fast Software Encryption, FSE 1993*, volume 809 of *Lecture Notes in Computer Science*, Springer, pages 18–32, 1994.

181. J. Daemen, R. Govaerts, and J. Vandewalle. Weak keys for IDEA. In D. Stinson, editor, *Advances in Cryptology - CRYPTO '93*, volume 773 of *Lecture Notes in Computer Science*, Springer, pages 224–231, 1994.

182. J. Daemen, R. Govaerts, and J. Vandewalle. Correlation matrices. In B. Preneel, editor, *Fast Software Encryption, FSE 1994*, volume 1008 of *Lecture Notes in Computer Science*, Springer, pages 275–285, 1995.

183. J. Daemen, L. R. Knudsen, and V. Rijmen. The block cipher Square. In E. Biham, editor, *Fast Software Encryption, FSE 1997*, volume 1267 of *Lecture Notes in Computer Science*, Springer, pages 149–165, 1997.

184. J. Daemen, L. R. Knudsen, and V. Rijmen. Linear frameworks for block ciphers. *Designs, Codes and Cryptography*, 22(1):65–87, 2001.

185. J. Daemen and V. Rijmen. AES proposal: Rijndael. Version 2.0. Available via www.nist.gov/aes.

186. J. Daemen and V. Rijmen. Answer to "new observations on Rijndael". August 11, 2000. Available via www.isg.rhul.ac.uk/~mrobshaw.

187. J. Daemen and V. Rijmen. The block cipher BKSQ. In J.-J. Quisquater and B. Schneier, editors, *Smart Card Research and Applications*, volume 1820 of *Lecture Notes in Computer Science*, pages 247–256. Springer, 2000.

188. J. Daemen and V. Rijmen. AES and the wide trail design strategy. In L.R. Knudsen, editor, *Advances in Cryptology - EUROCRYPT 2002*, volume 2332 of *Lecture Notes in Computer Science*, Springer, pages 108–109, 2002.

189. J. Daemen and V. Rijmen. *The Design of Rijndael. AES - The Advanced Encryption Standard*. Springer, 2002.

190. I. Damgård and L. R. Knudsen. The breaking of the AR hash function. In T. Helleseth, editor, *Advances in Cryptology - EUROCRYPT '93*, volume 765 of *Lecture Notes in Computer Science*, Springer, pages 286–292, 1994.

191. I. Damgård and L. R. Knudsen. Two-key triple encryption. *J. Cryptology*, 11(3):209–218, 1998.

192. A. Dandalis, V. K. Prasanna, and J. D. P. Rolim. A comparative study of performance of AES final candidates using FPGAs. In Ç.K. Koç and C. Paar, editors, *Cryptographic Hardware and Embedded Systems - CHES 2000*, volume 1965 of *Lecture Notes in Computer Science*, Springer, 2000, pages 125–140, 2000.

193. D. W. Davies. Some regular properties of the DES. In A. Gersho, editor, *Advances in Cryptology: A report on CRYPTO '81, IEEE Workshop on Communications Security, Santa Barbara, August 24-26, 1981. U.C. Santa Barbara, Dept. of Elec. and Computer Eng., ECE Report No 82-04*, page 41, 1982.

194. D. W. Davies. Some regular properties of the 'Data Encryption Standard' algorithm. In D. Chaum, R.L. Rivest, and A.T. Sherman, editors, *Advances in Cryptology: Proceedings of CRYPTO '82*, Plenum, pages 89–96, 1983.

195. D. W. Davies and S. Murph. Pairs and triplets of DES S-boxes. *J. Cryptology*, 8(1):1–25, 1995.

196. D. W. Davies and G. I. P. Parkin. The average cycle size of the key stream in output feedback encipherment. In D. Chaum, R.L. Rivest, and A.T. Sherman, editors, *Advances in Cryptology: Proceedings of CRYPTO '82*, Plenum, pages 97–98, 1983.

197. D. W. Davies and G. I. P. Parkin. The average cycle size of the key-stream in output feedback encipherment. In T. Beth, editor, *Cryptography, Proceedings of the Workshop on Cryptography, Burg Feuerstein, Germany, March 29 - April 2, 1982*. volume 149 of *Lecture Notes in Computer Science*, Springer, pages 263–279, 1983.

198. D. W. Davies and W. L. Price. *Security for Computer Networks*. John Wiley & Sons, 1989.

199. M. Davio, Y. Desmedt, M. Fosseprez, R. Govaerts, J. Hulsbosch, P. Neutjens, P. Piret, J.-J. Quisquater, J. Vandewalle, and P. Wouters. Analytical characteristics of the DES. In D. Chaum, editor, *Advances in Cryptology: Proceedings of CRYPTO '83*, Plenum, pages 171–202, 1984.

200. M. Davio, Y. Desmedt, J. Goubert, F. Hoornaert, and J.-J. Quisquater. Efficient hardware and software implementations for the DES. In G.R. Blakley and D. Chaum, editors, *Advances in Cryptology - CRYPTO '84*, volume 196 of *Lecture Notes in Computer Science*, Springer, pages 144–146, 1985.

201. M. Davio, Y. Desmedt, and J.-J. Quisquater. Propagation characteristics of the DES. In T. Beth, N. Cot, and I. Ingemarsson, editors, *Advances in Cryptology - EUROCRYPT '84*, volume 209 of *Lecture Notes in Computer Science*, Springer, pages 62–73, 1984.

202. M. H. Dawson and S. E. Tavares. An expanded set of S-box design criteria based on information theory and its relation to differential-like attacks. In D.W. Davies, editor, *Advances in Cryptology - EUROCRYPT '91*, volume 547 of *Lecture Notes in Computer Science*, Springer, pages 352–367, 1992.

203. H. Demirci. Square-like attacks on reduced rounds of IDEA. In K. Nyberg and H.M. Heys, editors, *Selected Areas in Cryptography, SAC 2002*, volume 2595 of *Lecture Notes in Computer Science*, Springer, pages 147–159, 2003.

204. H. Demirci and A. A. Selçuk. A meet-in-the-middle attack on 8-round AES. In K. Nyberg, editor, *Fast Software Encryption, FSE 2008*, volume 5086 of *Lecture Notes in Computer Science*, Springer, pages 116–126, 2008.

205. H. Demirci, A. A. Selçuk, and E. Türe. A new meet-in-the-middle attack on the IDEA block cipher. In M. Matsui and R.J. Zuccherato, editors, *Selected Areas in Cryptography, SAC 2003*, volume 3006 of *Lecture Notes in Computer Science*, Springer, pages 117–129, 2004.

206. B. Den Boer. Cryptanalysis of F.E.A.L. In C.G. Günther, editor, *Advances in Cryptology - EUROCRYPT '88*, volume 330 of *Lecture Notes in Computer Science*, Springer, pages 293–299, 1988.

207. A. Desai. New paradigms for constructing symmetric encryption schemes secure against chosen-ciphertext attack. In M. Bellare, editor, *Advances in Cryptology - CRYPTO 2000*, volume 1880 of *Lecture Notes in Computer Science*, Springer, pages 394–412, 2000.

208. Y. Desmedt, F. Hoornaert, and J.-J. Quisquater. Several exhaustive key search machines and DES. In I. Ingemarsson, editor, *Abstracts of Papers: EUROCRYPT 1986, Linköping, Sweden*, pages 17–19, 1986.

209. Y. Desmedt, J.-J. Quisquater, and M. Davio. Dependence of output on input in DES: Small avalanche characteristics. In G.R. Blakley and D. Chaum, editors, *Advances in Cryptology - CRYPTO '84*, volume 196 of *Lecture Notes in Computer Science*, Springer, pages 359–376, 1985.

210. J. Detombe and S. E. Tavares. Constructing large cryptographically strong S-boxes. In J. Seberry and Y. Zheng, editors, *Advances in Cryptology - ASIACRYPT '92*, volume 718 of *Lecture Notes in Computer Science*, Springer, pages 165–181, 1993.

211. C. D'Halluin, G. Bijnens, B. Preneel, and V. Rijmen. Equivalent keys of HPC. In K-Y. Lam, E. Okamoto and C. Xing, editors, *Advances in Cryptology - ASIACRYPT '99*, volume 1716 of *Lecture Notes in Computer Science*, Springer, pages 29–42, 1999.

212. C. D'Halluin, G. Bijnens, V. Rijmen, and B. Preneel. Attack on six rounds of Crypton. In L.R. Knudsen, editor, *Fast Software Encryption, FSE 1999*, volume 1636 of *Lecture Notes in Computer Science*, Springer, pages 46–59, 1999.

213. A. di Porto and W. Wolfowicz. VINO: A block cipher including variable permutations. In R. Anderson, editor, *Fast Software Encryption, FSE 1993*, volume 809 of *Lecture Notes in Computer Science*, Springer, pages 205–210, 1994.

214. C. Diem. The XL-algorithm and a conjecture from commutative algebra. In P.J. Lee, editor, *Advances in Cryptology - ASIACRYPT 2004*, volume 3329 of *Lecture Notes in Computer Science*, Springer, pages 323–337, 2004.

215. W. Diffie and M. E. Hellman. Exhaustive cryptanalysis of the NBS data encryption standard. *Computer*, pages 74–84, 1977.

216. W. Diffie and S. Landau. *Privacy on the Line*. MIT Press, 1998.

217. H. Dobbertin, L. R. Knudsen, and M. J. B. Robshaw. The cryptanalysis of the AES - a brief survey. In V. Rijmen, H. Dobbertin, and A. Sowa, editors, *Advanced Encryption Standard - AES, Fourth International Conference*, volume 3373 of *Lecture Notes in Computer Science*, Springer, pages 1–10, 2005.

218. Y. Dodis, K. Pietrzak, and P. Puniya. A new mode of operation for block ciphers and length-preserving MACs. In N. Smart, editor, *Advances in Cryptology - EUROCRYPT 2008*, volume 4965 of *Lecture Notes in Computer Science*, Springer, pages 198–219, 2007.

219. O. Dunkelman and N. Keller. An improved impossible differential attack on MISTY1. In J. Pieprzyk, editor, *Advances in Cryptology - ASIACRYPT 2008*, volume 5350 of *Lecture Notes in Computer Science*, Springer, pages 441–454, 2008.

220. O. Dunkelman and N. Keller. A new attack on the LEX stream cipher. In J. Pieprzyk, editor, *Advances in Cryptology - ASIACRYPT 2008*, volume 5350 of *Lecture Notes in Computer Science*, Springer, pages 539–556, 2008.

221. O. Dunkelman, N. Keller, and J. Kim. Related-key rectangle attack on the full SHACAL-1. In E. Biham and A.M. Youssef, editors, *Selected Areas in Cryptography, SAC 2006*, volume 4356 of *Lecture Notes in Computer Science*, Springer, pages 28–44, 2007.

222. O. Dunkelman, N. Keller, and A. Shamir. A practical-time related-key attack on the KASUMI cryptosystem used in GSM and 3G telephony. In T. Rabin, editor, *Advances in Cryptology - CRYPTO 2010*, volume 6223 of *Lecture Notes in Computer Science*, Springer, pages 393–410, 2010.

223. H. Eberle. A high-speed DES implementation for network applications. In E.F. Brickell, editor, *Advances in Cryptology - CRYPTO '92*, volume 740 of *Lecture Notes in Computer Science*, Springer, pages 521–539, 1993.

224. ECRYPT. ECRYPT yearly report on algorithms and keysizes. Available via www.ecrypt.eu.org.

225. T. Eisenbarth, T. Kasper, A. Moradi, C. Paar, M. Salmasizadeh, and M. T. M. Shalmani. On the power of power analysis in the real world: A complete break of the KeeLoqCode hopping scheme. In D. Wagner, editor, *Advances in Cryptology - CRYPTO 2008*, volume 5157 of *Lecture Notes in Computer Science*, Springer, pages 203–220, 2008.

226. Electronic Frontier Foundation. *Cracking DES: Secrets of Encryption Research, Wiretap Politics & Chip Design*. O'Reilly Press, 1998.

227. J. Etrog and M. J. B. Robshaw. The cryptanalysis of reduced-round SMS4: Refinements and improvements. In R.M. Avanzi, L. Keliher, and F. Sica, editors, *Selected Areas in Cryptography, SAC 2008*, volume 5381 of *Lecture Notes in Computer Science*, Springer, pages 51–65, 2009.

228. ETSI. TS 135 202 V7.0.0: Universal mobile telecommunications system (UMTS); specification of the 3GPP confidentiality and integrity algorithms; Document 2: Kasumi specification (3GPP TS 35.202 version 7.0.0 Release 7). Available via www.etsi.org.

229. S. Even and O. Goldreich. On the power of cascade ciphers. In D. Chaum, editor, *Advances in Cryptology: Proceedings of CRYPTO '83*, Plenum, pages 43–50, 1984.

230. S. Even and Y. Mansour. A construction of a cipher from a single pseudorandom permutation. *J. Cryptology*, 10(3):151–162, 1997.

231. J.-H. Evertse. Linear structures in blockciphers. In D. Chaum, W.L. Price, editors, *Advances in Cryptology - EUROCRYPT '87*, volume 304 of *Lecture Notes in Computer Science*, Springer, pages 249–266, 1988.

232. J.-C. Faugère. A new efficient algorithm for computing Gröbner bases (F_4). *Journal of Pure and Applied Algebra*, 139:61–88, 1999.

233. J.-C. Faugère. A new efficient algorithm for computing Gröbner bases without reduction to 0 (F_5). In T. Mora, editor, *Proceedings of ISSAC*, pages 75–83. ACM Press, 2002.

234. H. Feistel. Cryptography and computer privacy. *Scientific American*, 228(5):15–23, 1973.

235. H. Feistel, W. A. Notz, and J. L. Smith. Some cryptographic techniques for machine-to-machine data communications. *Proceedings of IEEE*, 63(11):1545–1554, 1975.

236. M. Feldhofer, S. Dominikus, and J. Wolkerstorfer. Strong authentication for RFID systems using the AES algorithm. In M. Joye and J.-J. Quisquater, editors, *Cryptographic Hardware and Embedded Systems - CHES 2004*, volume 3156 of *Lecture Notes in Computer Science*, Springer, 2004, pages 357–370, 2004.

237. M. Feldhofer, J. Wolkerstorfer, and V. Rijmen. AES implementation on a grain of sand. *Information Security, IEE Proceedings*, 152:13–20, 2005.

238. F. A. Feldman. Fast spectral tests for measuring nonrandomness and the DES. In C. Pomerance, editor, *Advances in Cryptology - CRYPTO '87*, volume 293 of *Lecture Notes in Computer Science*, Springer, pages 243–254, 1988.

239. W. Feller. *Probability Theory and Its Applications, Vol. I*. John Wiley & Sons, 1950.

240. N. Ferguson, J. Kelsey, S. Lucks, B. Schneier, M. Stay, D. Wagner, and D. Whiting. Improved cryptanalysis of Rijndael. In B. Schneier, editor, *Fast Software Encryption, FSE 2000*, volume 1978 of *Lecture Notes in Computer Science*, Springer, pages 213–230, 2000.

241. N. Ferguson, R. Schroeppel, and D. Whiting. A simple algebraic representation of Rijndael. In S. Vaudenay and A.M. Youssef, editors, *Selected Areas in Cryptography, SAC 2001*, volume 2259 of *Lecture Notes in Computer Science*, Springer, pages 103–111, 2001.

242. V. Fischer and M. Drutarovský. Two methods of Rijndael implementation in reconfigurable hardware. In Ç.K. Koç, D. Naccache and C. Paar, editors, *Cryptographic Hardware and Embedded Systems - CHES 2001*, volume 2162 of *Lecture Notes in Computer Science*, Springer, 2001, pages 77–92, 2001.

243. E. Fleischmann, M. Gorski, and S. Lucks. Attacking 9 and 10 rounds of AES-256. In C. Boyd and J.G. Nieto, editors, *Proceedings of ACISP 2009*, volume 5594 of *Lecture Notes in Computer Science*, Springer, pages 60–72, 2009.

244. S. R. Fluhrer. Cryptanalysis of the Mercy block cipher. In M. Matsui, editor, *Fast Software Encryption, FSE 2001*, volume 2355 of *Lecture Notes in Computer Science*, Springer, pages 28–36, 2001.

245. R. Forre. Methods and instruments for designing S-boxes. *J. Cryptology*, 2(3):115–130, 1990.

246. P.-A. Fouque, L. Granboulan, and J. Stern. Differential cryptanalysis for multivariate schemes. In R. Cramer, editor, *Advances in Cryptology - EUROCRYPT 2005*, volume 3494 of *Lecture Notes in Computer Science*, Springer, pages 341–353, 2005.

247. P.-A. Fouque, A. Joux, G. Martinet, and F. Valette. Authenticated on-line encryption. In M. Matsui and R.J. Zuccherato, editors, *Selected Areas in Cryptography, SAC 2003*, volume 3006 of *Lecture Notes in Computer Science*, Springer, pages 145–159, 2004.

248. P.-A. Fouque, A. Joux, and G. Poupard. Blockwise adversarial model for on-line ciphers and symmetric encryption schemes. In H. Handschuh and M.A. Hasan, editors, *Selected Areas in Cryptography, SAC 2004*, volume 3357 of *Lecture Notes in Computer Science*, Springer, pages 212–226, 2004.

249. P.-A. Fouque, G. Martinet, and G. Poupard. Practical symmetric on-line encryption. In T. Johansson, editor, *Fast Software Encryption, FSE 2003*, volume 2887 of *Lecture Notes in Computer Science*, Springer, pages 362–375, 2003.

250. J. Fuller and W. Millan. Linear redundancy in S-boxes. In T. Johansson, editor, *Fast Software Encryption, FSE 2003*, volume 2887 of *Lecture Notes in Computer Science*, Springer, pages 74–86, 2003.

251. W. Fumy. On the f-function of FEAL. In C. Pomerance, editor, *Advances in Cryptology - CRYPTO '87*, volume 293 of *Lecture Notes in Computer Science*, Springer, pages 434–437, 1988.

252. V. Furman. Differential cryptanalysis of Nimbus. In M. Matsui, editor, *Fast Software Encryption, FSE 2001*, volume 2355 of *Lecture Notes in Computer Science*, Springer, pages 187–195, 2001.

253. P. Gauravaram, L. R. Knudsen, K. Matusiewicz, F. Mendel, C. Rechberger, M. Schläffer, and S. S. Thomsen. Grøstl—a SHA-3 candidate. Available from [542].

254. C. Gentry and Z. Ramzan. Eliminating random permutation oracles in the Even-Mansour cipher. In P.J. Lee, editor, *Advances in Cryptology - ASIACRYPT 2004*, volume 3329 of *Lecture Notes in Computer Science*, Springer, pages 32–47, 2004.

255. H. Gilbert. The security of "one-block-to-many" modes of operation. In T. Johansson, editor, *Fast Software Encryption, FSE 2003*, volume 2887 of *Lecture Notes in Computer Science*, Springer, pages 376–395, 2003.

256. H. Gilbert and G. Chassé. A statistical attack of the FEAL-8 cryptosystem. In A.J. Menezes and S.A. Vanstone, editors, *Advances in Cryptology - CRYPTO '90*, volume 537 of *Lecture Notes in Computer Science*, Springer, pages 22–33, 1991.

257. H. Gilbert and P. Chauvaud. A chosen plaintext attack of the 16-round Khufu cryptosystem. In Y.G. Desmedt, editor, *Advances in Cryptology - CRYPTO '94*, volume 839 of *Lecture Notes in Computer Science*, Springer, pages 359–368, 1994.

258. H. Gilbert, H. Handschuh, A. Joux, and S. Vaudenay. A statistical attack on RC6. In B. Schneier, editor, *Fast Software Encryption, FSE 2000*, volume 1978 of *Lecture Notes in Computer Science*, Springer, pages 64–74, 2000.

259. H. Gilbert and M. Minier. A collision attack on 7 rounds of Rijndael. In 3rd *Advanced Encryption Standard Candidate Conference*, pages 230–241. National Institute of Standards and Technology, April 2000. Available via www.nist.gov/aes.

260. H. Gilbert and M. Minier. New results on the pseudorandomness of some blockcipher constructions. In M. Matsui, editor, *Fast Software Encryption, FSE 2001*, volume 2355 of *Lecture Notes in Computer Science*, Springer, pages 248–266, 2001.

261. V. D. Gligor and P. Donescu. Fast encryption and authentication: XCBC encryption and XECB authentication modes. In M. Matsui, editor, *Fast Software Encryption, FSE 2001*, volume 2355 of *Lecture Notes in Computer Science*, Springer, pages 92–108, 2001.

262. D. Goldenberg, S. Hohenberger, M. Liskov, E. C. Schwartz, and H. Seyalioglu. On tweaking Luby-Rackoff blockciphers. In K. Kurosawa, editor, *Advances in Cryptology - ASIACRYPT 2007*, volume 4833 of *Lecture Notes in Computer Science*, Springer, pages 342–356, 2007.

263. J. D. Golic and C. Tymen. Multiplicative masking and power analysis of AES. In B.S. Kaliski Jr., Ç.K. Koç and C. Paar, editors, *Cryptographic Hardware and Embedded Systems - CHES 2002*, volume 2523 of *Lecture Notes in Computer Science*, Springer, 2003, pages 198–212, 2003.

264. T. Good and M. Benaissa. AES on FPGA from the fastest to the smallest. In J.R. Rao and B. Sunar, editors, *Cryptographic Hardware and Embedded Systems - CHES 2005*, volume 3659 of *Lecture Notes in Computer Science*, Springer, 2005, pages 427–440, 2005.

265. J. A. Gordon and H. Retkin. Are big S-boxes best? In T. Beth, editor, *Cryptography, Proceedings of the Workshop on Cryptography, Burg Feuerstein, Germany, March 29 - April 2, 1982*. volume 149 of *Lecture Notes in Computer Science*, Springer, pages 257–262, 1983.

266. L. Goubin, J.-M. Masereel, and M. Quisquater. Cryptanalysis of white box DES implementations. In C.M. Adams, A. Miri, and M.J. Wiener, editors, *Selected Areas in Cryptography, SAC 2007*, volume 4876 of *Lecture Notes in Computer Science*, Springer, pages 278–295, 2007.

267. L. Goubin and J. Patarin. DES and differential power analysis (the "duplication" method). In Ç.K. Koç and C. Paar, editors, *Cryptographic Hardware and Embedded Systems, CHES '99*, volume 1717 of *Lecture Notes in Computer Science*, Springer, 1999, pages 158–172, 1999.

268. L. Granboulan. Flaws in differential cryptanalysis of Skipjack. In M. Matsui, editor, *Fast Software Encryption, FSE 2001*, volume 2355 of *Lecture Notes in Computer Science*, Springer, pages 328–335, 2001.

269. L. Granboulan, P. Q. Nguyen, F. Noilhan, and S. Vaudenay. DFCv2. In D.R. Stinson and S.E. Tavares, editors, *Selected Areas in Cryptography, SAC 2000*, volume 2012 of *Lecture Notes in Computer Science*, Springer, pages 57–71, 2001.

270. S. Gueron. Intel's advanced encryption standard (AES) instructions set. Intel Corporation White Paper, March 2009. Available via software.intel.com.

271. H. Gustafson, E. Dawson, and W. J. Caelli. Comparison of block ciphers. In J. Seberry and J. Pieprzyk, editors, *Advances in Cryptology - AUSCRYPT '90*, volume 453 of *Lecture Notes in Computer Science*, Springer, pages 208–220, 1990.

272. S. Halevi, W. Hall, and C. Jutla. The hash function Fugue. Available from [542].

273. S. Halevi and P. Rogaway. A tweakable enciphering mode. In D. Boneh, editor, *Advances in Cryptology - CRYPTO 2003*, volume 2729 of *Lecture Notes in Computer Science*, Springer, pages 482–499, 2003.

274. C. Hall, J. Kelsey, V. Rijmen, B. Schneier, and D. Wagner. Cryptanalysis of SPEED. In S.E. Tavares and H. Meijer, editors, *Selected Areas in Cryptography, SAC 1998*, volume 1556 of *Lecture Notes in Computer Science*, Springer, pages 319–338, 1999.

275. P. Hämäläinen, T. Alho, Hännikäinen, and T. Hämäläinen. Design and implementation of low-area and low-power AES encryption hardware core. In *9th EUROMICRO Conference on Digital System Design (DSD'06)*, pages 577–583, 2006.

276. H. Handschuh and H. M. Heys. A timing attack on RC5. In S.E. Tavares and H. Meijer, editors, *Selected Areas in Cryptography, SAC 1998*, volume 1556 of *Lecture Notes in Computer Science*, Springer, pages 306–318, 1999.

277. H. Handschuh, L. R. Knudsen, and M. J. B. Robshaw. Analysis of SHA-1 in encryption mode. In D. Naccache, editor, *Topics in Cryptology – CT-RSA 2001*, volume 2020 of *Lecture Notes in Computer Science*, pages 70–83. Springer, 2001.

278. H. Handschuh and D. Naccache. SHACAL. Primitive submitted to NESSIE. In Final report of European Project IST-1999-12324, *New European Schemes for Signatures, Integrity, and Encryption*, April 19, 2004. Available via www.cryptonessie.org.

279. H. Handschuh and B. Preneel. On the security of double and 2-key triple modes of operation. In L.R. Knudsen, editor, *Fast Software Encryption, FSE 1999*, volume 1636 of *Lecture Notes in Computer Science*, Springer, pages 215–230, 1999.

280. H. Handschuh and S. Vaudenay. A universal encryption standard. In H.M. Heys and C.M. Adams, editors, *Selected Areas in Cryptography, SAC 1999*, volume 1758 of *Lecture Notes in Computer Science*, Springer, pages 1–12, 2000.

281. C. Harpes, G. G. Kramer, and J. L. Massey. A generalization of linear cryptanalysis and the applicability of Matsui's piling-up lemma. In L. Guillou and J.-J. Quisquater, editors, *Advances in Cryptology - EUROCRYPT '95*, volume 921 of *Lecture Notes in Computer Science*, Springer, pages 24–38, 1995.

282. C. Harpes and J. L. Massey. Partitioning cryptanalysis. In E. Biham, editor, *Fast Software Encryption, FSE 1997*, volume 1267 of *Lecture Notes in Computer Science*, Springer, pages 13–27, 1997.

283. S. Harris and C. M. Adams. Key-dependent S-box manipulations. In S.E. Tavares and H. Meijer, editors, *Selected Areas in Cryptography, SAC 1998*, volume 1556 of *Lecture Notes in Computer Science*, Springer, pages 15–26, 1999.

284. Y. Hatano, H. Sekine, and T. Kaneko. Higher order differential attack of Camellia (II). In K. Nyberg and H.M. Heys, editors, *Selected Areas in Cryptography, SAC 2002*, volume 2595 of *Lecture Notes in Computer Science*, Springer, pages 129–146, 2003.

285. P. Hawkes. Differential-linear weak key classes of IDEA. In K. Nyberg, editor, *Advances in Cryptology - EUROCRYPT '98*, volume 1403 of *Lecture Notes in Computer Science*, Springer, pages 112–126, 1998.

286. P. Hawkes and L. O'Connor. On applying linear cryptanalysis to IDEA. In K. Kim and T. Matsumoto, editors, *Advances in Cryptology - ASIACRYPT '96*, volume 1163 of *Lecture Notes in Computer Science*, Springer, pages 105–115, 1996.

287. P. Hawkes and L. O'Connor. Xor and non-xor differential probabilities. In J. Stern, editor, *Advances in Cryptology - EUROCRYPT '99*, volume 1592 of *Lecture Notes in Computer Science*, Springer, pages 272–285, 1999.

288. M. E. Hellman. A cryptanalytic time-memory trade-off. *IEEE Trans. Information Theory*, 26(4):401–406, July 1980.

289. M. E. Hellman, R. Merkle, R. Schroeppel, L. Washington, W. Diffie, S. Pohlig, and P. Schweitzer. Results of an initial attempt to cryptanalyze the NBS Data Encryption Standard. Technical report, Stanford University, U.S.A., September 1976.

290. M. E. Hellman and J. M. Reyneri. Drainage and the DES. In D. Chaum, R.L. Rivest, and A.T. Sherman, editors, *Advances in Cryptology: Proceedings of CRYPTO '82*, Plenum, pages 129–131, 1983.

291. L. Hemme. A differential fault attack against early rounds of (triple-)DES. In M. Joye and J.-J. Quisquater, editors, *Cryptographic Hardware and Embedded Systems - CHES 2004*, volume 3156 of *Lecture Notes in Computer Science*, Springer, 2004, pages 254–267, 2004.

292. M. Henricksen and L. R. Knudsen. Cryptanalysis of the CRUSH hash function. In C.M. Adams, A. Miri, and M.J. Wiener, editors, *Selected Areas in Cryptography, SAC 2007*, volume 4876 of *Lecture Notes in Computer Science*, Springer, pages 74–83, 2007.

293. M. Hermelin, J. Y. Cho, and K. Nyberg. Multidimensional extension of Matsui's algorithm 2. In O. Dunkelman, editor, *Fast Software Encryption, FSE 2009,* volume 5665 of *Lecture Notes in Computer Science,* Springer, pages 209–227, 2009.

294. H. M. Heys. A tutorial on linear and differential cryptanalysis. *Cryptologia,* XXVI(3):189–221, 2002. Appears also as Technical Report CORR 2001-17, Centre for Applied Cryptographic Research, Department of Combinatorics and Optimization, University of Waterloo, Mar. 2001. Available via www.engr.mun.ca/~howard/PAPERS/ldc_tutorial.pdf.

295. H. M. Heys and S. E. Tavares. Substitution-permutation networks resistant to differential and linear cryptanalysis. *J. Cryptology,* 9(1):1–19, 1996.

296. F. H. Hinsley and A. Stripp. *Code Breakers: The Inside Story of Bletchley Park.* Oxford University Press, 1993.

297. S. Hirose. Some plausible constructions of double-block-length hash functions. In M.J.B. Robshaw, editor, *Fast Software Encryption, FSE 2006,* volume 4047 of *Lecture Notes in Computer Science,* Springer, pages 210–225, 2006.

298. S. Hirose, H. Kuwakado, and H. Yoshida. SHA-3 proposal: Lesamnta. Available from [542].

299. W. Hohl, X. Lai, T. Meier, and C. Waldvogel. Security of iterated hash functions based on block ciphers. In D. Stinson, editor, *Advances in Cryptology - CRYPTO '93,* volume 773 of *Lecture Notes in Computer Science,* Springer, pages 379–390, 1994.

300. D. Hong, S. Hong, W. Lee, S. Lee, J. Lim, J. Sung, and O. Yi. Known-IV, known-in-advance-IV, and replayed-and-known-IV attacks on multiple modes of operation of block ciphers. *J. Cryptology,* 19(4):441–462, 2006.

301. D. Hong, J. Sung, S. Hong, W. Lee, S. Lee, J. Lim, and O. Yi. Known-IV attacks on triple modes of operation of block ciphers. In C. Boyd, editor, *Advances in Cryptology - ASIACRYPT 2001,* volume 2248 of *Lecture Notes in Computer Science,* Springer, pages 208–221, 2001.

302. D. Hong, J. Sung, S. Hong, J. Lim, B. K. Sangjin Lee, C. Lee, D. Chang, J. Lee, K. Jeong, H. Kim, J. Kim, and S. Chee. HIGHT: A new block cipher suitable for low-resource device. In L. Goubin and M. Matsui, editors, *Cryptographic Hardware and Embedded Systems - CHES 2006,* volume 4249 of *Lecture Notes in Computer Science,* Springer, 2006, pages 46–59, 2006.

303. D. Hong, J. Sung, S. Moriai, S. Lee, and J. Lim. Impossible differential cryptanalysis of Zodiac. In M. Matsui, editor, *Fast Software Encryption, FSE 2001,* volume 2355 of *Lecture Notes in Computer Science,* Springer, pages 300–311, 2001.

304. E. Hong, J.-H. Chung, and C. H. Lim. Hardware design and performance estimation of the 128-bit block cipher Crypton. In Ç.K. Koç and C. Paar, editors, *Cryptographic Hardware and Embedded Systems, CHES '99,* volume 1717 of *Lecture Notes in Computer Science,* Springer, 1999, pages 49–60, 1999.

305. S. Hong, J. Kim, S. Lee, and B. Preneel. Related-key rectangle attacks on reduced versions of SHACAL-1 and AES-192. In H. Gilbert and H. Handschuh, editors, *Fast Software Encryption, FSE 2005,* volume 3557 of *Lecture Notes in Computer Science,* Springer, pages 368–383, 2005.

306. S. Hong, S. Lee, J. Lim, J. Sung, D. H. Cheon, and I. Cho. Provable security against differential and linear cryptanalysis for the SPN structure. In B. Schneier, editor, *Fast Software Encryption, FSE 2000,* volume 1978 of *Lecture Notes in Computer Science,* Springer, pages 273–283, 2000.

307. F. Hoornaert, J. Goubert, and Y. Desmedt. Efficient hardware implementation of the DES. In G.R. Blakley and D. Chaum, editors, *Advances in Cryptology - CRYPTO '84,* volume 196 of *Lecture Notes in Computer Science,* Springer, pages 147–173, 1985.

308. G. Hornauer, W. Stephan, and R. Wernsdorf. Markov ciphers and alternating groups. In T. Helleseth, editor, *Advances in Cryptology - EUROCRYPT '93,* volume 765 of *Lecture Notes in Computer Science,* Springer, pages 453–460, 1994.

309. K. Hwang, W. Lee, S. Lee, S. Lee, and J. Lim. Saturation attacks on reduced round Skipjack. In J. Daemen and V. Rijmen, editors, *Fast Software Encryption, FSE 2002, February 2002,* volume 2365 of *Lecture Notes in Computer Science,* Springer, pages 100–111, 2002.

310. IEEE Computer Society. Wireless LAN Medium Access Control and Physical Layer Specifications (802.11). Information available via standards.ieee.org, 1999.

311. H. Imai and A. Yamagishi. CRYPTREC project - cryptographic evaluation project for the Japanese electronic government. In T. Okamoto, editor, *Advances in Cryptology - ASIACRYPT 2000*, volume 1976 of *Lecture Notes in Computer Science*, Springer, pages 399–400, 2000.

312. S. Indesteege. The LANE hash function. Available from [542].

313. S. Indesteege, N. Keller, O. Dunkelman, E. Biham, and B. Preneel. A practical attack on KeeLoq. In N. Smart, editor, *Advances in Cryptology - EUROCRYPT 2008*, volume 4965 of *Lecture Notes in Computer Science*, Springer, pages 1–18, 2007.

314. ISO/IEC. Banking - key management (wholesale), ISO 8732. International Organization for Standardization, 1988.

315. ISO/IEC. Information technology - security techniques - data integrity mechanism using a cryptographic check function employing a block cipher algorithm, ISO/IEC 9797. International Organization for Standardization, 1994.

316. T. Iwata. New blockcipher modes of operation with beyond the birthday bound security. In M.J.B. Robshaw, editor, *Fast Software Encryption, FSE 2006*, volume 4047 of *Lecture Notes in Computer Science*, Springer, pages 310–327, 2006.

317. T. Iwata and T. Kohno. New security proofs for the 3GPP confidentiality and integrity algorithms. In W. Meier and B. Roy, editors, *Fast Software Encryption, FSE 2004*, volume 3017 of *Lecture Notes in Computer Science*, Springer, pages 427–445, 2004.

318. T. Iwata and K. Kurosawa. Probabilistic higher order differential attack and higher order bent functions. In K-Y. Lam, E. Okamoto and C. Xing, editors, *Advances in Cryptology - ASIACRYPT '99*, volume 1716 of *Lecture Notes in Computer Science*, Springer, pages 62–74, 1999.

319. T. Iwata and K. Kurosawa. On the pseudorandomness of the AES finalists - RC6 and Serpent. In B. Schneier, editor, *Fast Software Encryption, FSE 2000*, volume 1978 of *Lecture Notes in Computer Science*, Springer, pages 231–243, 2000.

320. T. Iwata and K. Kurosawa. OMAC: One-Key CBC MAC. In T. Johansson, editor, *Fast Software Encryption, FSE 2003*, volume 2887 of *Lecture Notes in Computer Science*, Springer, pages 129–153, 2003.

321. T. Iwata and K. Kurosawa. How to enhance the security of the 3GPP confidentiality and integrity algorithms. In H. Gilbert and H. Handschuh, editors, *Fast Software Encryption, FSE 2005*, volume 3557 of *Lecture Notes in Computer Science*, Springer, pages 268–283, 2005.

322. T. Iwata, T. Yoshino, and K. Kurosawa. Non-cryptographic primitive for pseudorandom permutation. In J. Daemen and V. Rijmen, editors, *Fast Software Encryption, FSE 2002, February 2002*, volume 2365 of *Lecture Notes in Computer Science*, Springer, pages 149–163, 2002.

323. T. Iwata, T. Yoshino, T. Yuasa, and K. Kurosawa. Round security and super pseudorandomness of MISTY type structure. In M. Matsui, editor, *Fast Software Encryption, FSE 2001*, volume 2355 of *Lecture Notes in Computer Science*, Springer, pages 233–247, 2001.

324. G. Jakimoski and Y. Desmedt. Related-key differential cryptanalysis of 192-bit key AES variants. In M. Matsui and R.J. Zuccherato, editors, *Selected Areas in Cryptography, SAC 2003*, volume 3006 of *Lecture Notes in Computer Science*, Springer, pages 208–221, 2004.

325. G. Jakimoski and K. P. Subbalakshmi. On efficient message authentication via block cipher design techniques. In K. Kurosawa, editor, *Advances in Cryptology - ASIACRYPT 2007*, volume 4833 of *Lecture Notes in Computer Science*, Springer, pages 232–248, 2007.

326. T. Jakobsen. Cryptanalysis of block ciphers with probabilistic non-linear relations of low degree. In H. Krawczyk, editor, *Advances in Cryptology - CRYPTO '98*, volume 1462 of *Lecture Notes in Computer Science*, Springer, pages 212–222, 1998.

327. T. Jakobsen and L. R. Knudsen. The interpolation attack on block ciphers. In E. Biham, editor, *Fast Software Encryption, FSE 1997*, volume 1267 of *Lecture Notes in Computer Science*, Springer, pages 28–40, 1997.

328. T. Jakobsen and L. R. Knudsen. Attacks on block ciphers of low algebraic degree. *J. Cryptology*, 14(3):197–210, 2001.

329. C. J. A. Jansen and D. E. Boekee. Modes of blockcipher algorithms and their protection against active eavesdropping. In D. Chaum, W.L. Price, editors, *Advances in Cryptology - EUROCRYPT '87*, volume 304 of *Lecture Notes in Computer Science*, Springer, pages 281–286, 1988.

330. E. Jaulmes, A. Joux, and F. Valette. On the security of randomized CBC-MAC beyond the birthday paradox limit: A new construction. In J. Daemen and V. Rijmen, editors, *Fast Software Encryption, FSE 2002, February 2002*, volume 2365 of *Lecture Notes in Computer Science*, Springer, pages 237–251, 2002.

331. J. Jonsson. On the security of CTR + CBC-MAC. In K. Nyberg and H.M. Heys, editors, *Selected Areas in Cryptography, SAC 2002*, volume 2595 of *Lecture Notes in Computer Science*, Springer, pages 76–93, 2003.

332. A. Joux. Cryptanalysis of the EMD mode of operation. In E. Biham, editor, *Advances in Cryptology - EUROCRYPT 2003*, volume 2656 of *Lecture Notes in Computer Science*, Springer, pages 1–16, 2003.

333. A. Joux, G. Martinet, and F. Valette. Blockwise-adaptive attackers: Revisiting the (in)security of some provably secure encryption models: CBC, GEM, IACBC. In M. Yung, editor, *Advances in Cryptology - CRYPTO 2002*, volume 2442 of *Lecture Notes in Computer Science*, Springer, pages 17–30, 2002.

334. R. R. Jueneman. Analysis of certain aspects of output feedback mode. In D. Chaum, R.L. Rivest, and A.T. Sherman, editors, *Advances in Cryptology: Proceedings of CRYPTO '82*, Plenum, pages 99–127, 1983.

335. P. Junod. On the complexity of Matsui's attack. In S. Vaudenay and A.M. Youssef, editors, *Selected Areas in Cryptography, SAC 2001*, volume 2259 of *Lecture Notes in Computer Science*, Springer, pages 199–211, 2001.

336. P. Junod. On the optimality of linear, differential, and sequential distinguishers. In E. Biham, editor, *Advances in Cryptology - EUROCRYPT 2003*, volume 2656 of *Lecture Notes in Computer Science*, Springer, pages 17–32, 2003.

337. P. Junod. New attacks against reduced-round versions of IDEA. In H. Gilbert and H. Handschuh, editors, *Fast Software Encryption, FSE 2005*, volume 3557 of *Lecture Notes in Computer Science*, Springer, pages 384–397, 2005.

338. P. Junod and M. Macchetti. Revisiting the IDEA philosophy. In O. Dunkelman, editor, *Fast Software Encryption, FSE 2009*, volume 5665 of *Lecture Notes in Computer Science*, Springer, pages 277–295, 2009.

339. P. Junod and S. Vaudenay. Optimal key ranking procedures in a statistical cryptanalysis. In T. Johansson, editor, *Fast Software Encryption, FSE 2003*, volume 2887 of *Lecture Notes in Computer Science*, Springer, pages 235–246, 2003.

340. P. Junod and S. Vaudenay. FOX : A new family of block ciphers. In H. Handschuh and M.A. Hasan, editors, *Selected Areas in Cryptography, SAC 2004*, volume 3357 of *Lecture Notes in Computer Science*, Springer, pages 114–129, 2004.

341. P. Junod and S. Vaudenay. Perfect diffusion primitives for block ciphers. In H. Handschuh and M.A. Hasan, editors, *Selected Areas in Cryptography, SAC 2004*, volume 3357 of *Lecture Notes in Computer Science*, Springer, pages 84–99, 2004.

342. C. S. Jutla. Generalized birthday attacks on unbalanced Feistel networks. In H. Krawczyk, editor, *Advances in Cryptology - CRYPTO '98*, volume 1462 of *Lecture Notes in Computer Science*, Springer, pages 186–199, 1998.

343. C. S. Jutla. Lower bound on linear authenticated encryption. In M. Matsui and R.J. Zuccherato, editors, *Selected Areas in Cryptography, SAC 2003*, volume 3006 of *Lecture Notes in Computer Science*, Springer, pages 348–360, 2004.

344. C. S. Jutla. Encryption modes with almost free message integrity. *J. Cryptology*, 21(4):547–578, 2008.

345. D. Kahn. *The Codebreakers: The Story of Secret Writing*. Macmillan Press, 1967.

346. B. S. Kaliski Jr., R. L. Rivest, and A. T. Sherman. Is the Data Encryption Standard a group? (preliminary abstract). In F. Pichler, editor, *Advances in Cryptology - EUROCRYPT '85*, volume 219 of *Lecture Notes in Computer Science*, Springer, pages 81–95, 1984.

347. B. S. Kaliski Jr., R. L. Rivest, and A. T. Sherman. Is DES a pure cipher? (results of more cycling experiments on DES). In H.C. Williams, editor, *Advances in Cryptology - CRYPTO '85*, volume 218 of *Lecture Notes in Computer Science*, Springer, pages 212–226, 1986.

348. B. S. Kaliski Jr., R. L. Rivest, and A. T. Sherman. Is the Data Encryption Standard a group? (results of cycling experiments on DES). *J. Cryptology*, 1(1):3–36, 1988.

349. B. S. Kaliski Jr. and M. J. B. Robshaw. Fast block cipher proposal. In R. Anderson, editor, *Fast Software Encryption, FSE 1993*, volume 809 of *Lecture Notes in Computer Science*, Springer, pages 33–40, 1994.

350. B. S. Kaliski Jr. and M. J. B. Robshaw. Linear cryptanalysis using multiple approximations. In Y.G. Desmedt, editor, *Advances in Cryptology - CRYPTO '94*, volume 839 of *Lecture Notes in Computer Science*, Springer, pages 26–39, 1994.

351. B. S. Kaliski Jr. and M. J. B. Robshaw. Linear cryptanalysis using multiple approximations and FEAL. In B. Preneel, editor, *Fast Software Encryption, FSE 1994*, volume 1008 of *Lecture Notes in Computer Science*, Springer, pages 249–264, 1995.

352. B. S. Kaliski Jr. and Y. L. Yin. On differential and linear cryptanalysis of the RC5 encryption algorithm. In D. Coppersmith, editor, *Advances in Cryptology - CRYPTO '95*, volume 963 of *Lecture Notes in Computer Science*, Springer, pages 171–184, 1995.

353. M. Kanda. Practical security evaluation against differential and linear cryptanalyses for Feistel ciphers with SPN round function. In D.R. Stinson and S.E. Tavares, editors, *Selected Areas in Cryptography, SAC 2000*, volume 2012 of *Lecture Notes in Computer Science*, Springer, pages 324–338, 2001.

354. M. Kanda and T. Matsumoto. Security of Camellia against truncated differential cryptanalysis. In M. Matsui, editor, *Fast Software Encryption, FSE 2001*, volume 2355 of *Lecture Notes in Computer Science*, Springer, pages 286–299, 2001.

355. M. Kanda, Y. Takashima, T. Matsumoto, K. Aoki, and K. Ohta. A strategy for constructing fast round functions with practical security against differential and linear cryptanalysis. In S.E. Tavares and H. Meijer, editors, *Selected Areas in Cryptography, SAC 1998*, volume 1556 of *Lecture Notes in Computer Science*, Springer, pages 264–279, 1999.

356. T. Kaneko. A known-plaintext attack of FEAL-4 based on the system of linear equations on difference. In H. Imai, R.L. Rivest and T. Matsumoto, editors, *Advances in Cryptology - ASIACRYPT '91*, volume 739 of *Lecture Notes in Computer Science*, Springer, pages 485–488, 1993.

357. J.-P. Kaps. Chai-tea, cryptographic hardware implementations of xTEA. In D. R. Chowdhury, V. Rijmen, and A. Das, editors, *Progress in Cryptology - INDOCRYPT 2008*, volume 5365 of *Lecture Notes in Computer Science*, pages 363–375. Springer, 2008.

358. J.-P. Kaps and C. Paar. Fast DES implementation for FPGAs and its application to a universal key-search machine. In S.E. Tavares and H. Meijer, editors, *Selected Areas in Cryptography, SAC 1998*, volume 1556 of *Lecture Notes in Computer Science*, Springer, pages 234–247, 1999.

359. L. Keliher. Refined analysis of bounds related to linear and differential cryptanalysis for the AES. In V. Rijmen, H. Dobbertin, and A. Sowa, editors, *Advanced Encryption Standard - AES, Fourth International Conference*, volume 3373 of *Lecture Notes in Computer Science, Springer*, pages 42–57, 2005.

360. L. Keliher, H. Meijer, and S. E. Tavares. Modeling linear characteristics of substitution-permutation networks. In H.M. Heys and C.M. Adams, editors, *Selected Areas in Cryptography, SAC 1999*, volume 1758 of *Lecture Notes in Computer Science*, Springer, pages 78–91, 2000.

361. L. Keliher, H. Meijer, and S. E. Tavares. Improving the upper bound on the maximum average linear hull probability for Rijndael. In S. Vaudenay and A.M. Youssef, editors, *Selected Areas in Cryptography, SAC 2001*, volume 2259 of *Lecture Notes in Computer Science*, Springer, pages 112–128, 2001.

362. L. Keliher, H. Meijer, and S. E. Tavares. New method for upper bounding the maximum average linear hull probability for SPNs. In B. Pfitzmann, editor, *Advances in Cryptology - EUROCRYPT 2001*, volume 2045 of *Lecture Notes in Computer Science*, Springer, pages 420–436, 2001.

363. J. Kelsey. NIST and the AES in 2004. In V. Rijmen, H. Dobbertin, and A. Sowa, editors, *Advanced Encryption Standard - AES, Fourth International Conference*, volume 3373 of *Lecture Notes in Computer Science, Springer*, 2005.

364. J. Kelsey, T. Kohno, and B. Schneier. Amplified boomerang attacks against reduced-round MARS and Serpent. In B. Schneier, editor, *Fast Software Encryption, FSE 2000*, volume 1978 of *Lecture Notes in Computer Science*, Springer, pages 75–93, 2000.

365. J. Kelsey and B. Schneier. Key-schedule cryptanalysis of DEAL. In H.M. Heys and C.M. Adams, editors, *Selected Areas in Cryptography, SAC 1999*, volume 1758 of *Lecture Notes in Computer Science*, Springer, pages 118–134, 2000.

366. J. Kelsey, B. Schneier, and D. Wagner. Key-schedule cryptanalysis of IDEA, G-DES, GOST, SAFER, and Triple-DES. In N. Koblitz, editor, *Advances in Cryptology - CRYPTO '96*, volume 1109 of *Lecture Notes in Computer Science*, Springer, pages 237–251, 1996.

367. J. Kelsey, B. Schneier, and D. Wagner. Mod n cryptanalysis, with applications against RC5P and M6. In L.R. Knudsen, editor, *Fast Software Encryption, FSE 1999*, volume 1636 of *Lecture Notes in Computer Science*, Springer, pages 139–155, 1999.

368. M. M. Kermani and A. Reyhani-Masoleh. A lightweight concurrent fault detection scheme for the AES S-boxes using normal basis. In E. Oswald and P. Rohatgi, editors, *Cryptographic Hardware and Embedded Systems - CHES 2008*, volume 5154 of *Lecture Notes in Computer Science*, Springer, 2008, pages 113–129, 2008.

369. D. Khovratovich, A. Biryukov, and I. Nikolić. The hash function Cheetah. Available from [542].

370. J. Kilian and P. Rogaway. How to protect DES against exhaustive key search. In N. Koblitz, editor, *Advances in Cryptology - CRYPTO '96*, volume 1109 of *Lecture Notes in Computer Science*, Springer, pages 252–267, 1996.

371. J. Kilian and P. Rogaway. How to protect DES against exhaustive key search (an analysis of DESX). *J. Cryptology*, 14(1):17–35, 2001.

372. J. Kim, S. Hong, and B. Preneel. Related-key rectangle attacks on reduced AES-192 and AES-256. In A. Biryukov, editor, *Fast Software Encryption, FSE 2007*, volume 4593 of *Lecture Notes in Computer Science*, Springer, pages 225–241, 2007.

373. J. Kim, D. Moon, W. Lee, S. Hong, S. Lee, and S. Jung. Amplified boomerang attack against reduced-round SHACAL. In Y. Zheng, editor, *Advances in Cryptology - ASIACRYPT 2002*, volume 2501 of *Lecture Notes in Computer Science*, Springer, pages 243–253, 2002.

374. K. Kim. Construction of DES-like S-boxes based on Boolean functions satisfying the SAC. In H. Imai, R.L. Rivest and T. Matsumoto, editors, *Advances in Cryptology - ASIACRYPT '91*, volume 739 of *Lecture Notes in Computer Science*, Springer, pages 59–72, 1993.

375. K. Kim, S. Lee, S. Park, and D. Lee. Securing DES S-boxes against three robust cryptanalysis. In *Workshop Record of SAC'95 (Selected Areas in Cryptology), May 18-19, 1995, Carleton Univ, Canada*, pages 145–157, 1995.

376. K. Kim, T. Matsumoto, and H. Imai. A recursive construction method of S-boxes satisfying strict avalanche criterion. In A.J. Menezes and S.A. Vanstone, editors, *Advances in Cryptology - CRYPTO '90*, volume 537 of *Lecture Notes in Computer Science*, Springer, pages 564–574, 1991.

377. K. Kim, S. Park, and S. Lee. The reconstruction of s^2-DES s-boxes and their immunity to differential cryptanalysis. In *Proceedings of the 1993 Japan-Korea Workshop on Information Security and Cryptography, Seoul, Korea*, pages 282–291, October 1993.

378. A. Kipnis and A. Shamir. Cryptanalysis of the HFE public key cryptosystem by relinearization. In M. Wiener, editor, *Advances in Cryptology - CRYPTO '99*, volume 1666 of *Lecture Notes in Computer Science*, Springer, pages 19–30, 1999.

379. L. R. Knudsen. Cryptanalysis of LOKI. In H. Imai, R.L. Rivest and T. Matsumoto, editors, *Advances in Cryptology - ASIACRYPT '91*, volume 739 of *Lecture Notes in Computer Science*, Springer, pages 22–35, 1993.

380. L. R. Knudsen. Cryptanalysis of LOKI91. In J. Seberry and Y. Zheng, editors, *Advances in Cryptology - ASIACRYPT '92*, volume 718 of *Lecture Notes in Computer Science*, Springer, pages 196–208, 1993.

381. L. R. Knudsen. Iterative characteristics of DES and s^2-DES. In E.F. Brickell, editor, *Advances in Cryptology - CRYPTO '92*, volume 740 of *Lecture Notes in Computer Science*, Springer, pages 497–511, 1993.

382. L. R. Knudsen. *Block Ciphers – Analysis, Design and Applications*. PhD thesis, Aarhus University, Denmark, 1994.

383. L. R. Knudsen. Practically secure Feistel ciphers. In R. Anderson, editor, *Fast Software Encryption, FSE 1993*, volume 809 of *Lecture Notes in Computer Science*, Springer, pages 211–221, 1994.

384. L. R. Knudsen. A key-schedule weakness in SAFER K-64. In D. Coppersmith, editor, *Advances in Cryptology - CRYPTO '95*, volume 963 of *Lecture Notes in Computer Science*, Springer, pages 274–286, 1995.

385. L. R. Knudsen. New potentially 'weak' keys for DES and LOKI (extended abstract). In A. De Santis, editor, *Advances in Cryptology - EUROCRYPT '94*, volume 950 of *Lecture Notes in Computer Science*, Springer, pages 419–424, 1995.

386. L. R. Knudsen. Truncated and higher order differentials. In B. Preneel, editor, *Fast Software Encryption, FSE 1994*, volume 1008 of *Lecture Notes in Computer Science*, Springer, pages 196–211, 1995.

387. L. R. Knudsen. A chosen-text attack on CBC-MAC. *Electronics Letters*, 33(1):48, 1997.

388. L. R. Knudsen. DEAL - a 128-bit block cipher. Technical Report 151, Department of Informatics, University of Bergen, Norway, February 1998. Submitted as an AES candidate by Richard Outerbridge. Available via www.nist.gov/aes.

389. L. R. Knudsen. A detailed analysis of SAFER K. *J. Cryptology*, 13(4):417–436, 2000.

390. L. R. Knudsen. The security of Feistel ciphers with six rounds or less. *J. Cryptology*, 15(3):207–222, 2002.

391. L. R. Knudsen and T. A. Berson. Truncated differentials of SAFER. In D. Gollmann, editor, *Fast Software Encryption, FSE 1996*, volume 1039 of *Lecture Notes in Computer Science*, Springer, pages 15–26, 1996.

392. L. R. Knudsen and X. Lai. New attacks on all double block length hash functions of hash rate 1, including the Parallel-DM. In A. De Santis, editor, *Advances in Cryptology - EUROCRYPT '94*, volume 950 of *Lecture Notes in Computer Science*, Springer, pages 410–418, 1995.

393. L. R. Knudsen, X. Lai, and B. Preneel. Attacks on fast double block length hash functions. *Journal of Cryptology*, 11(1):59–72, 1998.

394. L. R. Knudsen and J. E. Mathiassen. A chosen-plaintext linear attack on DES. In B. Schneier, editor, *Fast Software Encryption, FSE 2000*, volume 1978 of *Lecture Notes in Computer Science*, Springer, pages 262–272, 2000.

395. L. R. Knudsen and W. Meier. Improved differential attacks on RC5. In N. Koblitz, editor, *Advances in Cryptology - CRYPTO '96*, volume 1109 of *Lecture Notes in Computer Science*, Springer, pages 216–228, 1996.

396. L. R. Knudsen and W. Meier. Correlations in RC6 with a reduced number of rounds. In B. Schneier, editor, *Fast Software Encryption, FSE 2000*, volume 1978 of *Lecture Notes in Computer Science*, Springer, pages 94–108, 2000.

397. L. R. Knudsen, F. Mendel, C. Rechberger, and S. S. Thomsen. Cryptanalysis of MDC-2. In A. Joux, editor, *Advances in Cryptology - EUROCRYPT 2009*, volume 5479 of *Lecture Notes in Computer Science*, Springer, pages 106–120, 2009.

398. L. R. Knudsen and C. J. Mitchell. Partial key recovery attack against RMAC. *J. Cryptology*, 18(4):375–389, 2005.

399. L. R. Knudsen and B. Preneel. Hash functions based on block ciphers and quaternary codes. In K. Kim and T. Matsumoto, editors, *Advances in Cryptology - ASIACRYPT '96*, volume 1163 of *Lecture Notes in Computer Science*, Springer, pages 77–90, 1996.

400. L. R. Knudsen and H. Raddum. Distinguishing attack on five-round Feistel networks. *Electronics Letters*, 39(16):1175–1177, 2003.

401. L. R. Knudsen and V. Rijmen. On the decorrelated fast cipher (DFC) and its theory. In L.R. Knudsen, editor, *Fast Software Encryption, FSE 1999,* volume 1636 of *Lecture Notes in Computer Science,* Springer, pages 81–94, 1999.

402. L. R. Knudsen and V. Rijmen. Weaknesses in LOKI'97. Presented at the 2nd AES Candidate Conference, March, 1999.

403. L. R. Knudsen and V. Rijmen. Ciphertext-only attack on Akelarre. *Cryptologia,* XXIV (2):135–147, April 2000.

404. L. R. Knudsen and V. Rijmen. Known-key distinguishers for some block ciphers. In K. Kurosawa, editor, *Advances in Cryptology - ASIACRYPT 2007,* volume 4833 of *Lecture Notes in Computer Science,* Springer, pages 315–324, 2007.

405. L. R. Knudsen, V. Rijmen, R. L. Rivest, and M. J. B. Robshaw. On the design and security of RC2. In S. Vaudenay, editor, *Fast Software Encryption, FSE 1998,* volume 1372 of *Lecture Notes in Computer Science,* Springer, pages 206–221, 1998.

406. L. R. Knudsen and M. J. B. Robshaw. Non-linear approximations in linear cryptanalysis. In U. Maurer, editor, *Advances in Cryptology - EUROCRYPT '96,* volume 1070 of *Lecture Notes in Computer Science,* Springer, pages 224–236, 1996.

407. L. R. Knudsen, M. J. B. Robshaw, and D. Wagner. Truncated differentials and Skipjack. In M. Wiener, editor, *Advances in Cryptology - CRYPTO '99,* volume 1666 of *Lecture Notes in Computer Science,* Springer, pages 165–180, 1999.

408. L. R. Knudsen and D. Wagner. On the structure of Skipjack. *Discrete Applied Mathematics,* 11(1-2):103–116, 2001.

409. L. R. Knudsen and D. Wagner. Integral cryptanalysis. In J. Daemen and V. Rijmen, editors, *Fast Software Encryption, FSE 2002, February 2002,* volume 2365 of *Lecture Notes in Computer Science,* Springer, pages 112–127, 2002.

410. Y. Ko, S. Hong, W. Lee, S. Lee, and J.-S. Kang. Related key differential attacks on 27 rounds of XTEA and full-round GOST. In W. Meier and B. Roy, editors, *Fast Software Encryption, FSE 2004,* volume 3017 of *Lecture Notes in Computer Science,* Springer, pages 299–316, 2004.

411. P. C. Kocher. Timing attacks on implementations of Diffie-Hellman, RSA, DSS, and other systems. In N. Koblitz, editor, *Advances in Cryptology - CRYPTO '96,* volume 1109 of *Lecture Notes in Computer Science,* Springer, pages 104–113, 1996.

412. P. C. Kocher, J. Jaffe, and B. Jun. Differential power analysis. In M. Wiener, editor, *Advances in Cryptology - CRYPTO '99,* volume 1666 of *Lecture Notes in Computer Science,* Springer, pages 388–397, 1999.

413. T. Kohno, J. Viega, and D. Whiting. CWC: A high-performance conventional authenticated encryption mode. In W. Meier and B. Roy, editors, *Fast Software Encryption, FSE 2004,* volume 3017 of *Lecture Notes in Computer Science,* Springer, pages 408–426, 2004.

414. A. G. Konheim. *Cryptography: A Primer.* John Wiley & Sons, 1981.

415. M. Kounavis and S. Gueron. Vortex: A new family of one way hash functions based on Rijndael rounds and carry-less multiplication. Available from [542].

416. K. Koyama and R. Terada. How to strengthen DES-like cryptosystems against differential cryptanalysis. *IEICE Transactions on Fundamentals of Electronics, Communications and Computer Science,* E76-A(1):63–69, 1993.

417. U. Kühn. Cryptanalysis of reduced-round MISTY. In B. Pfitzmann, editor, *Advances in Cryptology - EUROCRYPT 2001,* volume 2045 of *Lecture Notes in Computer Science,* Springer, pages 325–339, 2001.

418. U. Kühn. Improved cryptanalysis of MISTY1. In J. Daemen and V. Rijmen, editors, *Fast Software Encryption, FSE 2002, February 2002,* volume 2365 of *Lecture Notes in Computer Science,* Springer, pages 61–75, 2002.

419. S. Kumar, C. Paar, J. Pelzl, G. Pfeiffer, and M. Schimmler. Breaking ciphers with COPACOBANA - a cost-optimized parallel code breaker. In L. Goubin and M. Matsui, editors, *Cryptographic Hardware and Embedded Systems - CHES 2006,* volume 4249 of *Lecture Notes in Computer Science,* Springer, 2006, pages 101–118, 2006.

420. S. Kunz-Jacques and F. Muller. New improvements of Davies-Murphy cryptanalysis. In B.K. Roy, editor, *Advances in Cryptology - ASIACRYPT 2005*, volume 3788 of *Lecture Notes in Computer Science*, Springer, pages 425–442, 2005.

421. S. Kunz-Jacques, F. Muller, and F. Valette. The Davies-Murphy power attack. In P.J. Lee, editor, *Advances in Cryptology - ASIACRYPT 2004*, volume 3329 of *Lecture Notes in Computer Science*, Springer, pages 451–467, 2004.

422. H. Kuo and I. Verbauwhede. Architectural optimization for a 1.82Gbits/sec VLSI implementation of the AES Rijndael algorithm. In Ç.K. Koç, D. Naccache and C. Paar, editors, *Cryptographic Hardware and Embedded Systems - CHES 2001*, volume 2162 of *Lecture Notes in Computer Science*, Springer, 2001, pages 51–64, 2001.

423. K. Kurosawa and T. Iwata. TMAC: Two-key CBC MAC. In M. Joye, editor, *Topics in Cryptology - CT-RSA 2003*, volume 2612 of *Lecture Notes in Computer Science*, pages 33–49. Springer, 2003.

424. K. Kurosawa, T. Iwata, and Q. V. Duong. Root finding interpolation attack. In D.R. Stinson and S.E. Tavares, editors, *Selected Areas in Cryptography, SAC 2000*, volume 2012 of *Lecture Notes in Computer Science*, Springer, pages 303–314, 2001.

425. K. Kurosawa and T. Satoh. Generalization of higher order SAC to vector output Boolean functions. In K. Kim and T. Matsumoto, editors, *Advances in Cryptology - ASIACRYPT '96*, volume 1163 of *Lecture Notes in Computer Science*, Springer, pages 218–231, 1996.

426. K. Kusuda and T. Matsumoto. Optimization of time-memory trade-off cryptanalysis and its application to DES, FEAL-32, and Skipjack. *IEICE Trans. Fundamentals*, E79-A(1), January 1996.

427. M. Kwan. Simultaneous attacks in differential cryptanalysis (getting more pairs per encryption). In H. Imai, R.L. Rivest and T. Matsumoto, editors, *Advances in Cryptology - ASIACRYPT '91*, volume 739 of *Lecture Notes in Computer Science*, Springer, pages 489–492, 1993.

428. M. Kwan. The design of the ICE encryption algorithm. In E. Biham, editor, *Fast Software Encryption, FSE 1997*, volume 1267 of *Lecture Notes in Computer Science*, Springer, pages 69–82, 1997.

429. X. Lai. On the design and security of block ciphers. In J. L. Massey, editor, *ETH Series in Information Processing*, volume 1. Hartung-Gorre Verlag, Konstanz, 1992.

430. X. Lai. Higher order derivatives and differential cryptanalysis. In R. Blahut, editor, *Communication and Cryptography, Two Sides of One Tapestry*. Kluwer Academic Publishers, 1994. ISBN 0-7923-9469-0.

431. X. Lai and J. L. Massey. A proposal for a new block encryption standard. In I.B. Damgård, editor, *Advances in Cryptology - EUROCRYPT '90*, volume 473 of *Lecture Notes in Computer Science*, Springer, pages 389–404, 1991.

432. X. Lai and J. L. Massey. Hash function based on block ciphers. In R. Rueppel, editor, *Advances in Cryptology - EUROCRYPT '92*, volume 658 of *Lecture Notes in Computer Science*, Springer, pages 55–70, 1993.

433. X. Lai, J. L. Massey, and S. Murphy. Markov ciphers and differential cryptanalysis. In D.W. Davies, editor, *Advances in Cryptology - EUROCRYPT '91*, volume 547 of *Lecture Notes in Computer Science*, Springer, pages 17–38, 1992.

434. X. Lai and R. A. Rueppel. Attacks on the HKM/HFX cryptosystem. In D. Gollmann, editor, *Fast Software Encryption, FSE 1996*, volume 1039 of *Lecture Notes in Computer Science*, Springer, pages 1–14, 1996.

435. S. Landau. Standing the test of time: the Data Encryption Standard. *Notices of the American Mathematical Society*, pages 341–349, March 2000.

436. S. Landau. Polynomials in the nation's service: Using algebra to design the Advanced Encryption Standard. *American Mathematical Monthly*, pages 89–117, February 2004.

437. S. K. Langford and M. E. Hellman. Differential-linear cryptanalysis. In Y.G. Desmedt, editor, *Advances in Cryptology - CRYPTO '94*, volume 839 of *Lecture Notes in Computer Science*, Springer, pages 17–25, 1994.

438. G. Leander, C. Paar, A. Poschmann, and K. Schramm. New lightweight DES variants. In A. Biryukov, editor, *Fast Software Encryption, FSE 2007*, volume 4593 of *Lecture Notes in Computer Science*, Springer, pages 196–210, 2007.

439. D. Lefranc, P. Painchault, V. Rouat, and E. Mayer. A generic method to design modes of operation beyond the birthday bound. In C.M. Adams, A. Miri, and M.J. Wiener, editors, *Selected Areas in Cryptography, SAC 2007*, volume 4876 of *Lecture Notes in Computer Science*, Springer, pages 328–343, 2007.

440. D. Lei, L. Chao, and K. Feng. New observation on Camellia. In B. Preneel and S.E. Tavares, editors, *Selected Areas in Cryptography, SAC 2005*, volume 3897 of *Lecture Notes in Computer Science*, Springer, pages 51–64, 2006.

441. H. Leitold, W. Mayerwieser, U. Payer, K. C. Posch, R. Posch, and J. Wolkerstorfer. A 155 mbps Triple-DES network encryptor. In Ç.K. Koç and C. Paar, editors, *Cryptographic Hardware and Embedded Systems - CHES 2000*, volume 1965 of *Lecture Notes in Computer Science*, Springer, 2000, pages 164–174, 2000.

442. K. Lemke, K. Schramm, and C. Paar. DPA on n-bit sized Boolean and arithmetic operations and its application to IDEA, RC6, and the HMAC-construction. In M. Joye and J.-J. Quisquater, editors, *Cryptographic Hardware and Embedded Systems - CHES 2004*, volume 3156 of *Lecture Notes in Computer Science*, Springer, 2004, pages 205–219, 2004.

443. N. Li and W.-F. Qi. Construction and analysis of Boolean functions of $2t+1$ variables with maximum algebraic immunity. In X. Lai and K. Chen, editors, *Advances in Cryptology - ASIACRYPT 2006*, volume 4284 of *Lecture Notes in Computer Science*, Springer, pages 84–98, 2006.

444. C. H. Lim. A revised version of Crypton - Crypton v1.0. In L.R. Knudsen, editor, *Fast Software Encryption, FSE 1999*, volume 1636 of *Lecture Notes in Computer Science*, Springer, pages 31–45, 1999.

445. C. H. Lim and H. S. Hwang. CRYPTON: A new 128-bit block cipher - specification and analysis (version 1.0). Submitted as candidate for AES. Available via www.nist.gov/aes, 1997.

446. C. H. Lim and T. Korkishko. mCrypton - a lightweight block cipher for security of low-cost RFID tags and sensors. In J. Song, T. Kwon, and M. Yung, editors, *Information Security Applications, 6th International Workshop, WISA 2005*, volume 3786 of *Lecture Notes in Computer Science*, pages 243–258. Springer, 2006.

447. C.-W. Lim and K. Khoo. An analysis of XSL applied to BES. In A. Biryukov, editor, *Fast Software Encryption, FSE 2007*, volume 4593 of *Lecture Notes in Computer Science*, Springer, pages 242–253, 2007.

448. H. Lipmaa. IDEA: A cipher for multimedia architectures. In S.E. Tavares and H. Meijer, editors, *Selected Areas in Cryptography, SAC 1998*, volume 1556 of *Lecture Notes in Computer Science*, Springer, pages 248–263, 1999.

449. M. Liskov, R. L. Rivest, and D. Wagner. Tweakable block ciphers. In M. Yung, editor, *Advances in Cryptology - CRYPTO 2002*, volume 2442 of *Lecture Notes in Computer Science*, Springer, pages 31–46, 2002.

450. F. Liu, W. Ji, L. Hu, J. Ding, S. Lv, A. Pyshkin, and R.-P. Weinmann. Analysis of the SMS4 block cipher. In J. Pieprzyk, H. Ghodosi, and E. Dawson, editors, *Proceedings of ACISP 2007*, volume 4586 of *Lecture Notes in Computer Science*, Springer, pages 158–170, 2007.

451. J. Lu. Attacking reduced-round versions of the SMS4 block cipher in the Chinese WAPI standard. In S. Qing, H. Imai, and G. Wang, editors, *Proceedings of ICICS 2007*, volume 4861 of *Lecture Notes in Computer Science*, Springer, pages 306–318, 2007.

452. Y. Lu, W. Meier, and S. Vaudenay. The conditional correlation attack: A practical attack on Bluetooth encryption. In V. Shoup, editor, *Advances in Cryptology - CRYPTO 2005*, volume 3621 of *Lecture Notes in Computer Science*, Springer, pages 97–117, 2005.

453. M. Luby and C. Rackoff. How to construct pseudorandom permutations from pseudorandom functions. *SIAM Journal of Computing*, 17(2):373–386, 1988.

454. S. Lucks. Faster Luby-Rackoff ciphers. In D. Gollmann, editor, *Fast Software Encryption, FSE 1996*, volume 1039 of *Lecture Notes in Computer Science*, Springer, pages 189–203, 1996.

455. S. Lucks. Attacking triple encryption. In S. Vaudenay, editor, *Fast Software Encryption, FSE 1998*, volume 1372 of *Lecture Notes in Computer Science*, Springer, pages 239–253, 1998.

456. S. Lucks. On security of the 128-bit block cipher DEAL. In L.R. Knudsen, editor, *Fast Software Encryption, FSE 1999*, volume 1636 of *Lecture Notes in Computer Science*, Springer, pages 60–70, 1999.

457. S. Lucks. Attacking seven rounds of Rijndael under 192-bit keys and 256-bit keys. In *Proceedings of the 3rd Advanced Encryption Standard Candidate Conference*, pages 215–229. National Institute of Standards and Technology, April 2000. Available via www.nist. gov/aes.

458. S. Lucks. The saturation attack - a bait for Twofish. In M. Matsui, editor, *Fast Software Encryption, FSE 2001*, volume 2355 of *Lecture Notes in Computer Science*, Springer, pages 1–15, 2001.

459. S. Lucks. Ciphers secure against related-key attacks. In W. Meier and B. Roy, editors, *Fast Software Encryption, FSE 2004*, volume 3017 of *Lecture Notes in Computer Science*, Springer, pages 359–370, 2004.

460. S. Lucks. Two-pass authenticated encryption faster than generic composition. In H. Gilbert and H. Handschuh, editors, *Fast Software Encryption, FSE 2005*, volume 3557 of *Lecture Notes in Computer Science*, Springer, pages 284–298, 2005.

461. S. Lucks and R. Weis. How to make DES-based smartcards fit for the 21st century. In J. Domingo-Ferrer, D. Chan, and A. Watson, editors, *CARDIS*, volume 180 of *IFIP Conference Proceedings*, pages 93–114. Kluwer, 2000.

462. A. K. Lutz, J. Treichler, F. K. Gürkaynak, H. Kaeslin, G. Basler, A. Erni, S. Reichmuth, P. Rommens, and W. F. Stephan Oetiker. 2gbit/s hardware realizations of Rijndael and Serpent: A comparative analysis. In B.S. Kaliski Jr., Ç.K. Koç and C. Paar, editors, *Cryptographic Hardware and Embedded Systems - CHES 2002*, volume 2523 of *Lecture Notes in Computer Science*, Springer, 2003, pages 144–158, 2003.

463. W. E. Madryga. A high performance encryption algorithm. *Computer Security: A Global Challenge*, pages 557–570, 1984.

464. S. S. Magliveras and N. D. Memon. Algebraic properties of cryptosystem PGM. *Journal of Cryptology*, 5(3):167–184, 1992.

465. S. Mangard, N. Pramstaller, and E. Oswald. Successfully attacking masked AES hardware implementations. In J.R. Rao and B. Sunar, editors, *Cryptographic Hardware and Embedded Systems - CHES 2005*, volume 3659 of *Lecture Notes in Computer Science*, Springer, 2005, pages 157–171, 2005.

466. G. A. Maranon, A. F. Sabater, D. G. Martnez, F. M. Vitini, and A. P. Dominguez. Akelarre: a new block cipher algorithm. In *Selected Areas in Cryptography, SAC 1996, Proceedings*, pages 1–14, 1996.

467. M. Masoumi, F. Raissi, and M. Ahmadian. NanoCMOS-molecular realization of Rijndael. In L. Goubin and M. Matsui, editors, *Cryptographic Hardware and Embedded Systems - CHES 2006*, volume 4249 of *Lecture Notes in Computer Science*, Springer, 2006, pages 285–297, 2006.

468. J. L. Massey. Strengthened key schedule for the cipher SAFER. Posted on USENET newsgroup sci.crypt, September 9, 1995.

469. J. L. Massey. Cryptography: Fundamentals and applications. Copies of transparencies, Advanced Technology Seminars, 1993.

470. J. L. Massey. SAFER K-64: A byte-oriented block-ciphering algorithm. In R. Anderson, editor, *Fast Software Encryption, FSE 1993*, volume 809 of *Lecture Notes in Computer Science*, Springer, pages 1–17, 1994.

471. J. L. Massey. SAFER K-64: One year later. In B. Preneel, editor, *Fast Software Encryption, FSE 1994*, volume 1008 of *Lecture Notes in Computer Science*, Springer, pages 212–241, 1995.

472. J. L. Massey, G. Khachatrian, and M. Kuregian. Nomination of SAFER+ as candidate algorithm for the Advanced Encryption Standard (AES). Submitted as candidate for AES. Available via www.nist.gov/aes.

473. J. L. Massey, G. Khachatrian, and M. Kuregian. Nomination of SAFER++ as candidate algorithm for the new European schemes for signatures, integrity, and encryption (NESSIE). Presented at the First Open NESSIE Workshop, November 2000.

474. J. L. Massey, U. M. Maurer, and M. Wang. Non-expanding, key-minimal, robustly-perfect, linear and bilinear ciphers. In D. Chaum, W.L. Price, editors, *Advances in Cryptology - EUROCRYPT '87*, volume 304 of *Lecture Notes in Computer Science*, Springer, pages 237–247, 1988.

475. M. Matsui. The first experimental cryptanalysis of the Data Encryption Standard. In Y.G. Desmedt, editor, *Advances in Cryptology - CRYPTO '94*, volume 839 of *Lecture Notes in Computer Science*, Springer, pages 1–11, 1994.

476. M. Matsui. Linear cryptanalysis method for DES cipher. In T. Helleseth, editor, *Advances in Cryptology - EUROCRYPT '93*, volume 765 of *Lecture Notes in Computer Science*, Springer, pages 386–397, 1994.

477. M. Matsui. On correlation between the order of S-boxes and the strength of DES. In A. De Santis, editor, *Advances in Cryptology - EUROCRYPT '94*, volume 950 of *Lecture Notes in Computer Science*, Springer, pages 366–375, 1995.

478. M. Matsui. New structure of block ciphers with provable security against differential and linear cryptanalysis. In D. Gollmann, editor, *Fast Software Encryption, FSE 1996*, volume 1039 of *Lecture Notes in Computer Science*, Springer, pages 205–218, 1996.

479. M. Matsui. New block encryption algorithm MISTY. In E. Biham, editor, *Fast Software Encryption, FSE 1997*, volume 1267 of *Lecture Notes in Computer Science*, Springer, pages 54–68, 1997.

480. M. Matsui. How far can we go on the x64 processors? In M.J.B. Robshaw, editor, *Fast Software Encryption, FSE 2006*, volume 4047 of *Lecture Notes in Computer Science*, Springer, pages 341–358, 2006.

481. M. Matsui and S. Fukuda. How to maximize software performance of symmetric primitives on Pentium III and 4 processors. In H. Gilbert and H. Handschuh, editors, *Fast Software Encryption, FSE 2005*, volume 3557 of *Lecture Notes in Computer Science*, Springer, pages 398–412, 2005.

482. M. Matsui and T. Tokita. Cryptanalysis of a reduced version of the block cipher E2. In L.R. Knudsen, editor, *Fast Software Encryption, FSE 1999*, volume 1636 of *Lecture Notes in Computer Science*, Springer, pages 71–80, 1999.

483. M. Matsui and A. Yamagishi. A new method for known plaintext attack of FEAL cipher. In R. Rueppel, editor, *Advances in Cryptology - EUROCRYPT '92*, volume 658 of *Lecture Notes in Computer Science*, Springer, pages 81–91, 1993.

484. S. M. Matyas, C. H. Meyer, and J. Oseas. Generating strong one-way functions with cryptographic algorithm. *IBM Technical Disclosure Bulletin*, 27:5658–5659, 1985.

485. U. M. Maurer. A simplified and generalized treatment of Luby-Rackoff pseudorandom permutation generator. In R. Rueppel, editor, *Advances in Cryptology - EUROCRYPT '92*, volume 658 of *Lecture Notes in Computer Science*, Springer, pages 239–255, 1993.

486. U. M. Maurer and J. L. Massey. Cascade ciphers: The importance of being first. *J. Cryptology*, 6(1):55–61, 1993.

487. U. M. Maurer, Y. A. Oswald, K. Pietrzak, and J. Sjödin. Luby-Rackoff ciphers from weak round functions? In S. Vaudenay, editor, *Advances in Cryptology - EUROCRYPT 2006*, volume 4004 of *Lecture Notes in Computer Science*, Springer, pages 391–408, 2006.

488. U. M. Maurer and K. Pietrzak. The security of many-round Luby-Rackoff pseudo-random permutations. In E. Biham, editor, *Advances in Cryptology - EUROCRYPT 2003*, volume 2656 of *Lecture Notes in Computer Science*, Springer, pages 544–561, 2003.

489. D. A. McGrew and S. R. Fluhrer. The security of the extended codebook (XCB) mode of operation. In C.M. Adams, A. Miri, and M.J. Wiener, editors, *Selected Areas in Cryptography, SAC 2007*, volume 4876 of *Lecture Notes in Computer Science*, Springer, pages 311–327, 2007.

490. M. McLoone and J. V. McCanny. High performance single-chip FPGA Rijndael algorithm implementations. In Ç.K. Koç, D. Naccache and C. Paar, editors, *Cryptographic Hardware*

and Embedded Systems - CHES 2001, volume 2162 of *Lecture Notes in Computer Science,* Springer, 2001, pages 65–76, 2001.

491. W. Meier. On the security of the IDEA block cipher. In T. Helleseth, editor, *Advances in Cryptology - EUROCRYPT '93,* volume 765 of *Lecture Notes in Computer Science,* Springer, pages 371–385, 1994.

492. H. Meijer and S. G. Akl. Two new secret key cryptosystems. In F. Pichler, editor, *Advances in Cryptology - EUROCRYPT '85,* volume 219 of *Lecture Notes in Computer Science,* Springer, pages 96–102, 1984.

493. A. J. Menezes, P. C. V. Oorschot, and S. A. Vanstone. *Handbook of Applied Cryptography.* CRC Press, 1997.

494. R. Merkle and M. E. Hellman. On the security of multiple encryption. *Communications of the ACM,* 24(7):465–467, 1981.

495. R. C. Merkle. One way hash functions and DES. In G. Brassard, editor, *Advances in Cryptology - CRYPTO '89,* volume 435 of *Lecture Notes in Computer Science,* Springer, pages 428–446, 1990.

496. R. C. Merkle. Fast software encryption functions. In A.J. Menezes and S.A. Vanstone, editors, *Advances in Cryptology - CRYPTO '90,* volume 537 of *Lecture Notes in Computer Science,* Springer, pages 476–501, 1991.

497. C. Meyer and M. Schilling. Secure program load with manipulation detection code. In *Proceedings of 6th Worldwide Congress on Computer and Communications Security and Protection SECURICOM 1988,* pages 111–130, 1988.

498. C. H. Meyer and S. M. Matyas. *A New Direction in Computer Data Security.* John Wiley & Sons, 1982.

499. K. Minematsu. Improved security analysis of XEX and LRW modes. In E. Biham and A.M. Youssef, editors, *Selected Areas in Cryptography, SAC 2006,* volume 4356 of *Lecture Notes in Computer Science,* Springer, pages 96–113, 2007.

500. K. Minematsu. Beyond-birthday-bound security based on tweakable block cipher. In O. Dunkelman, editor, *Fast Software Encryption, FSE 2009,* volume 5665 of *Lecture Notes in Computer Science,* Springer, pages 308–326, 2009.

501. M. Minier. A three rounds property of the AES. In V. Rijmen, H. Dobbertin, and A. Sowa, editors, *Advanced Encryption Standard - AES, Fourth International Conference,* volume 3373 of *Lecture Notes in Computer Science, Springer,* pages 16–26, 2005.

502. M. Minier and H. Gilbert. Stochastic cryptanalysis of Crypton. In B. Schneier, editor, *Fast Software Encryption, FSE 2000,* volume 1978 of *Lecture Notes in Computer Science,* Springer, pages 121–133, 2000.

503. S. Mister and R. J. Zuccherato. An attack on CFB mode encryption as used by OpenPGP. In B. Preneel and S.E. Tavares, editors, *Selected Areas in Cryptography, SAC 2005,* volume 3897 of *Lecture Notes in Computer Science,* Springer, pages 82–94, 2006.

504. S. Miyaguchi. The FEAL-8 cryptosystem and a call for attack. In G. Brassard, editor, *Advances in Cryptology - CRYPTO '89,* volume 435 of *Lecture Notes in Computer Science,* Springer, pages 624–627, 1990.

505. S. Miyaguchi. The FEAL cipher family. In A.J. Menezes and S.A. Vanstone, editors, *Advances in Cryptology - CRYPTO '90,* volume 537 of *Lecture Notes in Computer Science,* Springer, pages 627–638, 1991.

506. S. Miyaguchi, K. Ohta, and M. Iwata. 128-bit hash function (N-hash). *NTT Review,* 2:128–132, 1990.

507. H. Miyano. A method to estimate the number of ciphertext pairs for differential cryptanalysis. In H. Imai, R.L. Rivest and T. Matsumoto, editors, *Advances in Cryptology - ASIACRYPT '91,* volume 739 of *Lecture Notes in Computer Science,* Springer, pages 51–58, 1993.

508. A. A. Moldovyan and N. A. Moldovyan. A cipher based on data-dependent permutations. *J. Cryptology,* 15(1):61–72, 2002.

509. D. Moon, K. Hwang, W. Lee, S. Lee, and J. Lim. Impossible differential cryptanalysis of reduced round XTEA and TEA. In J. Daemen and V. Rijmen, editors, *Fast Software Encryption, FSE 2002, February 2002,* volume 2365 of *Lecture Notes in Computer Science,* Springer, pages 49–60, 2002.

510. G. E. Moore. Cramming more components onto integrated circuits. *Electronics*, 38(8), April 1965.
511. J. H. Moore and G. J. Simmons. Cycle structures of the DES with weak and semi-weak keys. In A.M. Odlyzko, editor, *Advances in Cryptology - CRYPTO '86*, volume 263 of *Lecture Notes in Computer Science*, Springer, pages 9–32, 1988.
512. T. E. Moore and S. E. Tavares. A layered approach to the design of private key cryptosystems. In H.C. Williams, editor, *Advances in Cryptology - CRYPTO '85*, volume 218 of *Lecture Notes in Computer Science*, Springer, pages 227–245, 1986.
513. A. Moradi, M. T. M. Shalmani, and M. Salmasizadeh. A generalized method of differential fault attack against AES cryptosystem. In L. Goubin and M. Matsui, editors, *Cryptographic Hardware and Embedded Systems - CHES 2006*, volume 4249 of *Lecture Notes in Computer Science*, Springer, 2006, pages 91–100, 2006.
514. S. Moriai, T. Shimoyama, and T. Kaneko. Higher order differential attak of CAST cipher. In S. Vaudenay, editor, *Fast Software Encryption, FSE 1998*, volume 1372 of *Lecture Notes in Computer Science*, Springer, pages 17–31, 1998.
515. S. Moriai, T. Shimoyama, and T. Kaneko. Higher order differential attack using chosen higher order differences. In S.E. Tavares and H. Meijer, editors, *Selected Areas in Cryptography, SAC 1998*, volume 1556 of *Lecture Notes in Computer Science*, Springer, pages 106–117, 1999.
516. S. Moriai, T. Shimoyama, and T. Kaneko. Interpolation attacks of the block cipher: SNAKE. In L.R. Knudsen, editor, *Fast Software Encryption, FSE 1999*, volume 1636 of *Lecture Notes in Computer Science*, Springer, pages 275–289, 1999.
517. S. Moriai, M. Sugita, K. Aoki, and M. Kanda. Security of E2 against truncated differential cryptanalysis. In H.M. Heys and C.M. Adams, editors, *Selected Areas in Cryptography, SAC 1999*, volume 1758 of *Lecture Notes in Computer Science*, Springer, pages 106–117, 2000.
518. S. Moriai and S. Vaudenay. On the pseudorandomness of top-level schemes of block ciphers. In T. Okamoto, editor, *Advances in Cryptology - ASIACRYPT 2000*, volume 1976 of *Lecture Notes in Computer Science*, Springer, pages 289–302, 2000.
519. S. Morioka and A. Satoh. An optimized S-box circuit architecture for low power AES design. In B.S. Kaliski Jr., Ç.K. Koç and C. Paar, editors, *Cryptographic Hardware and Embedded Systems - CHES 2002*, volume 2523 of *Lecture Notes in Computer Science*, Springer, 2003, pages 172–186, 2003.
520. H. Morita, K. Ohta, and S. Miyaguchi. Results of switching-closure-test on FEAL (extended abstract). In H. Imai, R.L. Rivest and T. Matsumoto, editors, *Advances in Cryptology - ASIACRYPT '91*, volume 739 of *Lecture Notes in Computer Science*, Springer, pages 247–252, 1993.
521. D. R. Morrison. Subtractive encryptors - alternatives to the DES. In A. Gersho, editor, *Advances in Cryptology: A report on CRYPTO '81, IEEE Workshop on Communications Security, Santa Barbara, August 24-26, 1981. U.C. Santa Barbara, Dept. of Elec. and Computer Eng., ECE Report No 82-04*, pages 42–52, 1982.
522. D. M'Raïhi, D. Naccache, J. Stern, and S. Vaudenay. XMX: A firmware-oriented block cipher based on modular multiplications. In E. Biham, editor, *Fast Software Encryption, FSE 1997*, volume 1267 of *Lecture Notes in Computer Science*, Springer, pages 166–171, 1997.
523. F. Muller. A new attack against Khazad. In C-S. Laih, editor, *Advances in Cryptology - ASIACRYPT 2003*, volume 2894 of *Lecture Notes in Computer Science*, Springer, pages 347–358, 2003.
524. S. Murphy. The cryptanalysis of FEAL-4 with 20 chosen plaintexts. *J. Cryptology*, 2(3):145–154, 1990.
525. S. Murphy. An analysis of SAFER. *J. Cryptology*, 11(4):235–251, 1998.
526. S. Murphy. The effectiveness of the linear hull effect. Technical Report RHUL-MA-2009-19, Royal Holloway, University of London, October 2009. Avaiable via www.ma.rhul.ac.uk/static/techrep/2009/.

527. S. Murphy. Overestimates for the gain of multiple linear approximations. Technical Report RHUL-MA-2009-21, Royal Holloway, University of London, October 2009. Avaiable via www.ma.rhul.ac.uk/static/techrep/2009/.

528. S. Murphy. The return of the boomerang. Technical Report RHUL-MA-2009-20, Royal Holloway, University of London, October 2009. Avaiable via www.ma.rhul.ac.uk/static/techrep/2009/.

529. S. Murphy, K. Paterson, and P. Wild. A weak cipher that generates the symmetric group. *Journal of Cryptology*, 7(1):61–65, 1994.

530. S. Murphy, F. Piper, M. Walker, and P. Wild. Maximum likelihood estimation for block cipher keys. Technical report, University of London, Royal Holloway, 1994. Available via www.isg.rhul.ac.uk/~sean/maxlik.pdf.

531. S. Murphy and M. Robshaw. Key-dependent s-boxes and differential cryptanalysis. *Designs, Codes and Cryptography*, 27(3):229–255, 2002.

532. S. Murphy and M. J. B. Robshaw. Further comments on the structure of Rijndael. August 17, 2000. Available via www.isg.rhul.ac.uk/~mrobshaw.

533. S. Murphy and M. J. B. Robshaw. New observations on Rijndael. August 7, 2000. Available via www.isg.rhul.ac.uk/~mrobshaw.

534. S. Murphy and M. J. B. Robshaw. Essential algebraic structure within the AES. In M. Yung, editor, *Advances in Cryptology - CRYPTO 2002*, volume 2442 of *Lecture Notes in Computer Science*, Springer, pages 1–16, 2002.

535. S. Murphy and M. J. B. Robshaw. Comments on the security of the AES and the XSL technique. *Electronics Letters*, 39(1):26–38, 2003.

536. J. Nakahara, P. Sepehrdad, B. Zhang, and M. Wang. Linear (hull) and algebraic cryptanalysis of the block cipher PRESENT. In J. A. Garay, A. Miyaji, and A. Otsuka, editors, *CANS 2009*, volume 5888 of *Lecture Notes in Computer Science*, pages 58–75. Springer, 2009.

537. J. Nakahara Jr., B. Preneel, and J. Vandewalle. Linear cryptanalysis of reduced-round versions of the SAFER block cipher family. In B. Schneier, editor, *Fast Software Encryption, FSE 2000*, volume 1978 of *Lecture Notes in Computer Science*, Springer, pages 244–261, 2000.

538. M. Naor and O. Reingold. On the construction of pseudorandom permutations: Luby-Rackoff revisited. *J. Cryptology*, 12(1):29–66, 1999.

539. National Bureau of Standards. Computer security and the data encryption standard. Proceedings of *Conference on Computer Security and the Data Encryption Standard, February 1977*, Special Publication 500-27, February 1978.

540. National Institute of Standards and Technology. Consultation paper on the selection of a block cipher based MAC algorithm. Available via crsc.nist.gov.

541. National Institute of Standards and Technology. Cryptographic hash algorithm competition. Available via www.nist.gov/hash-competition.

542. National Institute of Standards and Technology. First round candidates of the SHA-3 hash function competition. Available via csrc.nist.gov/groups/ST/hash/sha-3/Round1/submissions_md1.html.

543. National Institute of Standards and Technology. Modes development. Available via csrc.nist.gov/groups/ST/toolkit/BCM/modes_development.html.

544. National Institute of Standards and Technology. Rationale for the selection of the OMAC variation of XCBC. Available via crsc.nist.gov.

545. National Institute of Standards and Technology. Data encryption standard. Federal Information Processing Standard (FIPS), Publication 46, U.S. Department of Commerce, Washington D.C., January 1977.

546. National Institute of Standards and Technology. DES modes of operation. Federal Information Processing Standard (FIPS), Publication 81, U.S. Department of Commerce, Washington D.C., December 1980.

547. National Institute of Standards and Technology. Computer Data Authentication. Federal Information Processing Standard (FIPS), Publication 113, U.S. Department of Commerce, Washington D.C., May 1985.

548. National Institute of Standards and Technology. Secure hash standard. Federal Information Processing Standard (FIPS), Publication 180-1, U.S. Department of Commerce, Washington D.C., April 1995.

549. National Institute of Standards and Technology. Data encryption standard. Federal Information Processing Standard (FIPS), Publication 46-3, U.S. Department of Commerce, Washington D.C., October 1999.

550. National Institute of Standards and Technology. Report on the development of the Advanced Encryption Standard (AES). Available via crsc.nist.gov, October 2000.

551. National Institute of Standards and Technology. Advanced encryption standard. Federal Information Processing Standard (FIPS), Publication 197, U.S. Department of Commerce, Washington D.C., November 2001.

552. National Institute of Standards and Technology. Recommendation for block cipher modes of operation. Special Publication 800-38A, December 2001.

553. National Institute of Standards and Technology. Draft recommendation for block cipher modes of operation: The RMAC authentication mode. Draft Special Publication 800-38B, November 2002.

554. National Institute of Standards and Technology. Recommendation for block cipher modes of operation: The CCM mode for authentication and confidentiality. Special Publication 800-38C, May 2004.

555. National Institute of Standards and Technology. Recommendation for the triple data encryption algorithm (TDEA) block cipher. Special Publication 800-67, Version 1, May 2004.

556. National Institute of Standards and Technology. Recommendation for block cipher modes of operation: Three variants of ciphertext stealing for CBC mode. Addendum to special publication 800-38A. October 2010.

557. National Security Agency (NSA). SKIPJACK and KEA algorithm specifications. Available via csrc.ncsl.nist.gov/encryption/skipjack-1.pdf, May 1998.

558. NESSIE. New European schemes for signatures, integrity and encryption. Available via www.cryptonessie.org/.

559. I. Nikolić, A. Biryukov, and D. Khovratovich. Hash family LUX. Available from [542].

560. F. Noilhan. Software optimization of decorrelation module. In H.M. Heys and C.M. Adams, editors, *Selected Areas in Cryptography, SAC 1999,* volume 1758 of *Lecture Notes in Computer Science,* Springer, pages 175–183, 2000.

561. K. Nyberg. Constructions of bent functions and difference sets. In I.B. Damgård, editor, *Advances in Cryptology - EUROCRYPT '90,* volume 473 of *Lecture Notes in Computer Science,* Springer, pages 151–160, 1991.

562. K. Nyberg. Perfect nonlinear S-boxes. In D.W. Davies, editor, *Advances in Cryptology - EUROCRYPT '91,* volume 547 of *Lecture Notes in Computer Science,* Springer, pages 378–386, 1992.

563. K. Nyberg. On the construction of highly nonlinear permutations. In R. Rueppel, editor, *Advances in Cryptology - EUROCRYPT '92,* volume 658 of *Lecture Notes in Computer Science,* Springer, pages 92–98, 1993.

564. K. Nyberg. Differentially uniform mappings for cryptography. In T. Helleseth, editor, *Advances in Cryptology - EUROCRYPT '93,* volume 765 of *Lecture Notes in Computer Science,* Springer, pages 55–64, 1994.

565. K. Nyberg. Linear approximation of block ciphers. In A. De Santis, editor, *Advances in Cryptology - EUROCRYPT '94,* volume 950 of *Lecture Notes in Computer Science,* Springer, pages 439–444, 1995.

566. K. Nyberg. S-boxes and round functions with controllable linearity and differential uniformity. In B. Preneel, editor, *Fast Software Encryption, FSE 1994,* volume 1008 of *Lecture Notes in Computer Science,* Springer, pages 111–130, 1995.

567. K. Nyberg. Generalized Feistel networks. In K. Kim and T. Matsumoto, editors, *Advances in Cryptology - ASIACRYPT '96,* volume 1163 of *Lecture Notes in Computer Science,* Springer, pages 91–104, 1996.

568. K. Nyberg and L. R. Knudsen. Provable security against differential cryptanalysis. In E.F. Brickell, editor, *Advances in Cryptology - CRYPTO '92*, volume 740 of *Lecture Notes in Computer Science*, Springer, pages 566–574, 1993.

569. K. Nyberg and L. R. Knudsen. Provable security against a differential attack. *J. Cryptology*, 8(1):27–37, 1995.

570. L. O'Connor. Enumerating nondegenerate permutations. In D.W. Davies, editor, *Advances in Cryptology - EUROCRYPT '91*, volume 547 of *Lecture Notes in Computer Science*, Springer, pages 368–377, 1992.

571. L. O'Connor. An analysis of a class of algorithms for S-box construction. *J. Cryptology*, 7(3):133–151, 1994.

572. L. O'Connor. On the distribution of characteristics in bijective mappings. In T. Helleseth, editor, *Advances in Cryptology - EUROCRYPT '93*, volume 765 of *Lecture Notes in Computer Science*, Springer, pages 360–370, 1994.

573. L. O'Connor. On the distribution of characteristics in composite permutations. In D. Stinson, editor, *Advances in Cryptology - CRYPTO '93*, volume 773 of *Lecture Notes in Computer Science*, Springer, pages 403–412, 1994.

574. L. O'Connor. Convergence in differential distributions. In L. Guillou and J.-J. Quisquater, editors, *Advances in Cryptology - EUROCRYPT '95*, volume 921 of *Lecture Notes in Computer Science*, Springer, pages 13–23, 1995.

575. L. O'Connor. On the distribution of characteristics in bijective mappings. *J. Cryptology*, 8(2):67–86, 1995.

576. L. O'Connor. Properties of linear approximation tables. In B. Preneel, editor, *Fast Software Encryption, FSE 1994*, volume 1008 of *Lecture Notes in Computer Science*, Springer, pages 131–136, 1995.

577. L. O'Connor and J. D. Golic. A unified Markov approach to differential and linear cryptanalysis. In J. Pieprzyk and R. Safavi-Naini, editors, *Advances in Cryptology - ASIACRYPT '94*, volume 917 of *Lecture Notes in Computer Science*, Springer, pages 387–397, 1995.

578. P. Oechslin. Making a faster cryptanalytic time-memory trade-off. In D. Boneh, editor, *Advances in Cryptology - CRYPTO 2003*, volume 2729 of *Lecture Notes in Computer Science*, Springer, pages 617–630, 2003.

579. K. Ohkuma. Weak keys of reduced-round PRESENT for linear cryptanalysis. In M.J. Jacobson Jr., V. Rijmen, and R. Safavi-Naini, editors, *Selected Areas in Cryptography, SAC 2009*, volume 5867 of *Lecture Notes in Computer Science*, Springer, pages 249–265, 2009.

580. K. Ohkuma, H. Muratani, F. Sano, and S. Kawamura. The block cipher Hierocrypt. In D.R. Stinson and S.E. Tavares, editors, *Selected Areas in Cryptography, SAC 2000*, volume 2012 of *Lecture Notes in Computer Science*, Springer, pages 72–88, 2001.

581. K. Ohta and K. Aoki. Linear cryptanalysis of the fast data encipherment algorithm. In Y.G. Desmedt, editor, *Advances in Cryptology - CRYPTO '94*, volume 839 of *Lecture Notes in Computer Science*, Springer, pages 12–16, 1994.

582. P. C. V. Oorschot and M. J. Wiener. A known plaintext attack on two-key triple encryption. In I.B. Damgård, editor, *Advances in Cryptology - EUROCRYPT '90*, volume 473 of *Lecture Notes in Computer Science*, Springer, pages 318–325, 1991.

583. P. C. V. Oorschot and M. J. Wiener. Improving implementable meet-in-the-middle attacks by orders of magnitude. In N. Koblitz, editor, *Advances in Cryptology - CRYPTO '96*, volume 1109 of *Lecture Notes in Computer Science*, Springer, pages 229–236, 1996.

584. D. A. Osvik, A. Shamir, and E. Tromer. Cache attacks and countermeasures - the case of the AES. In D. Pointcheval, editor, *Topics in Cryptology - CT-RSA 2006*, volume 3860 of *Lecture Notes in Computer Science*, pages 1–20. Springer, 2006.

585. E. Oswald, S. Mangard, N. Pramstaller, and V. Rijmen. A side-channel analysis resistant description of the AES S-box. In H. Gilbert and H. Handschuh, editors, *Fast Software Encryption, FSE 2005*, volume 3557 of *Lecture Notes in Computer Science*, Springer, pages 413–423, 2005.

586. O. Özen, K. Varici, C. Tezcan, and Çelebi Kocair. Lightweight block ciphers revisited: Cryptanalysis of reduced round PRESENT and HIGHT. In C. Boyd and J.G. Nieto, editors, *Proceedings of ACISP 2009*, volume 5594 of *Lecture Notes in Computer Science*, Springer, pages 90–107, 2009.

587. C. Paar, A. Poschmann, and M. J. B. Robshaw. New designs in lightweight symmetric encryption. In P. Kitsos and Y. Zhang, editors, *RFID Security: Techniques, Protocols and System-on-Chip Design*, pages 349–372. Springer, 2008.

588. S. Park, S. H. Sung, S. Chee, E.-J. Yoon, and J. Lim. On the security of Rijndael-like structures against differential and linear cryptanalysis. In Y. Zheng, editor, *Advances in Cryptology - ASIACRYPT 2002*, volume 2501 of *Lecture Notes in Computer Science*, Springer, pages 176–191, 2002.

589. S. Park, S. H. Sung, S. Lee, and J. Lim. Improving the upper bound on the maximum differential and the maximum linear hull probability for SPN structures and AES. In T. Johansson, editor, *Fast Software Encryption, FSE 2003*, volume 2887 of *Lecture Notes in Computer Science*, Springer, pages 247–260, 2003.

590. J. Patarin. New results on pseudorandom permutation generators based on the DES scheme. In J. Feigenbaum, editor, *Advances in Cryptology - CRYPTO '91*, volume 576 of *Lecture Notes in Computer Science*, Springer, pages 301–312, 1992.

591. J. Patarin. How to construct pseudorandom and super pseudorandom permutations from one single pseudorandom function. In R. Rueppel, editor, *Advances in Cryptology - EUROCRYPT '92*, volume 658 of *Lecture Notes in Computer Science*, Springer, pages 256–266, 1993.

592. J. Patarin. About Feistel schemes with six (or more) rounds. In S. Vaudenay, editor, *Fast Software Encryption, FSE 1998*, volume 1372 of *Lecture Notes in Computer Science*, Springer, pages 103–121, 1998.

593. J. Patarin. Generic attacks on Feistel schemes. In C. Boyd, editor, *Advances in Cryptology - ASIACRYPT 2001*, volume 2248 of *Lecture Notes in Computer Science*, Springer, pages 222–238, 2001.

594. J. Patarin. Luby-Rackoff: 7 rounds are enough for $2^{n(1-epsilon)}$ security. In D. Boneh, editor, *Advances in Cryptology - CRYPTO 2003*, volume 2729 of *Lecture Notes in Computer Science*, Springer, pages 513–529, 2003.

595. J. Patarin. Security of random Feistel schemes with 5 or more rounds. In M. Franklin, editor, *Advances in Cryptology - CRYPTO 2004*, volume 3152 of *Lecture Notes in Computer Science*, Springer, pages 106–122, 2004.

596. J. Patarin, V. Nachef, and C. Berbain. Generic attacks on unbalanced Feistel schemes with contracting functions. In X. Lai and K. Chen, editors, *Advances in Cryptology - ASIACRYPT 2006*, volume 4284 of *Lecture Notes in Computer Science*, Springer, pages 396–411, 2006.

597. J. Patarin, V. Nachef, and C. Berbain. Generic attacks on unbalanced Feistel schemes with expanding functions. In K. Kurosawa, editor, *Advances in Cryptology - ASIACRYPT 2007*, volume 4833 of *Lecture Notes in Computer Science*, Springer, pages 325–341, 2007.

598. S. Patel, Z. Ramzan, and G. S. Sundaram. Luby-Rackoff ciphers: Why xor is not so exclusive. In K. Nyberg and H.M. Heys, editors, *Selected Areas in Cryptography, SAC 2002*, volume 2595 of *Lecture Notes in Computer Science*, Springer, pages 271–290, 2003.

599. S. Patel, Z. Ramzan, and G. S. Sundaram. Efficient constructions of variable-input-length block ciphers. In H. Handschuh and M.A. Hasan, editors, *Selected Areas in Cryptography, SAC 2004*, volume 3357 of *Lecture Notes in Computer Science*, Springer, pages 326–340, 2004.

600. K. G. Paterson and A. Yau. Padding oracle attacks on the ISO CBC mode encryption standard. In T. Okamoto, editor, *Topics in Cryptology – CT-RSA 2004*, volume 2964 of *Lecture Notes in Computer Science*, pages 305–323. Springer, 2004.

601. C. Patterson. A dynamic FPGA implementation of the Serpent block cipher. In Ç.K. Koç and C. Paar, editors, *Cryptographic Hardware and Embedded Systems - CHES 2000*, volume 1965 of *Lecture Notes in Computer Science*, Springer, 2000, pages 141–155, 2000.

602. E. Petrank and C. Rackoff. CBC MAC for real-time data sources. *J. Cryptology*, 13(3):315–338, 2000.

603. D. H. Phan and D. Pointcheval. About the security of ciphers (semantic security and pseudorandom permutations). In H. Handschuh and M.A. Hasan, editors, *Selected Areas in Cryptography, SAC 2004*, volume 3357 of *Lecture Notes in Computer Science*, Springer, pages 182–197, 2004.

604. R. C. W. Phan. Classes of impossible differentials of the Advanced Encryption Standard. *Electronics Letters*, 38(11):508–510, 2002.

605. R. C. W. Phan. Impossible differential cryptanalysis of 7-round Advanced Encryption Standard. *Information Processing Letters*, 91:33–38, 2004.

606. R. C. W. Phan and M. U. Siddiqi. Generalised impossible differentials of the Advanced Encryption Standard. *Electronics Letters*, 37(14):896–898, 2001.

607. J. Pieprzyk. How to construct pseudorandom permutations from single pseudorandom functions. In I.B. Damgård, editor, *Advances in Cryptology - EUROCRYPT '90*, volume 473 of *Lecture Notes in Computer Science*, Springer, pages 140–150, 1991.

608. G. Piret and J.-J. Quisquater. A differential fault attack technique against SPN structures, with application to the AES and KHAZAD. In C.D. Walter, Ç.K. Koç and C. Paar, editors, *Cryptographic Hardware and Embedded Systems - CHES 2003*, volume 2779 of *Lecture Notes in Computer Science*, Springer, 2003, pages 77–88, 2003.

609. G. Piret and J.-J. Quisquater. Security of the MISTY structure in the Luby-Rackoff model: Improved results. In H. Handschuh and M.A. Hasan, editors, *Selected Areas in Cryptography, SAC 2004*, volume 3357 of *Lecture Notes in Computer Science*, Springer, pages 100–113, 2004.

610. J. O. Pliam. A polynomial-time universal security amplifier in the class of block ciphers. In D.R. Stinson and S.E. Tavares, editors, *Selected Areas in Cryptography, SAC 2000*, volume 2012 of *Lecture Notes in Computer Science*, Springer, pages 169–188, 2001.

611. T. Pornin. Optimal resistance against the Davies and Murphy attack. In K. Ohta and D. Pei, editors, *Advances in Cryptology - ASIACRYPT '98*, volume 1514 of *Lecture Notes in Computer Science*, Springer, pages 148–159, 1998.

612. A. Poschmann. *Lightweight Cryptography–Cryptographic Engineering for a Pervasive World. Number 8 in IT Security. Europäischer Universitätsverlag*. PhD thesis, Ruhr University Bochum, 2009.

613. B. Preneel. *Analysis and Design of Cryptographic Hash Functions*. PhD thesis, Katholieke Universiteit Leuven, January 1993.

614. B. Preneel, R. Govaerts, and J. Vandewalle. Boolean functions satisfying higher order propagation criteria. In D.W. Davies, editor, *Advances in Cryptology - EUROCRYPT '91*, volume 547 of *Lecture Notes in Computer Science*, Springer, pages 141–152, 1992.

615. B. Preneel, R. Govaerts, and J. Vandewalle. Hash functions based on block ciphers: A synthetic approach. In D. Stinson, editor, *Advances in Cryptology - CRYPTO '93*, volume 773 of *Lecture Notes in Computer Science*, Springer, pages 368–378, 1994.

616. B. Preneel, W. V. Leekwijck, L. V. Linden, R. Govaerts, and J. Vandewalle. Propagation characteristics of Boolean functions. In I.B. Damgård, editor, *Advances in Cryptology - EUROCRYPT '90*, volume 473 of *Lecture Notes in Computer Science*, Springer, pages 161–173, 1991.

617. B. Preneel, M. Nuttin, V. Rijmen, and J. Buelens. Cryptanalysis of the CFB mode of the DES with a reduced number of rounds. In D. Stinson, editor, *Advances in Cryptology - CRYPTO '93*, volume 773 of *Lecture Notes in Computer Science*, Springer, pages 212–223, 1994.

618. E. Prouff. DPA attacks and S-boxes. In H. Gilbert and H. Handschuh, editors, *Fast Software Encryption, FSE 2005*, volume 3557 of *Lecture Notes in Computer Science*, Springer, pages 424–441, 2005.

619. E. Prouff, C. Giraud, and S. Aumônier. Provably secure S-box implementation based on Fourier transform. In L. Goubin and M. Matsui, editors, *Cryptographic Hardware and Embedded Systems - CHES 2006*, volume 4249 of *Lecture Notes in Computer Science*, Springer, 2006, pages 216–230, 2006.

620. J.-J. Quisquater and J.-P. Delescaille. Other cycling tests for DES (abstract). In C. Pomerance, editor, *Advances in Cryptology - CRYPTO '87*, volume 293 of *Lecture Notes in Computer Science*, Springer, pages 255–256, 1988.

621. J.-J. Quisquater and J.-P. Delescaille. How easy is collision search? application to DES (extended summary). In J.-J. Quisquater and J. Vandewalle, editors, *Advances in Cryptology - EUROCRYPT '89*, volume 434 of *Lecture Notes in Computer Science*, Springer, pages 429–434, 1990.

622. J.-J. Quisquater and J.-P. Delescaille. How easy is collision search? New results and applications to DES. In G. Brassard, editor, *Advances in Cryptology - CRYPTO '89*, volume 435 of *Lecture Notes in Computer Science*, Springer, pages 408–413, 1990.

623. J.-J. Quisquater, Y. Desmedt, and M. Davio. The importance of "good" key scheduling schemes (how to make a secure DES scheme with $<= 48$ bit keys). In H.C. Williams, editor, *Advances in Cryptology - CRYPTO '85*, volume 218 of *Lecture Notes in Computer Science*, Springer, pages 537–542, 1986.

624. J.-J. Quisquater and F.-X. Standaert. Exhaustive key search of the DES: Updates and refinements. In *Proceedings of SHARCS 2005 (Special-purpose Hardware for Attacking Cryptographic Systems)*, February 2005. Available via www.dice.ucl.ac.be/~fstandae/

625. H. Raddum. Cryptanalysis of IDEA-X/2. In T. Johansson, editor, *Fast Software Encryption, FSE 2003*, volume 2887 of *Lecture Notes in Computer Science*, Springer, pages 1–8, 2003.

626. H. Raddum. More dual Rijndaels. In V. Rijmen, H. Dobbertin, and A. Sowa, editors, *Advanced Encryption Standard - AES, Fourth International Conference*, volume 3373 of *Lecture Notes in Computer Science, Springer*, pages 142–147, 2005.

627. H. Raddum and L. R. Knudsen. A differential attack on reduced-round SC2000. In S. Vaudenay and A.M. Youssef, editors, *Selected Areas in Cryptography, SAC 2001*, volume 2259 of *Lecture Notes in Computer Science*, Springer, pages 190–198, 2001.

628. Z. Ramzan and L. Reyzin. On the round security of symmetric-key cryptographic primitives. In M. Bellare, editor, *Advances in Cryptology - CRYPTO 2000*, volume 1880 of *Lecture Notes in Computer Science*, Springer, pages 376–393, 2000.

629. J. A. Reeds and J. L. Manferdelli. DES has no per round linear factors. In G.R. Blakley and D. Chaum, editors, *Advances in Cryptology - CRYPTO '84*, volume 196 of *Lecture Notes in Computer Science*, Springer, pages 377–389, 1985.

630. B. Reichardt and D. Wagner. Markov truncated differential cryptanalysis of Skipjack. In K. Nyberg and H.M. Heys, editors, *Selected Areas in Cryptography, SAC 2002*, volume 2595 of *Lecture Notes in Computer Science*, Springer, pages 110–128, 2003.

631. V. Rijmen. *Cryptanalysis and Design of Iterated Block Ciphers*. PhD thesis, Katholieke Universiteit Leuven, October 1997.

632. V. Rijmen, J. Daemen, B. Preneel, A. Bosselaers, and E. De Win. The cipher SHARK. In D. Gollmann, editor, *Fast Software Encryption, FSE 1996*, volume 1039 of *Lecture Notes in Computer Science*, Springer, pages 99–111, 1996.

633. V. Rijmen and B. Preneel. Cryptanalysis of McGuffin. In B. Preneel, editor, *Fast Software Encryption, FSE 1994*, volume 1008 of *Lecture Notes in Computer Science*, Springer, pages 353–358, 1995.

634. V. Rijmen and B. Preneel. Improved characteristics for differential cryptanalysis of hash functions based on block ciphers. In B. Preneel, editor, *Fast Software Encryption, FSE 1994*, volume 1008 of *Lecture Notes in Computer Science*, Springer, pages 242–248, 1995.

635. V. Rijmen and B. Preneel. A family of trapdoor ciphers. In E. Biham, editor, *Fast Software Encryption, FSE 1997*, volume 1267 of *Lecture Notes in Computer Science*, Springer, pages 139–148, 1997.

636. R. L. Rivest. Cryptography. In Chapter 13 of *Handbook of Theoretical Computer Science*, Elsevier Science Publishers B.V., 1990.

637. R. L. Rivest. The MD5 message-digest algorithm. Request for Comments (RFC) 1321, Internet Activities Board, Internet Privacy Task Force, April 1992.

638. R. L. Rivest. The RC5 encryption algorithm. In B. Preneel, editor, *Fast Software Encryption, FSE 1994*, volume 1008 of *Lecture Notes in Computer Science*, Springer, pages 86–96, 1995.

639. R. L. Rivest, 1996. Attributed to Rivest in [370].

640. R. L. Rivest, 1996. Attributed to Rivest in *Cryptography and Data Security* by D.E. Denning, Addison-Wesley, 1982.

641. R. L. Rivest. A description of the RC2(r) encryption algorithm. Request for Comments (RFC) 2268, Internet Activities Board, Internet Privacy Task Force, March 1998.

642. R. L. Rivest, M. J. B. Robshaw, R. Sidney, and Y. L. Yin. The RC6 block cipher. Submitted as candidate for AES. Available via www.nist.gov/aes.

643. M. Robshaw and O. Billet. *New Stream Cipher Designs: The eSTREAM Finalists.* Springer, 2008.

644. P. Rogaway. Efficient instantiations of tweakable blockciphers and refinements to modes OCB and PMAC. In P.J. Lee, editor, *Advances in Cryptology - ASIACRYPT 2004,* volume 3329 of *Lecture Notes in Computer Science,* Springer, pages 16–31, 2004.

645. P. Rogaway. Nonce-based symmetric encryption. In W. Meier and B. Roy, editors, *Fast Software Encryption, FSE 2004,* volume 3017 of *Lecture Notes in Computer Science,* Springer, pages 348–359, 2004.

646. P. Rogaway and T. Shrimpton. A provable-security treatment of the key-wrap problem. In S. Vaudenay, editor, *Advances in Cryptology - EUROCRYPT 2006,* volume 4004 of *Lecture Notes in Computer Science,* Springer, pages 373–390, 2006.

647. P. Rogaway and J. Steinberger. Security/efficiency tradeoffs for permutation-based hashing. In N. Smart, editor, *Advances in Cryptology - EUROCRYPT 2008,* volume 4965 of *Lecture Notes in Computer Science,* Springer, pages 220–236, 2007.

648. C. Rolfes, A. Poschmann, G. Leander, and C. Paar. Ultra-lightweight implementations for smart devices - security for 1000 gate equivalents. In G. Grimaud and F.-X. Standaert, editors, *CARDIS 2008,* volume 5189 of *Lecture Notes in Computer Science,* pages 89–103. Springer, 2008.

649. B. V. Rompay, L. R. Knudsen, and V. Rijmen. Differential cryptanalysis of the ICE encryption algorithm. In S. Vaudenay, editor, *Fast Software Encryption, FSE 1998,* volume 1372 of *Lecture Notes in Computer Science,* Springer, pages 270–283, 1998.

650. RSA Laboratories. Secret key challenge. Available via www.rsasecurity.com/rsalabs.

651. M.-J. O. Saarinen. Cryptanalysis of block ciphers based on SHA-1 and MD5. In T. Johansson, editor, *Fast Software Encryption, FSE 2003,* volume 2887 of *Lecture Notes in Computer Science,* Springer, pages 36–44, 2003.

652. B. Sadeghiyan and J. Pieprzyk. A construction for super pseudorandom permutations from a single pseudorandom function. In R. Rueppel, editor, *Advances in Cryptology - EUROCRYPT '92,* volume 658 of *Lecture Notes in Computer Science,* Springer, pages 267–284, 1993.

653. B. Sadeghiyan and J. Pieprzyk. On necessary and sufficient conditions for the construction of super pseudorandom permutations. In H. Imai, R.L. Rivest and T. Matsumoto, editors, *Advances in Cryptology - ASIACRYPT '91,* volume 739 of *Lecture Notes in Computer Science,* Springer, pages 194–209, 1993.

654. K. Sakurai and S. Furuya. Improving linear cryptanalysis of LOKI91 by probabilistic counting method. In E. Biham, editor, *Fast Software Encryption, FSE 1997,* volume 1267 of *Lecture Notes in Computer Science,* Springer, pages 114–133, 1997.

655. A. Satoh and S. Morioka. Unified hardware architecture for 128-bit block ciphers AES and Camellia. In C.D. Walter, Ç.K. Koç and C. Paar, editors, *Cryptographic Hardware and Embedded Systems - CHES 2003,* volume 2779 of *Lecture Notes in Computer Science,* Springer, 2003, pages 304–318, 2003.

656. A. Satoh, S. Morioka, K. Takano, and S. Munetoh. A compact Rijndael hardware architecture with S-box optimization. In C. Boyd, editor, *Advances in Cryptology - ASIACRYPT 2001,* volume 2248 of *Lecture Notes in Computer Science,* Springer, pages 239–254, 2001.

657. A. Satoh, T. Sugawara, N. Homma, and T. Aoki. High-performance concurrent error detection scheme for AES hardware. In E. Oswald and P. Rohatgi, editors, *Cryptographic Hardware and Embedded Systems - CHES 2008,* volume 5154 of *Lecture Notes in Computer Science,* Springer, 2008, pages 100–112, 2008.

658. T. Satoh, T. Iwata, and K. Kurosawa. On cryptographically secure vectorial Boolean functions. In K-Y. Lam, E. Okamoto and C. Xing, editors, *Advances in Cryptology - ASIACRYPT '99,* volume 1716 of *Lecture Notes in Computer Science,* Springer, pages 20–28, 1999.

659. I. Schaumüller-Bichl. *Zur Analyse des Data Encryption Standard und Synthese Verwandter Chiffriersysteme.* PhD thesis, Linz University, May 1981.

660. I. Schaumüller-Bichl. Cryptanalysis of the Data Encryption Standard by the method of formal coding. In T. Beth, editor, *Cryptography, Proceedings of the Workshop on Cryptography,*

Burg Feuerstein, Germany, March 29 - April 2, 1982. volume 149 of *Lecture Notes in Computer Science,* Springer, pages 235–255, 1983.

661. I. Schaumüller-Bichl. On the design and analysis of new cipher systems related to the DES. Technical report, Linz University, 1983.

662. B. Schneier. Description of a new variable-length key, 64-bit block cipher (Blowfish). In R. Anderson, editor, *Fast Software Encryption, FSE 1993,* volume 809 of *Lecture Notes in Computer Science,* Springer, pages 191–204, 1994.

663. B. Schneier. *Applied Cryptography.* John Wiley & Sons, 1996.

664. B. Schneier and D. Banisar. *The Electronic Privacy Papers.* John Wiley & Sons, 1997.

665. B. Schneier and J. Kelsey. Unbalanced Feistel networks and block cipher design. In D. Gollmann, editor, *Fast Software Encryption, FSE 1996,* volume 1039 of *Lecture Notes in Computer Science,* Springer, pages 121–144, 1996.

666. B. Schneier, J. Kelsey, D. Whiting, D. Wagner, and C. Hall. On the Twofish key schedule. In S.E. Tavares and H. Meijer, editors, *Selected Areas in Cryptography, SAC 1998,* volume 1556 of *Lecture Notes in Computer Science,* Springer, pages 27–42, 1999.

667. B. Schneier, J. Kelsey, D. Whiting, D. Wagner, C. Hall, and N. Ferguson. Twofish: A 128-bit block cipher. Submitted as candidate for AES. Available via www.nist.gov/aes.

668. K. Schramm, G. Leander, P. Felke, and C. Paar. A collision-attack on AES: Combining side channel- and differential-attack. In M. Joye and J.-J. Quisquater, editors, *Cryptographic Hardware and Embedded Systems - CHES 2004,* volume 3156 of *Lecture Notes in Computer Science,* Springer, 2004, pages 163–175, 2004.

669. K. Schramm, T. J. Wollinger, and C. Paar. A new class of collision attacks and its application to DES. In T. Johansson, editor, *Fast Software Encryption, FSE 2003,* volume 2887 of *Lecture Notes in Computer Science,* Springer, pages 206–222, 2003.

670. R. C. Schroeppel and A. Shamir. A $t = o(2^{\frac{n}{2}}), s = o(2^{\frac{n}{4}})$ algorithm for certain NP-complete problems. *SIAM Journal of Computing,* 10:456–464, 1981.

671. J. Seberry and X.-M. Zhang. Highly nonlinear 0-1 balanced Boolean functions satisfying strict avalanche criterion. In J. Seberry and Y. Zheng, editors, *Advances in Cryptology - ASIACRYPT '92,* volume 718 of *Lecture Notes in Computer Science,* Springer, pages 145–155, 1993.

672. J. Seberry, X.-M. Zhang, and Y. Zheng. Pitfalls in designing substitution boxes (extended abstract). In Y.G. Desmedt, editor, *Advances in Cryptology - CRYPTO '94,* volume 839 of *Lecture Notes in Computer Science,* Springer, pages 383–396, 1994.

673. J. Seberry, X.-M. Zhang, and Y. Zheng. Relationships among nonlinear criteria (extended abstract). In A. De Santis, editor, *Advances in Cryptology - EUROCRYPT '94,* volume 950 of *Lecture Notes in Computer Science,* Springer, pages 376–388, 1995.

674. J. Seberry, X.-M. Zhang, and Y. Zheng. Structures of cryptographic functions with strong avalanche characteristics (extended abstract). In J. Pieprzyk and R. Safavi-Naini, editors, *Advances in Cryptology - ASIACRYPT '94,* volume 917 of *Lecture Notes in Computer Science,* Springer, pages 119–132, 1995.

675. H. Seki and T. Kaneko. Cryptanalysis of five rounds of CRYPTON using impossible differentials. In K-Y. Lam, E. Okamoto and C. Xing, editors, *Advances in Cryptology - ASIACRYPT '99,* volume 1716 of *Lecture Notes in Computer Science,* Springer, pages 43–51, 1999.

676. H. Seki and T. Kaneko. Differential cryptanalysis of reduced rounds of GOST. In D.R. Stinson and S.E. Tavares, editors, *Selected Areas in Cryptography, SAC 2000,* volume 2012 of *Lecture Notes in Computer Science,* Springer, pages 315–323, 2001.

677. A. A. Selçuk. New results in linear cryptanalysis of RC5. In S. Vaudenay, editor, *Fast Software Encryption, FSE 1998,* volume 1372 of *Lecture Notes in Computer Science,* Springer, pages 1–16, 1998.

678. A. A. Selçuk. On probability of success in linear and differential cryptanalysis. *J. Cryptology,* 21(1):131–147, 2008.

679. A. Shamir. On the security of DES. In H.C. Williams, editor, *Advances in Cryptology - CRYPTO '85,* volume 218 of *Lecture Notes in Computer Science,* Springer, pages 280–281, 1986.

680. C. E. Shannon. Communication theory of secrecy systems. *Bell System Technical Journal*, 28:656–715, 1949.

681. A. Shimizu and S. Miyaguchi. Fast data encipherment algorithm FEAL. In D. Chaum, W.L. Price, editors, *Advances in Cryptology - EUROCRYPT '87*, volume 304 of *Lecture Notes in Computer Science*, Springer, pages 267–278, 1988.

682. T. Shimoyama and T. Kaneko. Quadratic relation of S-box and its application to the linear attack of full round DES. In H. Krawczyk, editor, *Advances in Cryptology - CRYPTO '98*, volume 1462 of *Lecture Notes in Computer Science*, Springer, pages 200–211, 1998.

683. T. Shimoyama, M. Takenaka, and T. Koshiba. Multiple linear cryptanalysis of a reduced round RC6. In J. Daemen and V. Rijmen, editors, *Fast Software Encryption, FSE 2002, February 2002*, volume 2365 of *Lecture Notes in Computer Science*, Springer, pages 76–88, 2002.

684. T. Shimoyama, H. Yanami, K. Yokoyama, M. Takenaka, K. Itoh, J. Yajima, N. Torii, and H. Tanaka. The block cipher SC2000. In M. Matsui, editor, *Fast Software Encryption, FSE 2001*, volume 2355 of *Lecture Notes in Computer Science*, Springer, pages 312–327, 2001.

685. T. Shirai, S. Kanamaru, and G. Abe. Improved upper bounds of differential and linear characteristic probability for Camellia. In J. Daemen and V. Rijmen, editors, *Fast Software Encryption, FSE 2002, February 2002*, volume 2365 of *Lecture Notes in Computer Science*, Springer, pages 128–142, 2002.

686. T. Shirai and B. Preneel. On Feistel ciphers using optimal diffusion mappings across multiple rounds. In P.J. Lee, editor, *Advances in Cryptology - ASIACRYPT 2004*, volume 3329 of *Lecture Notes in Computer Science*, Springer, pages 1–15, 2004.

687. T. Shirai and K. Shibutani. Improving immunity of Feistel ciphers against differential cryptanalysis by using multiple MDS matrices. In W. Meier and B. Roy, editors, *Fast Software Encryption, FSE 2004*, volume 3017 of *Lecture Notes in Computer Science*, Springer, pages 260–278, 2004.

688. T. Shirai and K. Shibutani. On Feistel structures using a diffusion switching mechanism. In M.J.B. Robshaw, editor, *Fast Software Encryption, FSE 2006*, volume 4047 of *Lecture Notes in Computer Science*, Springer, pages 41–56, 2006.

689. T. Shirai, K. Shibutani, T. Akishita, S. Moriai, and T. Iwata. The 128-bit blockcipher CLEFIA. In A. Biryukov, editor, *Fast Software Encryption, FSE 2007*, volume 4593 of *Lecture Notes in Computer Science*, Springer, pages 181–195, 2007.

690. M. Sivabalan, S. E. Tavares, and L. E. Peppard. On the design of SP networks from an information theoretic point of view. In E.F. Brickell, editor, *Advances in Cryptology - CRYPTO '92*, volume 740 of *Lecture Notes in Computer Science*, Springer, pages 260–279, 1993.

691. M. E. Smid. DES '81: An update. In A. Gersho, editor, *Advances in Cryptology: A report on CRYPTO '81, IEEE Workshop on Communications Security, Santa Barbara, August 24-26, 1981. U.C. Santa Barbara, Dept. of Elec. and Computer Eng., ECE Report No 82-04*, pages 39–40, 1982.

692. B. Song and J. Seberry. Further observations on the structure of the AES algorithm. In T. Johansson, editor, *Fast Software Encryption, FSE 2003*, volume 2887 of *Lecture Notes in Computer Science*, Springer, pages 223–234, 2003.

693. F.-X. Standaert, S. B. Örs, and B. Preneel. Power analysis of an FPGA: Implementation of Rijndael: Is pipelining a DPA countermeasure. In M. Joye and J.-J. Quisquater, editors, *Cryptographic Hardware and Embedded Systems - CHES 2004*, volume 3156 of *Lecture Notes in Computer Science*, Springer, 2004, pages 30–44, 2004.

694. F.-X. Standaert, G. Piret, N. Gershenfeld, and J.-J. Quisquater. SEA: A scalable encryption algorithm for small embedded applications. In J. Domingo-Ferrer, J. Posegga, and D. Schreckling, editors, *CARDIS 2006*, volume 3928 of *Lecture Notes in Computer Science*, pages 222–236. Springer, 2006.

695. F.-X. Standaert, G. Piret, G. Rouvroy, J.-J. Quisquater, and J.-D. Legat. Iceberg: An involutional cipher efficient for block encryption in reconfigurable hardware. In W. Meier and B. Roy, editors, *Fast Software Encryption, FSE 2004*, volume 3017 of *Lecture Notes in Computer Science*, Springer, pages 279–299, 2004.

696. F.-X. Standaert, G. Rouvroy, J.-J. Quisquater, and J.-D. Legat. Efficient implementation of Rijndael encryption in reconfigurable hardware: Improvements and design tradeoffs. In C.D. Walter, Ç.K. Koç and C. Paar, editors, *Cryptographic Hardware and Embedded Systems - CHES 2003*, volume 2779 of *Lecture Notes in Computer Science*, Springer, 2003, pages 334–350, 2003.

697. F.-X. Standaert, G. Rouvroy, J.-J. Quisquater, and J.-D. Legat. A time-memory tradeoff using distinguished points: New analysis & FPGA results. In B.S. Kaliski Jr., Ç.K. Koç and C. Paar, editors, *Cryptographic Hardware and Embedded Systems - CHES 2002*, volume 2523 of *Lecture Notes in Computer Science*, Springer, 2003, pages 593–609, 2003.

698. J. Steinberger. The collision intractability of MDC-2 in the ideal-cipher model. In M. Naor, editor, *Advances in Cryptology - EUROCRYPT 2007*, volume 4515 of *Lecture Notes in Computer Science*, Springer, pages 34–51, 2007.

699. J. Stern and S. Vaudenay. CS-cipher. In S. Vaudenay, editor, *Fast Software Encryption, FSE 1998*, volume 1372 of *Lecture Notes in Computer Science*, Springer, pages 189–205, 1998.

700. D. R. Stinson. *Cryptography - Theory and Practice*. CRC Press, Inc., 2006. 3rd Edition.

701. M. Sugita, K. Kobara, and H. Imai. Security of reduced version of the block cipher Camellia against truncated and impossible differential cryptanalysis. In C. Boyd, editor, *Advances in Cryptology - ASIACRYPT 2001*, volume 2248 of *Lecture Notes in Computer Science*, Springer, pages 193–207, 2001.

702. B. Sun, L. Qu, and C. Li. New cryptanalysis of block ciphers with low algebraic degree. In O. Dunkelman, editor, *Fast Software Encryption, FSE 2009*, volume 5665 of *Lecture Notes in Computer Science*, Springer, pages 180–192, 2009.

703. J. Sung, S. Lee, J. I. Lim, S. Hong, and S. Park. Provable security for the Skipjack-like structure against differential cryptanalysis and linear cryptanalysis. In T. Okamoto, editor, *Advances in Cryptology - ASIACRYPT 2000*, volume 1976 of *Lecture Notes in Computer Science*, Springer, pages 274–288, 2000.

704. H. Tanaka, C. Ishii, and T. Kaneko. On the strength of KASUMI without FL functions against higher order differential attack. In D. Won, editor, 3rd *International Conference on Information Security and Cryptology (ICISC 2000)*, volume 2015 of *Lecture Notes in Computer Science*, pages 14–21. Springer, 2001.

705. A. Tardy-Corfdir and H. Gilbert. A known plaintext attack of FEAL-4 and FEAL-6. In J. Feigenbaum, editor, *Advances in Cryptology - CRYPTO '91*, volume 576 of *Lecture Notes in Computer Science*, Springer, pages 172–181, 1992.

706. S. Tillich and J. Großschädl. Instruction set extensions for efficient AES implementation on 32-bit processors. In L. Goubin and M. Matsui, editors, *Cryptographic Hardware and Embedded Systems - CHES 2006*, volume 4249 of *Lecture Notes in Computer Science*, Springer, 2006, pages 270–284, 2006.

707. S. Tillich and C. Herbst. Attacking state-of-the-art software countermeasures-a case study for AES. In E. Oswald and P. Rohatgi, editors, *Cryptographic Hardware and Embedded Systems - CHES 2008*, volume 5154 of *Lecture Notes in Computer Science*, Springer, 2008, pages 228–243, 2008.

708. T. Tokita, T. Sorimachi, and M. Matsui. Linear cryptanalysis of LOKI and s^2DES. In J. Pieprzyk and R. Safavi-Naini, editors, *Advances in Cryptology - ASIACRYPT '94*, volume 917 of *Lecture Notes in Computer Science*, Springer, pages 293–303, 1995.

709. Toshiba Corporation. The Hierocrypt block cipher. Available via www.cryptonessie.org, September 2001.

710. E. Trichina, D. De Seta, and L. Germani. Simplified adaptive multiplicative masking for AES. In B.S. Kaliski Jr., Ç.K. Koç and C. Paar, editors, *Cryptographic Hardware and Embedded Systems - CHES 2002*, volume 2523 of *Lecture Notes in Computer Science*, Springer, 2003, pages 187–197, 2003.

711. S. Trimberger, R. Pang, and A. Singh. A 12 gbps DES encryptor/decryptor core in an FPGA. In Ç.K. Koç and C. Paar, editors, *Cryptographic Hardware and Embedded Systems - CHES 2000*, volume 1965 of *Lecture Notes in Computer Science*, Springer, 2000, pages 156–163, 2000.

712. Y. Tsunoo, T. Saito, T. Suzaki, M. Shigeri, and H. Miyauchi. Cryptanalysis of DES implemented on computers with cache. In C.D. Walter, Ç.K. Koç and C. Paar, editors, *Cryptographic Hardware and Embedded Systems - CHES 2003*, volume 2779 of *Lecture Notes in Computer Science*, Springer, 2003, pages 62–76, 2003.

713. Y. Tsunoo, E. Tsujihara, M. Shigeri, T. Saito, T. Suzaki, and H. Kubo. Impossible differential cryptanalysis of CLEFIA. In K. Nyberg, editor, *Fast Software Encryption, FSE 2008*, volume 5086 of *Lecture Notes in Computer Science*, Springer, pages 398–411, 2008.

714. T.U. Graz. AES Lounge. Available via www.iaik.tugraz.at.

715. S. Vaudenay. On the need for multipermutations: Cryptanalysis of MD4 and SAFER. In B. Preneel, editor, *Fast Software Encryption, FSE 1994*, volume 1008 of *Lecture Notes in Computer Science*, Springer, pages 286–297, 1995.

716. S. Vaudenay. On the weak keys of Blowfish. In D. Gollmann, editor, *Fast Software Encryption, FSE 1996*, volume 1039 of *Lecture Notes in Computer Science*, Springer, pages 27–32, 1996.

717. S. Vaudenay. Provable security for block ciphers by decorrelation. In *STACS'98*, volume 1373 of *Lecture Notes in Computer Science*, pages 249–275. Springer, 1998.

718. S. Vaudenay. Feistel ciphers with L2-decorrelation. In S.E. Tavares and H. Meijer, editors, *Selected Areas in Cryptography, SAC 1998*, volume 1556 of *Lecture Notes in Computer Science*, Springer, pages 1–14, 1999.

719. S. Vaudenay. On the Lai-Massey scheme. In K-Y. Lam, E. Okamoto and C. Xing, editors, *Advances in Cryptology - ASIACRYPT '99*, volume 1716 of *Lecture Notes in Computer Science*, Springer, pages 8–19, 1999.

720. S. Vaudenay. On the security of CS-cipher. In L.R. Knudsen, editor, *Fast Software Encryption, FSE 1999*, volume 1636 of *Lecture Notes in Computer Science*, Springer, pages 260–274, 1999.

721. S. Vaudenay. Resistance against general iterated attacks. In J. Stern, editor, *Advances in Cryptology - EUROCRYPT '99*, volume 1592 of *Lecture Notes in Computer Science*, Springer, pages 255–271, 1999.

722. S. Vaudenay. Adaptive-attack norm for decorrelation and super-pseudorandomness. In H.M. Heys and C.M. Adams, editors, *Selected Areas in Cryptography, SAC 1999*, volume 1758 of *Lecture Notes in Computer Science*, Springer, pages 49–61, 2000.

723. S. Vaudenay. Decorrelation over infinite domains: The encrypted CBC-MAC case. In D.R. Stinson and S.E. Tavares, editors, *Selected Areas in Cryptography, SAC 2000*, volume 2012 of *Lecture Notes in Computer Science*, Springer, pages 189–201, 2001.

724. S. Vaudenay. Security flaws induced by CBC padding - applications to SSL, IPSEC, WTLS ... In L.R. Knudsen, editor, *Advances in Cryptology - EUROCRYPT 2002*, volume 2332 of *Lecture Notes in Computer Science*, Springer, pages 534–546, 2002.

725. S. Vaudenay. Decorrelation: A theory for block cipher security. *J. Cryptology*, 16(4):249–286, 2003.

726. I. Verbauwhede, F. Hoornaert, J. Vandewalle, and H. De Man. Security and performance optimization of a new DES data encryption chip. *IEEE Journal of Solid-State Circuits*, SC-23(3):647–656, June 1988.

727. I. Verbauwhede, F. Hoornaert, J. Vandewalle, and H. De Man. Security considerations in the design and implementation of a new DES chip. In D. Chaum, W.L. Price, editors, *Advances in Cryptology - EUROCRYPT '87*, volume 304 of *Lecture Notes in Computer Science*, Springer, pages 287–300, 1988.

728. D. Wagner. Cryptanalysis of some recently-proposed multiple modes of operation. In S. Vaudenay, editor, *Fast Software Encryption, FSE 1998*, volume 1372 of *Lecture Notes in Computer Science*, Springer, pages 254–269, 1998.

729. D. Wagner. The boomerang attack. In L.R. Knudsen, editor, *Fast Software Encryption, FSE 1999*, volume 1636 of *Lecture Notes in Computer Science*, Springer, pages 156–170, 1999.

730. D. Wagner. Towards a unifying view of block cipher cryptanalysis. In W. Meier and B. Roy, editors, *Fast Software Encryption, FSE 2004*, volume 3017 of *Lecture Notes in Computer Science*, Springer, pages 16–33, 2004.

731. D. Wagner, B. Schneier, and J. Kelsey. Cryptanalysis of the cellular encryption algorithm. In B.S. Kaliski Jr., editor, *Advances in Cryptology - CRYPTO '97*, volume 1294 of *Lecture Notes in Computer Science*, Springer, pages 526–537, 1997.

732. J. Wallén. Linear approximations of addition modulo 2^n. In T. Johansson, editor, *Fast Software Encryption, FSE 2003*, volume 2887 of *Lecture Notes in Computer Science*, Springer, pages 261–273, 2003.

733. M. Wang. Differential cryptanalysis of reduced-round PRESENT. In S. Vaudenay, editor, *Progress in Cryptology - AFRICACRYPT 2008*, volume 5023 of *Lecture Notes in Computer Science*, pages 40–49. Springer, 2008.

734. X. Wang, Y. L. Yin, and H. Yu. Finding collisions in the full SHA-1. In V. Shoup, editor, *Advances in Cryptology - CRYPTO 2005*, volume 3621 of *Lecture Notes in Computer Science*, Springer, pages 17–36, 2005.

735. X. Wang and H. Yu. How to break MD5 and other hash functions. In R. Cramer, editor, *Advances in Cryptology - EUROCRYPT 2005*, volume 3494 of *Lecture Notes in Computer Science*, Springer, pages 19–35, 2005.

736. P. C. Wayner. Content-addressable search engines and DES-like systems. In E.F. Brickell, editor, *Advances in Cryptology - CRYPTO '92*, volume 740 of *Lecture Notes in Computer Science*, Springer, pages 575–586, 1993.

737. A. F. Webster and S. E. Tavares. On the design of S-boxes. In H.C. Williams, editor, *Advances in Cryptology - CRYPTO '85*, volume 218 of *Lecture Notes in Computer Science*, Springer, pages 523–534, 1986.

738. R. Wernsdorf. The one-round functions of the DES generate the alternating group. In R. Rueppel, editor, *Advances in Cryptology - EUROCRYPT '92*, volume 658 of *Lecture Notes in Computer Science*, Springer, pages 99–112, 1993.

739. R. Wernsdorf. The round functions of Rijndael generate the alternating group. In J. Daemen and V. Rijmen, editors, *Fast Software Encryption, FSE 2002, February 2002*, volume 2365 of *Lecture Notes in Computer Science*, Springer, pages 143–148, 2002.

740. D. Wheeler. A bulk data encryption algorithm. In R. Anderson, editor, *Fast Software Encryption, FSE 1993*, volume 809 of *Lecture Notes in Computer Science*, Springer, pages 127–134, 1994.

741. D. Wheeler and R. Needham. Tea extensions. Available via www.ftp.cl.cam.ac.uk/ftp/users/djw3/. (Also Correction to XTEA. October, 1998.), October 1997.

742. D. J. Wheeler and R. M. Needham. TEA, a tiny encryption algorithm. In B. Preneel, editor, *Fast Software Encryption, FSE 1994*, volume 1008 of *Lecture Notes in Computer Science*, Springer, pages 363–366, 1995.

743. D. Whiting, R. Housley, and N. Ferguson. Counter with CBC-MAC (CCM). IETF Internet Draft with file name draft-housley-ccm-mode-01.txt, Available via www.ietf.org/internet-drafts/, September 2002.

744. M. J. Wiener. Efficient DES key search. Technical Report TR-244, School of Computer Science, Carleton University, Ottawa, Canada, May 1994. Presented at the Rump Session of CRYPTO'93.

745. M. J. Wiener. Efficient DES key search - an update. *CryptoBytes*, 3(2):6–8, 1998.

746. D. C. Wilcox, L. G. Pierson, P. J. Robertson, E. L. Witzke, and K. Gass. A DES ASIC suitable for network encryption at 10 Gbps and beyond. In Ç.K. Koç and C. Paar, editors, *Cryptographic Hardware and Embedded Systems, CHES '99*, volume 1717 of *Lecture Notes in Computer Science*, Springer, 1999, pages 37–48, 1999.

747. R. S. Winternitz. Producing a one-way hash function from DES. In D. Chaum, editor, *Advances in Cryptology: Proceedings of CRYPTO '83*, Plenum, pages 203–207, 1984.

748. H. Wu, F. Bao, R. H. Deng, and Q.-Z. Ye. Cryptanalysis of Rijmen-Preneel trapdoor ciphers. In K. Ohta and D. Pei, editors, *Advances in Cryptology - ASIACRYPT '98*, volume 1514 of *Lecture Notes in Computer Science*, Springer, pages 126–132, 1998.

749. H. Wu, F. Bao, R. H. Deng, and Q.-Z. Ye. Improved truncated differential attacks on SAFER. In K. Ohta and D. Pei, editors, *Advances in Cryptology - ASIACRYPT '98*, volume 1514 of *Lecture Notes in Computer Science*, Springer, pages 133–147, 1998.

750. W. Wu, D. Feng, and H. Chen. Collision attack and pseudorandomness of reduced-round Camellia. In H. Handschuh and M.A. Hasan, editors, *Selected Areas in Cryptography, SAC 2004*, volume 3357 of *Lecture Notes in Computer Science*, Springer, pages 252–266, 2004.
751. B. Wyseur, W. Michiels, P. Gorissen, and B. Preneel. Cryptanalysis of white-box DES implementations with arbitrary external encodings. In C.M. Adams, A. Miri, and M.J. Wiener, editors, *Selected Areas in Cryptography, SAC 2007*, volume 4876 of *Lecture Notes in Computer Science*, Springer, pages 264–277, 2007.
752. D. Yamamoto, J. Yajima, and K. Itoh. A very compact hardware implementation of the MISTY1 block cipher. In E. Oswald and P. Rohatgi, editors, *Cryptographic Hardware and Embedded Systems - CHES 2008*, volume 5154 of *Lecture Notes in Computer Science*, Springer, 2008, pages 315–330, 2008.
753. H. Yanami, T. Shimoyama, and O. Dunkelman. Differential and linear cryptanalysis of a reduced-round SC2000. In J. Daemen and V. Rijmen, editors, *Fast Software Encryption, FSE 2002, February 2002*, volume 2365 of *Lecture Notes in Computer Science*, Springer, pages 34–48, 2002.
754. A. K. L. Yau, K. G. Paterson, and C. J. Mitchell. Padding oracle attacks on CBC-mode encryption with secret and random IVs. In H. Gilbert and H. Handschuh, editors, *Fast Software Encryption, FSE 2005*, volume 3557 of *Lecture Notes in Computer Science*, Springer, pages 299–319, 2005.
755. Y. Yeom, S. Park, and I. Kim. On the security of CAMELLIA against the Square attack. In J. Daemen and V. Rijmen, editors, *Fast Software Encryption, FSE 2002, February 2002*, volume 2365 of *Lecture Notes in Computer Science*, Springer, pages 89–99, 2002.
756. A. Young and M. Yung. A subliminal channel in secret block ciphers. In H. Handschuh and M.A. Hasan, editors, *Selected Areas in Cryptography, SAC 2004*, volume 3357 of *Lecture Notes in Computer Science*, Springer, pages 198–211, 2004.
757. A. M. Youssef and G. Gong. On the interpolation attacks on block ciphers. In B. Schneier, editor, *Fast Software Encryption, FSE 2000*, volume 1978 of *Lecture Notes in Computer Science*, Springer, pages 109–120, 2000.
758. Y. Yu, Y. Yang, Y. Fan, and H. Min. Security scheme for RFID tag. Auto-ID Labs white paper WP-HARDWARE-022. Available via www.autoidlabs.org/.
759. M. R. Z'aba, H. Raddum, M. Henricksen, and E. Dawson. Bit-pattern based integral attack. In K. Nyberg, editor, *Fast Software Encryption, FSE 2008*, volume 5086 of *Lecture Notes in Computer Science*, Springer, pages 363–381, 2008.
760. I. A. Zabotin, G. P. Glazkov, and V. B. Isaeva. Cryptographic protection for information processing systems: Cryptographic transformation algorithm. Technical Report, Government Standard of the USSR, GOST 28147-89, 1989.
761. K. Zeng, J.-H. Yang, and Z. T. Dai. Patterns of entropy drop of the key in an S-box of the DES. In C. Pomerance, editor, *Advances in Cryptology - CRYPTO '87*, volume 293 of *Lecture Notes in Computer Science*, Springer, pages 438–444, 1988.
762. L. Zhang, W. Zhang, and W. Wu. Cryptanalysis of reduced-round SMS4 block cipher. In Y. Mi, W. Susilo, and J. Seberry, editors, *Proceedings of ACISP 2008*, volume 5107 of *Lecture Notes in Computer Science*, Springer, pages 216–229, 2008.
763. W. Zhang, W. Wu, L. Zhang, and D. Feng. Improved related-key impossible differential attacks on reduced-round AES-192. In E. Biham and A.M. Youssef, editors, *Selected Areas in Cryptography, SAC 2006*, volume 4356 of *Lecture Notes in Computer Science*, Springer, pages 15–27, 2007.
764. X.-M. Zhang and Y. Zheng. The nonhomomorphicity of Boolean functions. In S.E. Tavares and H. Meijer, editors, *Selected Areas in Cryptography, SAC 1998*, volume 1556 of *Lecture Notes in Computer Science*, Springer, pages 280–295, 1999.
765. Y. Zheng. The SPEED cipher. In R. Hirschfeld, editor, *Financial Cryptography*, volume 1318 of *Lecture Notes in Computer Science*, pages 71–90. Springer, 1997.
766. Y. Zheng, T. Matsumoto, and H. Imai. On the construction of block ciphers provably secure and not relying on any unproved hypotheses. In G. Brassard, editor, *Advances in Cryptology - CRYPTO '89*, volume 435 of *Lecture Notes in Computer Science*, Springer, pages 461–480, 1990.

767. Y. Zheng and X.-M. Zhang. On relationships among avalanche, nonlinearity, and correlation immunity. In T. Okamoto, editor, *Advances in Cryptology - ASIACRYPT 2000,* volume 1976 of *Lecture Notes in Computer Science,* Springer, pages 470–482, 2000.
768. Y. Zheng and X.-M. Zhang. Strong linear dependence and unbiased distribution of non-propagative vectors. In H.M. Heys and C.M. Adams, editors, *Selected Areas in Cryptography, SAC 1999,* volume 1758 of *Lecture Notes in Computer Science,* Springer, pages 92–105, 2000.
769. T. Zieschang. Combinatorial properties of basic encryption operations (extended abstract). In W. Fumy, editor, *Advances in Cryptology - EUROCRYPT '97,* volume 1233 of *Lecture Notes in Computer Science,* Springer, pages 14–26, 1997.